Lecture Notes in Computer Science 9941

Commenced Publication in 1973
Founding and Former Series Editors:
Gerhard Goos, Juris Hartmanis, and Jan van Leeuwen

Advanced Research in Computing and Software Science
Subline of Lecture Notes in Computer Science

More information about this series at http://www.springer.com/series/7407

Pinar Heggernes (Ed.)

Graph-Theoretic Concepts in Computer Science

42nd International Workshop, WG 2016
Istanbul, Turkey, June 22–24, 2016
Revised Selected Papers

 Springer

Editor
Pinar Heggernes
Department of Informatics
University of Bergen
Bergen
Norway

ISSN 0302-9743   ISSN 1611-3349 (electronic)
Lecture Notes in Computer Science
ISBN 978-3-662-53535-6   ISBN 978-3-662-53536-3 (eBook)
DOI 10.1007/978-3-662-53536-3

Library of Congress Control Number: 2016947500

LNCS Sublibrary: SL1 – Theoretical Computer Science and General Issues

Printed on acid-free paper

This Springer imprint is published by Springer Nature
The registered company is Springer-Verlag GmbH Germany
The registered company address is: Heidelberger Platz 3, 14197 Berlin, Germany

Preface

The 42nd International Workshop on Graph-Theoretic Concepts in Computer Science (WG 2016) took place in Istanbul, Turkey, on the south campus of Boğaziçi University during June 22–24, 2016. This volume contains the papers presented at the conference.

The WG conference series has a long tradition. Since 1975, it has taken place 23 times in Germany, four times in The Netherlands, three times in France, twice in Austria and the Czech Republic, as well as once in Greece, Israel, Italy, Norway, Slovakia, Switzerland, Turkey, and the UK. The WG conferences aim to connect theory and practice by demonstrating how graph-theoretic concepts can be applied to various areas of computer science and by extracting new graph problems from applications. Their goal is to present new research results and to identify and explore directions of future research.

WG 2016 received 74 submissions. Each submission was carefully reviewed by at least three members of the Program Committee. The Program Committee accepted 25 papers for presentation at WG 2016. The WG 2016 Best Paper Award was given to Emile Ziedan, Deepak Rajendraprasad, Rogers Mathew, Martin Charles Golumbic, and Jeremie Dusart for their paper on "Induced Separation Dimension". The WG 2016 Best Student Paper Award was given to Linda Kleist for her paper on "Drawing Planar Graphs with Prescribed Face Areas". The program included three inspiring invited talks: Saket Saurabh gave a talk on "Lossy Kernelization", Kavitha Telikepalli on "Popular Matchings", and Dominique de Werra on "Minimal Reinforcement of Edges: A Problem of Reliability in Graphs".

I would like to thank the authors of all the papers submitted for possible presentation at WG 2016, the speakers of the 25 talks presenting the accepted papers, the three invited speakers, the members of the Program Committee, and the external reviewers. Special thanks to the local Organizing Committee at Boğaziçi University; their outstanding performance made WG 2016 a great success.

WG 2016 is grateful to the Research Council of Norway for financial support; the University of Bergen, Norway, for financial and administrative support; Boğaziçi University, Turkey, for administrative and practical support; and Springer, Germany, for funding the Best Paper Award. The work of the Program Committee was conducted and the proceedings were prepared using the EasyChair conference management system, which provided an excellent working environment.

August 2016   Pinar Heggernes

Organization

Program Committee

Isolde Adler	University of Leeds, UK
Manu Basavaraju	National Institute of Technology, India
Cristina Bazgan	LAMSADE, Université Paris-Dauphine, France
Hans Bodlaender	Utrecht University, The Netherlands
Christophe Crespelle	Université Claude Bernard Lyon 1, France
Celina de Figueiredo	UFRJ, Brazil
Tınaz Ekim	Boğaziçi University, Turkey
Petr Golovach	University of Bergen, Norway
Gregory Gutin	Royal Holloway, University of London, UK
Pinar Heggernes	University of Bergen, Norway
Mamadou Kante	Université Blaise Pascal - Limos, France
Jan Kratochvil	Charles University, Czech Republic
Martin Milanic	UP IAM and UP FAMNIT, University of Primorska, Slovenia
Rolf Niedermeier	TU Berlin, Germany
Naomi Nishimura	University of Waterloo, Canada
Yota Otachi	Advanced Institute of Science and Technology, Japan
Daniel Paulusma	Durham University, UK
Michał Pilipczuk	University of Warsaw, Poland
Hadas Shachnai	The Technion, Israel
Erik Jan van Leeuwen	MPI Saarbrücken, Germany

Additional Reviewers

Aristotelis, Giannakos	Chechik, Shiri	Fomin, Fedor
Babu, Jasine	Chitnis, Rajesh	Francis, Mathew
Beaudou, Laurent	Corrêa, Ricardo	Ganian, Robert
Belmonte, Rémy	Dabrowski,	Gaspers, Serge
Ben Basat, Ran	Konrad Kazimierz	Gould, Ronald
Bevern, René Van	de Freitas, Rosiane	Grigoriev, Alexander
Biedl, Therese	De Montgolfier, Fabien	Groshaus, Marina
Bonomo, Flavia	Demange, Marc	Guillaume, Jean-Loup
Bornstein, Claudson	Deniz, Zakir	Guo, Jiong
Bresar, Bostjan	Eppstein, David	Gutierrez, Marisa
Busson, Anthony	Fernau, Henning	Habib, Michel
Casel, Katrin	Fluschnik, Till	Hartinger, Tatiana Romina

Havet, Frederic
Hermelin, Danny
Ishii, Toshimasa
Jansen, Bart M.P.
Jeż, Łukasz
Jones, Mark
Kamiyama, Naoyuki
Klavík, Pavel
Kolay, Sudeshna
Kothari, Nishad
Krause, Philipp Klaus
Krivine, Jean
Kwon, O-Joung
Leveque, Benjamin
Levit, Vadim
Lima, Leonardo
Lin, Min Chih
Liotta, Giuseppe
Liu, Chun-Hung
Lo, Allan
Lokshtanov, Daniel
M.S., Ramanujan
Madan, Vivek
Maia, Ana Karolinna
Manne, Fredrik
Mares, Martin
Martin, Daniel M.
Marx, Dániel

Mathew, Rogers
Mertzios, George
Miculan, Marino
Misra, Neeldhara
Misra, Pranabendu
Mnich, Matthias
Molter, Hendrik
Mouawad, Amer
Narayanaswamy, N.S.
Nasre, Meghana
Nichterlein, André
Obdrzalek, David
Onn, Shmuel
Ordyniak, Sebastian
Ossona de Mendez,
    Patrice
Paixao, Joao
Panolan, Fahad
Pap, Julia
Picouleau, Christophe
Pilipczuk, Marcin
Pizaña, Miguel
Rajendraprasad, Deepak
Raman, Venkatesh
Rawitz, Dror
Reidl, Felix
Romero Gonzalez,
    Heidi Jazmin

Safe, Martín Darío
Salazar, Gelasio
Salfelder, Felix
Sau, Ignasi
Saurabh, Saket
Schauer, Joachim
Schiermeyer, Ingo
Siebertz, Sebastian
Sinaimeri, Blerina
Sorge, Manuel
Soulignac, Francisco
Spieksma, Frits
Strozecki, Yann
Subramanya, Vijay
Suchy, Ondrej
Syed Mohammad,
    Meesum
Thierens, Dirk
Trotignon, Nicolas
Vaidyanathan, Krishna
Watrigant, Rémi
Weimann, Oren
Weller, Mathias
Wiese, Andreas
Wrochna, Marcin
Zaker, Manouchehr
Zehavi, Meirav

Contents

Sequences of Radius k for Complete Bipartite Graphs   1
   Michał Dębski, Zbigniew Lonc, and Paweł Rzążewski

Approximate Association via Dissociation   13
   Jie You, Jianxin Wang, and Yixin Cao

Geodetic Convexity Parameters for Graphs with Few Short Induced Paths . . .   25
   Mitre C. Dourado, Lucia D. Penso, and Dieter Rautenbach

Weighted Efficient Domination for P_6-Free and for P_5-Free Graphs   38
   Andreas Brandstädt and Raffaele Mosca

Saving Colors and Max Coloring: Some Fixed-Parameter
Tractability Results   50
   Bruno Escoffier

Finding Two Edge-Disjoint Paths with Length Constraints   62
   Leizhen Cai and Junjie Ye

Packing and Covering Immersion Models of Planar Subcubic Graphs   74
   *Archontia C. Giannopoulou, O-joung Kwon, Jean-Florent Raymond,
   and Dimitrios M. Thilikos*

The Maximum Weight Stable Set Problem in $(P_6,$ bull)-Free Graphs   85
   Frédéric Maffray and Lucas Pastor

Parameterized Power Vertex Cover   97
   Eric Angel, Evripidis Bampis, Bruno Escoffier, and Michael Lampis

Exhaustive Generation of k-Critical $\mathcal{H}$-Free Graphs   109
   Jan Goedgebeur and Oliver Schaudt

Induced Separation Dimension   121
   *Emile Ziedan, Deepak Rajendraprasad, Rogers Mathew,
   Martin Charles Golumbic, and Jérémie Dusart*

Tight Bounds for Gomory-Hu-like Cut Counting   133
   Rajesh Chitnis, Lior Kamma, and Robert Krauthgamer

Eccentricity Approximating Trees: Extended Abstract   145
   Feodor F. Dragan, Ekkehard Köhler, and Hend Alrasheed

Drawing Planar Graphs with Prescribed Face Areas. 158
Linda Kleist

Vertex Cover Structural Parameterization Revisited 171
Fedor V. Fomin and Torstein J.F. Strømme

On Distance-*d* Independent Set and Other Problems in Graphs
with "few" Minimal Separators. 183
Pedro Montealegre and Ioan Todinca

Parameterized Complexity of the MINCCA Problem on Graphs
of Bounded Decomposability 195
*Didem Gözüpek, Sibel Özkan, Christophe Paul, Ignasi Sau,
and Mordechai Shalom*

On Edge Intersection Graphs of Paths with 2 Bends 207
Martin Pergel and Paweł Rzążewski

Almost Induced Matching: Linear Kernels and Parameterized Algorithms . . . 220
Mingyu Xiao and Shaowei Kou

Parameterized Vertex Deletion Problems for Hereditary Graph Classes
with a Block Property 233
Édouard Bonnet, Nick Brettell, O-joung Kwon, and Dániel Marx

Harmonious Coloring: Parameterized Algorithms and Upper Bounds. 245
*Sudeshna Kolay, Ragukumar Pandurangan, Fahad Panolan,
Venkatesh Raman, and Prafullkumar Tale*

On Directed Steiner Trees with Multiple Roots. 257
Ondřej Suchý

A Faster Parameterized Algorithm for GROUP FEEDBACK EDGE SET 269
M.S. Ramanujan

Sequence Hypergraphs. 282
*Kateřina Böhmová, Jérémie Chalopin, Matúš Mihalák, Guido Proietti,
and Peter Widmayer*

On Subgraphs of Bounded Degeneracy in Hypergraphs 295
Kunal Dutta and Arijit Ghosh

Erratum to: Packing and Covering Immersion Models of Planar
Subcubic Graphs. E1
*Archontia C. Giannopoulou, O-joung Kwon, Jean-Florent Raymond,
and Dimitrios M. Thilikos*

Author Index 307

Sequences of Radius k for Complete Bipartite Graphs

Michał Dębski[1], Zbigniew Lonc[2], and Paweł Rzążewski[2(✉)]

[1] Faculty of Mathematics, Informatics and Mechanics,
University of Warsaw, Warsaw, Poland
michal.debski87@gmail.com
[2] Faculty of Mathematics and Information Science,
Warsaw University of Technology, Warsaw, Poland
{zblonc,p.rzazewski}@mini.pw.edu.pl

Abstract. Let G be a graph. A *k-radius sequence for* G is a sequence of vertices of G such that for every edge uv of G vertices u and v appear at least once within distance k in the sequence. The length of a shortest k-radius sequence for G is denoted by $f_k(G)$.

Such sequences appear in a problem related to computing values of some 2-argument functions. Suppose we have a set V of large objects, stored in an external database, and our cache can accommodate at most $k + 1$ objects from V at one time. If we are given a set E of pairs of objects for which we want to compute the value of some 2-argument function, and assume that our cache is managed in FIFO manner, then $f_k(G)$ (where $G = (V, E)$) is the minimum number of times we need to copy an object from the database to the cache.

We give an asymptotically tight estimation on $f_k(G)$ for complete bipartite graphs. We show that for every $\epsilon > 0$ we have $f_k(K_{m,n}) \le (1 + \epsilon)d_k \frac{mn}{k}$, provided that both m and n are sufficiently large – where d_k depends only on k. This upper bound asymptotically coincides with the lower bound $f_k(G) \ge d_k \frac{e(G)}{k}$, valid for all bipartite graphs.

We also show that determining $f_k(G)$ for an arbitrary graph G is NP-hard for every constant $k > 1$.

1 Introduction

1.1 k-radius Sequences

Suppose we need to compute values of a two-argument function, say H, for all pairs of large objects. The objects are stored in a remote database. To compute the values of the function, we need to place these objects in our cache before carrying out the computations. The cache is limited in size – it can hold up to $k + 1$ objects at one time. Our task is to provide a shortest possible sequence of (costly) read operations to ensure that each pair of objects will at some point

Research supported by the Polish National Science Center, decision nr DEC-2012/05/B/ST1/00652.

P. Heggernes (Ed.): WG 2016, LNCS 9941, pp. 1–12, 2016.
DOI: 10.1007/978-3-662-53536-3_1

reside in the cache together so that we can compute the values of H for all pairs of objects. This problem appeared in practice in processing large medical images (see Jaromczyk and Lonc [11]).

The read operation assumes that, if the cache is full, the next object takes the place of one of the objects currently residing in the cache. So far most of the research related to this problem has been concentrated on a special case when we assume that the replacement of objects is based on the first-in first-out strategy. This leads to the concept of a k-radius sequence. Let k and n be positive integers and let V be an n-element set (of objects). We say that a sequence (with possible repetitions) of elements of V is a *k-radius sequence* (or has a *k-radius property*) if every two elements in V are at distance at most k somewhere in the sequence. Observe that short k-radius sequences correspond to efficient caching strategies for our problem. Indeed, if $x_1, x_2, \ldots, x_m$ is a k-radius sequence, then at time t we load the element x_t and after this loading (for $t \geq k+1$) the cache holds the elements $x_{t-k}, x_{t-k+1}, \ldots, x_t$. The k-radius property guarantees that any pair of elements of V resides in the cache together at some point. We denote by $f_k(n)$ the length of a shortest k-radius sequence over an n-element set of objects.

The problem of constructing short k-radius sequences has been considered by several researchers (see Blackburn [3], Blackburn and McKee [4], Chee *et al.* [5], Dębski and Lonc [7], Jaromczyk and Lonc [11], Jaromczyk *et al.* [12]).

1.2 k-radius Sequences for Graphs

In this paper we consider a more general problem – we assume that the values of the function H need not be computed for all pairs of objects but only for some of them. Let V be a set of objects and let $G = (V, E)$ be a graph. We ask: what is the smallest number $c_k(G)$ of read operations that guarantees that each pair of vertices adjacent in G resides in the $(k+1)$-element cache together at some point?

If we assume additionally that the replacement of objects in the cache is based on the first-in first-out strategy, then we get the following generalization of k-radius sequences. A sequence of vertices of a graph $G = (V, E)$ is called a *k-radius sequence for G* (or alternatively, it has a *k-radius property with respect to G*) if each pair of adjacent vertices of G appears at distance at most k in the sequence. More precisely, a sequence $x_1, x_2, \ldots, x_m$ of vertices of a graph $G = (V, E)$ is called a k-radius sequence if for each two vertices u and v adjacent in G there are i, j, $1 \leq i, j \leq m$, such that $u = x_i$, $v = x_j$ and $|j - i| \leq k$. We denote by $f_k(G)$ the length of a shortest k-radius sequence for the graph G. Clearly, assuming the first-in first-out strategy, $f_k(G)$ is equal to the least number of read operations that guarantees that each pair of vertices adjacent in G resides in the cache together at some point. Thus $f_k(G) \geq c_k(G)$.

We shall always assume that G has more than $k+1$ non-isolated vertices because otherwise the problems of finding $c_k(G)$ and $f_k(G)$ are trivial. If G satisfies this condition, then there is an obvious lower bound for both numbers $c_k(G)$ and $f_k(G)$:

$$f_k(G) \geq c_k(G) \geq \frac{e(G)}{k} + \frac{k+1}{2}, \tag{1}$$

where $e(G)$ is the number of edges in G.

Indeed, consider a strategy that requires $m = c_k(G)$ read operations only and guarantees that each pair of vertices adjacent in G resides in the cache together at some point. Observe that if after loading a vertex the cache stores j vertices, then it contains at most $j - 1$ pairs of adjacent vertices which were not together in the cache before. Thus, as we start from an empty cache, after m read operations at most $0 + 1 + \ldots + (k-1) + (m-k)k = mk - \binom{k+1}{2}$ pairs of adjacent vertices have been in the cache together at some point. Consequently, $e(G) \leq mk - \binom{k+1}{2}$, which is equivalent to (1).

Using the terminology we have introduced, the initial problem mentioned in Sect. 1.1 is a special case of our generalization, where $G = K_n$ (the complete graph on n vertices). Blackburn [3] gave a simple replacement strategy that shows that, for a fixed k, $c_k(K_n)$ is asymptotically equal to the lower bound (1). Moreover, he proved using a non-constructive method that imposing the restriction to a first-in first-out strategy does not affect the asymptotic efficiency, i.e. that also the number $f_k(K_n)$ is asymptotically equal to the lower bound (1). Currently, the best known upper bound for $f_k(K_n)$ is $f_k(K_n) = \frac{n^2}{2k} + O(n^{1+\varepsilon})$. It was proved by a constructive method by Jaromczyk et al. [12].

Now, consider the case when G is a complete bipartite graph $K_{m,n}$. In terms of the initial motivation it means that we want to compute the values of a two-argument function H whose domain is a Cartesian product $X \times Y$, where X and Y are the sets that form the bipartition in G.

If k is fixed and both m and n are large, then $c_k(K_{m,n})$ is asymptotically equal to the lower bound (1) – more precisely, we have $c_k(K_{m,n}) = \frac{mn}{k} + O(m + n)$. This bound is attained by the following replacement strategy: pick k vertices from X and keep them in the cache while cycling through all vertices from Y, and repeat the process a total of $\left\lceil \frac{|X|}{k} \right\rceil$ times, each time picking k – or possibly less than k in the last iteration – different vertices from X.

1.3 Our Contributions

The main result of this paper is that for every k there is a constant d_k such that $f_k(K_{m,n})$ is roughly equal to $d_k \frac{mn}{k}$, in case when m and n are sufficiently large – that is, a shortest k-radius sequence for a complete bipartite graph is roughly d_k times longer than the trivial lower bound (1) would imply. We have $1 \leq d_k < \frac{1}{2-\sqrt{2}} \approx 1.71$ (and d_k is close to $\frac{1}{2-\sqrt{2}}$ for large k). Here is a precise statement of this result (see the end of Sect. 2 for the proof).

Theorem 1. *Let k be a positive integer. For every $\epsilon > 0$ if m and n are sufficiently large, then*

$$d_k \frac{mn}{k} \leq f_k(K_{m,n}) \leq (1 + \epsilon) d_k \frac{mn}{k},$$

where $\frac{k}{2k - \sqrt{2k(k-1)}} \leq d_k \leq \frac{k+1}{2k - \sqrt{2k(k-1)}}.$

It is worth highlighting that the lower bound in Theorem 1 generalizes to all bipartite graphs. The following result is a reformulation of our Corollary 7.

Theorem 2. *Let k be a positive integer. For every bipartite graph G we have $f_k(G) \geq d_k \frac{e(G)}{k}$, where d_k is the constant from Theorem 1.*

1.4 Related Problems

An additional motivation of our study comes from its relationship to maximum cuts in some graphs. A *maximum cut* in a graph G is a bipartition of the set of vertices of G maximizing the *size* of the cut, i.e. the number of edges that join vertices of the two sets of the bipartition; the size of the maximum cut in G is denoted $\mathrm{mc}(G)$. Finding a maximum cut in a graph is a widely studied problem which is important in both graph theory and combinatorial optimization (see Newman [13] and a survey by Poljak and Tuza [14]).

Let C_n^k denote a circulant graph obtained from the cycle C_n on n vertices by joining with edges all vertices at distance at most k. Our considerations yield an estimation on the size of a maximum cut in C_n^k (see the end of Sect. 2 for a proof).

Corollary 3. *For a fixed k, we have $\mathrm{mc}(C_n^k) = \frac{kn}{d_k}(1 - o(1))$, where d_k is the constant from Theorem 1.*

The implications go both ways – given the size of a maximum cut in a graph G we can derive a lower bound on $f_k(G)$. Note that a k-radius sequence for G must be also a k-radius sequence for every subgraph of G (in particular, the bipartite subgraph induced by the maximum cut). It gives an immediate consequence of Theorem 2.

Corollary 4. *For every graph G, we have $f_k(G) \geq d_k \frac{\mathrm{mc}(G)}{k}$, where d_k is the constant from Theorem 1.*

The problem of finding a shortest k-radius sequence for a graph is also related to the *bandwidth problem*. The *bandwidth* of a graph $G = (V, E)$ is the minimum of the values $\max\{|i - j| : v_i v_j \in E\}$ over all orderings $(v_1, v_2, \ldots, v_n)$ of V. Let us call such an ordering *bw-optimal*. Informally speaking, we want to place the vertices of G in integer points of a line in such a way, that the longest edge is as short as possible (see for example Chinn *et al.* [6]).

Consider a graph G with bandwidth k and the bw-optimal ordering of its vertices. It is easy to observe that it is a k-radius sequence for G. Thus the graph (with no isolated vertices) has a k-radius sequence containing each vertex exactly once if and only if its bandwidth is at most k. Since determining the bandwidth is NP-hard, even for subcubic graphs (see Garey *et al.* [10]), the problem of determining the existence of a k-radius sequence of length n is NP-hard as well (if k is a part of the instance).

In Sect. 3 we give stronger complexity results. We show the problem of determining $f_k(G)$ for an arbitrary graph G is NP-hard even if k is a constant greater than 1. Moreover, determining $c_k(G)$ for an arbitrary graph G is NP-hard for every constant $k \geq 1$.

2 Asymptotically Shortest k-radius Sequences for Complete Bipartite Graphs

For technical reasons it will be convenient to assume in this section that binary sequences we consider are *cyclic*. In other words we shall assume that the terms of a binary sequence $b_1 b_2 \ldots b_s$ are arranged in a "cyclic way", i.e. b_1 is a successor of b_s. Consequently, we redefine the notion of the distance for cyclic sequences to $d_c(b_i, b_j) = \min(|i - j|, s - |i - j|)$. Moreover, we shall omit commas in binary sequences to simplify the notation.

When we construct a k-radius sequence for a bipartite graph G, we have to jump from one bipartition class of vertices to the other many times. Let X and Y be the bipartition classes in G, $|X| = m$ and $|Y| = n$ and let $\boldsymbol{a} = a_1, a_2, \ldots, a_s$ be a sequence with the k-radius property for the graph G. We define the binary sequence $\boldsymbol{b}(\boldsymbol{a}) = b_1 b_2 \ldots b_s$ such that $b_i = 0$ whenever $a_i \in X$ and $b_i = 1$ whenever $a_i \in Y$. Every appearance of two identical symbols at distance at most k in $\boldsymbol{b}(\boldsymbol{a})$ corresponds to a pair of vertices of G which are at the same distance in $\boldsymbol{a}$ but do not form an edge in G. Therefore, we will call the pair of indices of such a pair of terms in $\boldsymbol{b}(\boldsymbol{a})$ a *bad pair*.

Formally, an unordered pair ij, $i \neq j$, is a k-*bad pair* (resp. a k-*good pair*) in a cyclic binary sequence $\boldsymbol{b}$, if $d_c(b_i, b_j) = \min(|i - j|, s - |i - j|) \leq k$ and $b_i = b_j$ (resp. $b_i \neq b_j$). For every k and s we will be interested in constructing a cyclic binary sequence $\boldsymbol{b}$ of length s with the least possible number of k-bad pairs. Let $w_k(s)$ be this number.

The number of all pairs of terms at distance at most k in a cyclic binary sequence of length s is equal to ks. Let M be the length of a shortest k-radius sequence for a bipartite graph G. Then, we obtain that $kM \geq e(G) + w_k(M)$. So, if prove that $w_k(s) \geq \alpha s$, for some $\alpha < k$, then we will get

$$f_k(G) \geq \frac{e(G)}{k - \alpha}. \tag{2}$$

Clearly, $w_1(s) = 0$ if s is even because the cyclic sequence $0101 \ldots 01$ has no 1-bad pairs. For a similar reason $w_1(s) = 1$ when s is odd.

Let B_k be the de Bruijn graph, i.e. the directed graph, whose vertices are all k-term binary sequences and an ordered pair of vertices vu is an edge if the $(k - 1)$-term suffix of v is the $(k - 1)$-term prefix of u. We identify each edge with the $(k + 1)$-term binary sequence which starts with the first term of v and is followed by all the terms of u.

Clearly, every binary cyclic sequence of length s corresponds to a directed closed walk of length s in B_k (both vertices and edges can appear in a walk an arbitrary number of times). We assign to every edge e in B_k the weight $t_k(e)$ which is equal to the number of appearances of the first term of e on the remaining k positions of e. For instance, if $e = 010001$ (here $k = 5$), then $t_5(e) = 3$. The weight $t_k(C)$ of a closed walk C in B_k is just the sum of weights of its edges (we count each edge as many times as it appears in the walk).

Proposition 5. *The number of k-bad pairs in a cyclic binary sequence is equal to the weight of the corresponding closed walk in the de Bruijn graph B_k.*

Proof. To see this, it suffices to observe that every k-bad pair contributes to the weight of exactly one edge of the corresponding closed walk – the edge starting with the element of the pair, which appears first in the sequence.   □

The *normalized weight* of a closed walk C in B_k is the ratio $\frac{t_k(C)}{|C|}$ (where $|C|$ is the number of edges in C - again we count each edge as many times as it appears in C).

Let a_k be the least possible normalized weight of a cycle in B_k, i.e.

$$a_k = \min\left\{\frac{t_k(C)}{|C|} : C \text{ is a cycle in } B_k\right\}$$

(we allow neither multiple appearances of vertices nor edges in cycles).

Proposition 6. *For integers $k, s > 0$, it holds $a_k s \le w_k(s) < a_k s + k(2^k + k)$.*

Proof. By Proposition 5, $w_k(s)$ is equal to the least possible weight of a closed walk, say C, of length s in the de Bruijn graph B_k. Clearly, the multiset of edges of the closed walk C can be split into sets of edges of cycles, say $C_1, C_2, \ldots, C_p$, in B_k.

By the definition of a_k, we have $t_k(C_i) \ge a_k|C_i|$, for $i = 1, \ldots, p$. Hence,

$$w_k(s) = t_k(C) = t_k(C_1) + \ldots + t_k(C_p) \ge a_k(|C_1| + \ldots + |C_p|) = a_k|C| = a_k s.$$

To complete the proof we need to construct a binary sequence of length s with less than $a_k s + k(2^k + k)$ bad pairs. Let ℓ be the length of a cycle C in B_k with the normalized weight equal to a_k. Moreover, let $q = \lfloor \frac{s}{\ell} \rfloor$ and $r = s - q\ell \le \ell - 1 < |V(B_k)| = 2^k$. We define C' to be the closed walk in B_k obtained by traversing the cycle C q times. Clearly, $t_k(C') = qt_k(C) = q\ell a_k \le sa_k$.

We insert anywhere in the cyclic sequence corresponding to the closed walk C' a sequence of r consecutive 0's. The number of bad pairs in the resulting binary sequence is not larger than $t_k(C') + (k + r)k < a_k s + k(2^k + k)$.   □

It follows from the proof of Proposition 6 that if s is divisible by the length ℓ of a cycle in B_k of minimum normalized weight (equal to a_k), then $w_k(s) = a_k s$ and there is a cyclic binary sequence with exactly $a_k s$ bad pairs which is periodic with the period equal to ℓ.

Moreover, by Proposition 6,

$$\lim_{s\to\infty} \frac{w_k(s)}{s} = a_k. \tag{3}$$

Clearly, the cyclic sequence $0101\ldots 01$ of length $2s$ proves that $w_k(2s) \le ks$, so $a_k \le \frac{k}{2} < k$. Thus, Proposition 6 and the inequality (2) give now the following statement.

Corollary 7. *Let k be a positive integer. For every bipartite graph G, we have $f_k(G) \geq \frac{e(G)}{k-a_k}$.* □

It turns out that this lower bound is asymptotically tight for $K_{m,n}$.

Theorem 8. *For every integer k and real $\varepsilon > 0$, if m and n are sufficiently large, then $f_k(K_{m,n}) \leq \frac{mn}{k-a_k}(1+\varepsilon)$.*

Proof (sketch). We shall use the following theorem by Frankl and Rödl [8] from hypergraph theory. Recall that a hypergraph is r-*uniform* if all its edges have cardinality r. It is d-*regular* if each of its vertices is contained in exactly d edges. By the *codegree* $\mathrm{codeg}_H(v,u)$ of a pair of distinct vertices v and u in a hypergraph H we mean the number of edges containing both v and u. Finally, a *covering* of H is a set of edges whose union is equal to the set of all vertices of H.

> **Theorem (Frankl, Rödl [8])[1].** *Let $r \in \mathbb{N}$ and $\delta > 0$ be fixed. There exist $d_0 \in \mathbb{N}$ and $\delta' > 0$ such that for every $N \geq d \geq d_0$ the following holds. If H is an r-uniform hypergraph with N vertices satisfying the conditions:*
> 1. *H is d-regular,*
> 2. *$\mathrm{codeg}_H(v,u) \leq \delta' \cdot d$ for any vertices $v,u,v \neq u$,*
> *then H has a covering by at most $(1+\delta)\frac{N}{r}$ edges.*

Let C_k be a cycle in B_k with the normalized weight equal to a_k. We denote by ℓ the length of C_k. Let qC_k be the closed walk in B_k obtained by traversing the cycle C_k q times, where $q = \left\lceil \frac{1+\varepsilon}{\varepsilon} \cdot \frac{k(k+1)}{\ell(k-a_k)} \right\rceil$. Clearly, the number of k-good pairs in the cyclic sequence c' corresponding to the closed walk qC_k is $(k - a_k)q\ell$. Let c be the (non-cyclic) binary sequence of length $q\ell$ obtained from c' by cutting it at some point and let r be the number of k-good pairs in c. Observe that a k-good pair in c' is either still k-good in c, or it is no longer at distance at most k. Since c' has $\frac{k(k+1)}{2}$ fewer pairs at distance at most k than c, we get $r \geq (k - a_k)q\ell - \frac{k(k+1)}{2}$. We denote by c_0 the number of 0's and by $c_1 = q\ell - c_0$ the number of 1's in c.

Let X and Y be the bipartition classes in $K_{m,n}$, with $|X| = m$ and $|Y| = n$. We denote by H be the hypergraph whose vertices are all ordered pairs xy such that $x \in X$, $y \in Y$. For every sequence a of $q\ell$ distinct vertices in $K_{m,n}$ such that $b(a) = c$ we define an edge e_a in H. The edge e_a consists of such vertices xy of H that x and y are at distance at most k in a.

It can be checked that H satisfies the assumptions of the Frankl-Rödl Theorem and thus there is a covering of the vertex set of H by at most $(1+\frac{\varepsilon}{2})\frac{mn}{r}$ edges. Let us consider a sequence obtained by concatenation of the sequences corresponding to these edges. In this sequence every two vertices forming an edge in $K_{m,n}$ are at distance at most k. Using the definition of q it can be shown that the length of this sequence is at most $\frac{mn}{(k-a_k)}(1+\varepsilon)$ which completes the proof our theorem. □

---

[1] This is a special case of a version of the original theorem that appears in Alon and Spencer [1, Theorem 4.7.1].

In view of Theorem 8 and Corollary 7 it would be interesting to find the exact values of a_k. The values of a_k for $k \leq 5$ as well as the optimal cycles in B_k (i.e. the cycles for which the normalized weight is equal to a_k) are shown in Table 1. We denote by $(b_1, b_2, \ldots, b_p)^*$ the cycle in B_k whose consecutive edges are $b_1 b_2 \ldots b_k$, $b_2 b_3 \ldots b_{k+1}, \ldots, b_p b_1 \ldots b_{k-1}$ (indices are computed modulo p).

It is routine to show that the normalized weight of the cycle $(0^t 1^t)^*$ (t 0's followed by t 1's) in B_k is equal to $\frac{\binom{t}{2} + \binom{k-t+1}{2}}{t}$, for $\frac{k}{2} \leq t \leq k+1$. Let

$$b_k = \min_{\frac{k}{2} \leq t \leq k+1} \frac{\binom{t}{2} + \binom{k-t+1}{2}}{t} = \min_{\frac{k}{2} \leq t \leq k+1} \left(t + \frac{(k+1)k}{2t} - k - 1 \right). \tag{4}$$

Obviously, $a_k \leq b_k$. We conjecture that $a_k = b_k$ for all positive integers k. The values of a_k in Table 1 show that the conjecture is true for $k \leq 5$.

Table 1. The values of a_k and optimal cycles in B_k for small k

k	a_k	Optimal cycles in B_k
1	0	$(01)^*$
2	1/2	$(0011)^*$
3	1	$(01)^*, (0011)^*, (000111)^*, (00011)^*, (00111)^*$
4	4/3	$(000111)^*$
5	7/4	$(00001111)^*$

Our next theorem gives a lower bound for a_k which is "very close" to b_k.

Theorem 9. *For every positive integer k, we have $a_k \geq \sqrt{2k(k-1)} - k$.*

Proof (sketch). Clearly, the theorem holds for $k = 1$, so we assume from now on that $k \geq 2$.

Consider a cyclic binary sequence $\boldsymbol{b} = b_0 b_1 \ldots b_{s-1}$ of length s with minimum possible number $w_k(s)$ of k-bad pairs. For $i = 0, 1, \ldots, s-1$ let $\bar{\ell}_i$ (resp. ℓ_i) denote the number of k-bad (resp. k-good) pairs containing the term b_i. Clearly $\bar{\ell}_i + \ell_i = 2k$ for all i and $w_k(s) = \frac{1}{2} \sum_{i=0}^{s-1} \bar{\ell}_i$. For a maximal segment $\boldsymbol{b}' = b_h b_{h+1} \ldots b_{h+t-1}$ in $\boldsymbol{b}$ of terms of the same value we define the *score* $Sc(\boldsymbol{b}') = \frac{1}{t} \sum_{j=0}^{t-1} \bar{\ell}_{h+j}$.

By an analysis of the structure of $\boldsymbol{b}'$ and considering the cases $t < \frac{2k-1}{3}$; $\frac{2k-1}{3} \leq t \leq k-1$; and $k \leq t \leq k+1$ separately, we can prove that $Sc(\boldsymbol{b}') \geq u(t)$, where

$$u(t) = \begin{cases} k - 2 - \frac{(t-1)^2}{4t} & \text{for } t < \frac{2k-1}{3}, \\ k - 2 - \frac{1}{t}(k - t - 1)(2t - k) & \text{for } \frac{2k-1}{3} \leq t \leq k-1, \\ t - 3 & \text{for } k \leq t \leq k+1 \end{cases}$$

is a real-valued function. It is routine to check that for $k \geq 2$, $u(t)$ reaches its minimum value in the interval $[1, k+1]$ at $t = \sqrt{\frac{k(k-1)}{2}}$.

Thus we obtain $Sc(\boldsymbol{b'}) \geq u\left(\sqrt{\frac{k(k-1)}{2}}\right) = 2\sqrt{2k(k-1)} - 2k$.

We divide $\boldsymbol{b}$ to maximal segments $\boldsymbol{b_1}, \boldsymbol{b_2}, \ldots, \boldsymbol{b_r}$ of terms of the same value. Let t_i denote the length of $\boldsymbol{b_i}$. Clearly, the number of k-bad pairs in $\boldsymbol{b}$ is $\frac{1}{2}\sum_{i=0}^{s-1}\ell_i = \frac{1}{2}\sum_{i=1}^{r}Sc(\boldsymbol{b_i})t_i$. Thus,

$$w_k(s) = \frac{1}{2}\sum_{i=1}^{r}Sc(\boldsymbol{b_i})t_i \geq \sum_{i=1}^{r}(\sqrt{2k(k-1)} - k)t_i = (\sqrt{2k(k-1)} - k)s.$$

Hence, by (3), $a_k = \lim_{s \to \infty}\frac{w_k(s)}{s} \geq \sqrt{2k(k-1)} - k$. $\qquad\square$

We have shown that

$$\sqrt{2k(k-1)} - k \leq a_k \leq b_k. \tag{5}$$

We shall see now that these bounds for a_k are very close to each other.

First observe that there exists a positive integer t such that

$$t + \frac{(k+1)k}{2t} \leq \sqrt{2(k+1)k+1}. \tag{6}$$

Indeed, consider the function $f(x) = x + \frac{(k+1)k}{2x}$ and let $x_1 = \sqrt{\frac{(k+1)k}{2} + \frac{1}{4}} - \frac{1}{2}$. It is easy to verify that for $x \in [x_1, x_1 + 1]$, $f(x) \leq f(x_1) = f(x_1 + 1) = \sqrt{2(k+1)k+1}$, so we define t to be the unique integer in the interval $[x_1, x_1+1)$.

By (4) and (6) we get

$$b_k = \min_{\frac{k}{2} \leq t \leq k+1}\left(t + \frac{(k+1)k}{2t} - k - 1\right) \leq \sqrt{2(k+1)k+1} - k - 1.$$

Using the inequality above one can readily verify that the difference between the upper and the lower bound for a_k given in (5) is smaller than 0.5 for $k \geq 5$ (and it tends to $\sqrt{2} - 1$ as k tends to infinity). Thus, since the actual value of a_4 differs from the lower bound in (5) by less than 0.5 too (see Table 1), we have the following statement.

Corollary 10. *For all $k \geq 4$ it holds that*

$$\sqrt{2k(k-1)} - k \leq a_k < \sqrt{2k(k-1)} - k + 1/2.$$

Proof of Theorem 1. Theorem 1 follows immediately from Theorem 8 and Corollary 7 by defining $d_k = \frac{k}{k - a_k}$. The bounds for d_k given in Theorem 1 can be easily obtained from Corollary 10 for $k \geq 4$ and by direct computations (using Table 1) for $k < 4$. $\qquad\square$

Proof of Corollary 3. Let $v_1, \ldots, v_n$ be the vertices of C_n^k, in a natural order. Any cyclic binary sequence $\boldsymbol{b} = b_1 b_2 \ldots b_n$ defines a bipartition (V_0, V_1) of the vertex set of C_n^k: a vertex v_i goes to V_0 if $b_i = 0$ and it goes to V_1 otherwise. One can readily verify that the number of good pairs in $\boldsymbol{b}$ is equal to the size of the bipartition (V_0, V_1), i.e. the number of edges joining vertices of the two sets V_0 and V_1. Consequently, the size of a maximum cut in C_n^k is equal to the maximum number $kn - w_k(n)$ of good pairs in a cyclic binary sequence of length n. Therefore, the proof is complete by the equalities (3) and $d_k = \frac{k}{k - a_k}$. $\qquad\square$

3 Complexity Results for Arbitrary Graphs

In this section we consider problems of finding the numbers $f_k(G)$ and $c_k(G)$ for arbitrary connected graphs G.

Let us first make the definition of $c_k(G)$ a bit more precise. We define for a graph G and $k < |V(G)|$ a *k-cover sequence* $\mathbf{c} = c_1, \ldots, c_m$ to be a sequence of $(k+1)$-subsets of $V(G)$ such that every two consecutive sets in $\mathbf{c}$ differ by one element (that is, $|c_i \setminus c_{i+1}| = |c_{i+1} \setminus c_i| = 1$ for $i = 1, \ldots, m-1$) and for every edge $e \in E(G)$ we have $e \subseteq c_i$ for some i. Clearly, a k-cover sequence describes replacements of objects in the cache; if we assume that at time 0 the cache holds the set c_1, then at time t (for $1 \le t \le m-1$) we replace the only object of $c_t \setminus c_{t+1}$ by the only object of $c_{t+1} \setminus c_t$. The number of read operations in this scenario is equal to $m + k$ (because we have to read the $k+1$ elements of c_1 at start). Consequently, $c_k(G)$ is equal to the sum of k and the length of a shortest k-cover sequence for G.

We shall discuss the computational complexities of the following two decision problems.

Problem R_k
Instance: a connected graph G and an integer M.
Question: Is there a k-radius sequence of length M for the graph G?

Problem B_k
Instance: a connected graph G and an integer M.
Question: Is there a k-cover sequence of length M for the graph G?

Let us consider the problem R_k first. We start with a simple lower bound for the value of $f_k(G)$.

Proposition 11. *For any graph G it holds that $f_k(G) \ge \sum_{v \in V} \left\lceil \frac{\deg v}{2k} \right\rceil$.*

Proof. Consider a shortest k-radius sequence $\mathbf{x}$ for G and for each vertex v let $m(v)$ denote the number of appearances of v in $\mathbf{x}$. For every appearance of v in $\mathbf{x}$, at most $2k$ neighbors of v appear at distance at most k in $\mathbf{x}$. Thus $m(v) \ge \left\lceil \frac{\deg v}{2k} \right\rceil$ and $f_k(G) = \sum_{v \in V} m(v) \ge \sum_{v \in V} \left\lceil \frac{\deg v}{2k} \right\rceil$. $\square$

First consider the case $k = 1$. The problem R_1 asks for a sequence of vertices of G, in which the endvertices of every edge appear as consecutive elements. Let G be a connected graph with n vertices and m edges. Denote by n_o the number of vertices of odd degree in G. We add $n_0/2$ edges between them, creating a multigraph G', which has an Euler circuit. This Euler circuit corresponds to a 1-radius sequence in G of length $m + n_0/2$, which matches the lower bound given in Proposition 11. Thus the problem R_1 is polynomially solvable.

Theorem 12. *For $k \ge 2$ the problem R_k is NP-complete.*

Proof (main idea). To prove NP-hardness, we show a reduction from the problem of determining existence of a Hamiltonian path in a cubic triangle-free graph (see Garey and Johnson [9, p. 199]).

Let F be a cubic triangle-free graph with n vertices. For every vertex v in F we add $k - 2$ pendant edges $e_v^1, \ldots, e_v^{k-2}$ incident with v and call the resulting graph F'. We define G to be the line graph of F'. Then we prove that there exists a k-radius sequence for G of length $\frac{2}{k+1}e(G) + 1$ if and only if F has a Hamiltonian path. $\qquad\square$

We proceed to the problem B_k.

Theorem 13. *For all $k \geq 1$ the problem B_k is NP-complete.*

Proof (main idea). Clearly the problem is in NP. To show NP-hardness for $k = 1$ we observe that a graph H has a 1-cover sequence of length $e(H)$ if and only if its line graph $L(H)$ has a Hamiltonian path. Determining the existence of a Hamiltonian path in a line graph in NP-complete [2].

For $k \geq 2$ we show NP-hardness by a reduction from B_1. Let H be a graph with m edges and let $N = \binom{k}{2}(m - 1) + \binom{k+1}{2} + 3$. We define a graph G in which every edge uv of H is replaced by the "edge gadget" G_{uv}, depicted in Fig. 1.

We prove that there exists a k-cover sequence of length $mN + (m - 1)(k - 1)$ for G if and only if there exists a 1-cover sequence of length m for H. $\qquad\square$

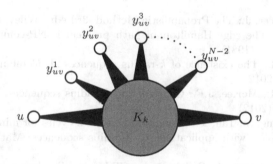

Fig. 1. The "edge gadget" G_{uv} used in the reduction from B_1 to B_k, for $k \geq 2$ (K_k denotes the complete graph on k vertices; the remaining N vertices of G_{uv} are joined by edges with all vertices of K_k)

4 Concluding Remarks and Open Problems

Theorem 1 gives a good estimate of $f_k(K_{m,n})$ when both m and n are large (tending to infinity), which leaves the following question open: what happens when only one of these parameters goes to infinity? This problem probably cannot be solved using the same proof technique (in particular, the assumption (2) of Frankl-Rödl theorem will not be satisfied) and it is not clear what the precise answer should be.

Problem 1. Give an asymptotically tight estimate on $f_k(K_{m,n})$ when m is constant and n tends to infinity.

We would like to see an analog of Theorem 1 for other classes of graphs – in particular, for tripartite complete t-partite graphs $K_{n_1,n_2,...,n_t}$ (for fixed t and large $n_1, n_2, \ldots, n_t$). Proofs of Proposition 6, Corollary 7 and Theorem 8 can be adapted to the t-partite case, but the main difficulty is determining (the analog of) the constant a_k.

Since this problem seems to be interesting on its own, we will make it more precise. An unordered pair ij, where $i \neq j$, is a k-bad pair in a t-ary sequence $\mathbf{r}$ if $|j - i| \leq k$ and $r_i = r_j$. Let $w_{k,t}(s)$ be the minimum number of k-bad pairs in a t-ary sequence of length s and set $a_{k,t} = \lim_{s \to \infty} \frac{w_{k,t}(s)}{s}$. (The existence of this limit follows from an analog of Proposition 6 for t-ary sequences.)

Problem 2. Give a good estimate on $a_{k,t}$.

Note that Corollary 10 gives a value of $a_{k,2}$ that is accurate only up to $\frac{1}{2}$, which means that we do not know the exact value of the constant d_k in Theorem 1. We know the values of $a_{k,2}$ for $k \leq 5$ (see Table 1) and it would be interesting to find a precise formula for $k > 5$.

References

1. Alon, N., Spencer, J.: The Probabilistic Method, 3rd edn. Wiley, Hoboken (2008)
2. Bertossi, A.A.: The edge Hamiltonian path problem is NP-complete. Inf. Proc. Lett. **13**, 157–159 (1981)
3. Blackburn, S.R.: The existence of k-radius sequences. J. Combin. Theor. Ser. A **119**, 212–217 (2012)
4. Blackburn, S.R., McKee, J.F.: Constructing k-radius sequences. Math. Comput. **81**, 2439–2459 (2012)
5. Chee, Y.M., Ling, S., Tan, Y., Zhang, X.: Universal cycles for minimum coverings of pairs by triples, with applications to 2-radius sequences. Math. Comput. **81**, 585–603 (2012)
6. Chinn, P., Chvátalová, J., Dewdney, A., Gibbs, N.: The bandwidth problem for graphs and matrices - a survey. J. Graph Theor. **6**, 223–254 (1982)
7. Dębski, M., Lonc, Z.: Sequences of large radius. Eur. J. Comb. **41**, 197–204 (2014)
8. Frankl, P., Rödl, V.: Near perfect coverings in graphs and hypergraphs. Eur. J. Comb. **6**, 317–326 (1985)
9. Garey, M.R., Johnson, D.S.: Computers and Intractability, A Guide to the Theory of NP-Completeness. Freeman, New York (1979)
10. Garey, M.R., Graham, R.L., Johnson, D.S., Knuth, D.E.: Complexity results for bandwidth minimization. SIAM J. Appl. Math. **34**, 477–495 (1978)
11. Jaromczyk, J.W., Lonc, Z.: Sequences of radius k: how to fetch many huge objects into small memory for pairwise computations. In: Fleischer, R., Trippen, G. (eds.) ISAAC 2004. LNCS, vol. 3341, pp. 594–605. Springer, Heidelberg (2004)
12. Jaromczyk, J., Lonc, Z., Truszczyński, M.: Constructions of asymptotically shortest k-radius sequences. J. Combin. Theor. Ser. A **119**, 731–746 (2012)
13. Newman, A.: Max-cut. Encycl. Algorithms **1**, 489–492 (2008)
14. Poljak, S., Tuza, Z.: Maximum cuts and large bipartite subgraphs. DIMACS Ser. Discrete Math. Theoret. Comput. Sci. **20**, 181–244 (1995)

Approximate Association via Dissociation

Jie You[1,2], Jianxin Wang[1], and Yixin Cao[2](✉)

[1] School of Information Science and Engineering,
Central South University, Changsha, China
jxwang@mail.csu.edu.cn
[2] Department of Computing, Hong Kong Polytechnic University,
Hong Kong, China
yixin.cao@polyu.edu.hk

Abstract. A vertex set X of a graph G is an association set if each component of $G - X$ is a clique, or a dissociation set if each component of $G - X$ is a single vertex or a single edge. Interestingly, $G - X$ is then precisely a graph containing no induced P_3's or containing no P_3's, respectively. We observe some special structures and show that if none of them exists, then the minimum association set problem can be reduced to the minimum (weighted) dissociation set problem. This yields the first nontrivial approximation algorithm for the association set problem, with approximation ratio is 2.5. The reduction is based on a combinatorial study of modular decomposition of graphs free of these special structures. Further, a novel algorithmic use of modular decomposition enables us to implement this approach in $O(mn + n^2)$ time.

1 Introduction

A *cluster graph* comprises a family of disjoint cliques, each an *association*. Cluster graphs have been an important model in the study of clustering objects based on their pairwise similarities, particularly in computational biology and machine learning [3]. If we represent each object with a vertex, and add an edge between two objects that are similar, we would expect a cluster graph. If this fails, a natural problem is then to find and exclude a minimum number of vertices such that the rest forms a cluster graph; this is the *association set* problem. This problem has recently received significant interest from the community of parameterized computation, where it is more commonly called *cluster vertex deletion* [4,15]. The cardinality of a minimum association set of a graph is also known as its *distance to clusters*. It is one of the few structural parameters for dense graphs [9,10], in contrast with a multitude of structural parameters for sparse graphs, thereby providing another motivation for this line of research. For example, Bruhn et al. [5] recently showed that the boxicity problem (of deciding the minimum d such that a graph G can be represented as an intersection graph

Supported in part by NSFC under grants 61572414 and 61420106009, and RGC under grant 252026/15E.

P. Heggernes (Ed.): WG 2016, LNCS 9941, pp. 13–24, 2016.
DOI: 10.1007/978-3-662-53536-3_2

of axis-aligned boxes in the d-dimension Euclidean space) is fixed-parameter tractable parameterized by the distance to clusters.

The association set problem belongs to the family of vertex deletion problems studied by Yannakakis et al. [16,18]. The task in these problems is to delete the minimum number of vertices from a graph so that the remaining subgraph satisfies a hereditary property; recall that a graph property is *hereditary* if it is closed under taking induced subgraphs [16]. It is known that a hereditary property can be characterized by a (possibly infinite) set of forbidden induced subgraphs. In our case, the property is "being a cluster graph," and the forbidden induced subgraphs are P_3's (i.e., paths on three vertices). A trivial approximation algorithm of ratio 3 can be derived as follows. We search for induced P_3's, and we delete all its three vertices if one is found. This trivial upper bound is hitherto the best known. Indeed, this is a simple application of Lund and Yannakakis's observation [18], which applies to all graph classes with finite forbidden induced subgraphs.

Closely related is the cluster editing problem, which allows us to use, instead of vertex deletions, both edge additions and deletions [3]. Approximation algorithms of the cluster editing problem have been intensively studied, and the current best approximation ratio is 2.5 [1,2,8]. Our main result is the first nontrivial approximation algorithm for the association set problem, with a ratio matching the best ratio of the closely related cluster editing problem. As usual, n and m denote the numbers of vertices and edges respectively in the input graph. Without loss of generality, we assume throughout the paper that the input graph contains no isolated vertices (vertices of degree 0), hence $n = O(m)$.

Theorem 1. *There is an $O(mn)$-time approximation algorithm of ratio 2.5 for the association set problem.*

Our approach is to reduce the association set problem to the weighted dissociation set problem. Given a vertex-weighted graph, the *weighted dissociation set* problem asks for a set of vertices with the minimum weight such that its deletion breaks all P_3's, thereby leaving a graph of maximum degree 1 or 0. This problem was first studied by Yannakakis [25], who proved that its unweighted version is already NP-hard on bipartite graphs. Note that a P_3 that is not induced must be in a triangle. Thus, in triangle-free graphs, the weighted version of the association set problem is equivalent to the weighted dissociation set problem. It is easy to observe that for the association set problem, vertices in a twin class (i.e., whose vertices have the same closed neighborhood) are either fully contained in or disjoint from a minimum solution. This observation inspires us to transform the input graph G into a vertex-weighted graph Q by identifying each twin class of G with a vertex of Q whose weight is the size of the corresponding twin class. We further observe that there are five small graphs such that if G has none of them as an induced subgraph, then Q either has a simple structure, hence trivially solvable, or is triangle-free, and can be solved using the ratio-2 approximation algorithm for the weighted dissociation set problem [21,22]. From the obtained solution for Q we can easily retrieve a solution for the original graph

G. Since each of these five graphs has at most five vertices and at least two of them need to be deleted to make it free of induced P_3's, the approximation ratio 2.5 follows.

The main idea of this paper appears in the argument justifying the reduction from the (unweighted) association set problem to the weighted dissociation set problem. Indeed, we are able to provide a stronger algorithmic result that implies the aforementioned combinatorial result. We develop an efficient algorithm that detects one of the five graphs in G, solves the problem completely, or determines that Q is already triangle-free. Our principal tool is modular decomposition. A similar use of modular decomposition was recently invented by the authors [17] in parameterized algorithms. It is worth noting that the basic observation on vertex deletion problems to graph properties with finite forbidden induced subgraphs has been used on both approximation and parameterized algorithms, by Lund and Yannakakis [18] and by Cai [6] respectively.

After a preliminary version of this work appeared in arxiv, Fiorini et al. [11] managed to further improve the ratio to 7/3. The first part of their algorithm is similar as ours, with more small induced subgraphs taken into consideration, while their analysis, using the "local ratio" technique, is quite different from ours.

As a final remark, cluster editing has a $2k$-vertex kernel [7], while it remains an open problem to find a linear-vertex kernel for the association set (cluster vertex deletion) problem.

2 Preliminaries

This paper will be only concerned with undirected and simple graphs. The vertex set and edge set of a graph G are denoted by $V(G)$ and $E(G)$ respectively. For $\ell \geq 3$, let P_ℓ and C_ℓ denote respectively an induced path and an induced cycle on ℓ vertices. A C_3 is also called a *triangle*. For a given set $\mathcal{F}$ of graphs, a graph G is $\mathcal{F}$-*free* if it contains no graph in $\mathcal{F}$ as an induced subgraph. When $\mathcal{F}$ consists of a single graph F, we use also F-*free* for short. For each vertex v in $V(G)$, its *neighborhood* and *closed neighborhood* are denoted by $N_G(v)$ and $N_G[v]$ respectively.

A subset M of vertices forms a *module* of G if all vertices in M have the same neighborhood outside M. In other words, for every pair of vertices $u, v \in M$, a vertex $x \notin M$ is adjacent to u if and only if it is adjacent to v as well. The set $V(G)$ and all singleton vertex sets are modules, called *trivial*. A graph on at least four vertices is *prime* if it contains only trivial modules, e.g., a P_4 and a C_5. Given any partition $\{M_1, \ldots, M_p\}$ of $V(G)$ such that M_i for every $1 \leq i \leq p$ is a module of G, we can derive a p-vertex *quotient graph* Q such that for any pair of distinct i, j with $1 \leq i, j \leq p$, the ith and jth vertices of Q are adjacent if and only if M_i and M_j are adjacent in G (every vertex in M_i is adjacent to every vertex in M_j). It should be noted that a single-vertex graph and G itself are both trivial quotient graphs of G, defined by the trivial module partitions $\{V(G)\}$ and $\{\{v_1\}, \ldots, \{v_n\}\}$ respectively.

A module M is *strong* if for every other module M' that intersects M, one of M and M' is a proper subset of the other. All trivial modules are clearly strong. We say that a strong module M different from $V(G)$ is *maximal* if the only strong module properly containing M is $V(G)$. (It can be contained by non-strong modules, e.g., in a graph that is a clique, the maximal strong modules are simply the singletons, while every subset of vertices is a module.) The set of maximal strong modules of G partitions $V(G)$, and defines a special quotient graph of G, denoted by $\widetilde{Q}(G)$.[1] The reader who is unfamiliar with modular decomposition is referred to the survey of Habib and Paul [13] for more information. The following proposition will be crucial for our algorithm.

Proposition 1. [12,20] *If a graph G is connected, then $\widetilde{Q}(G)$ is either a clique or prime. Any prime graph contains an induced P_4.*

Let Q be a quotient graph of G, and let M be a module of G in the module partition defining Q. By abuse of notation, we will also use M to denote the corresponding node of Q; hence $M \in V(Q)$ and $M \subseteq V(G)$, and its meaning will be clear from context. Accordingly, by $N_G(M)$ we mean those vertices of G adjacent to M in G, and by $N_Q(M)$ we mean those nodes of Q adjacent to M in Q—note that the union of those vertices of G represented by $N_Q(M)$ is exactly $N_G(M)$. Sets $N_G[M]$ and $N_Q[M]$ are understood analogously.

The weighted versions of the associated set problem and the dissociation set problem are formally defined as follows.

Associated set
Input: A vertex-weighted graph G.
Task: find a subset $X \subset V(G)$ of the minimum weight such that every component of $G - X$ is a clique.

Dissociation set
Input: A vertex-weighted graph G.
Task: find a subset $X \subset V(G)$ of the minimum weight such that every component of $G - X$ is a single vertex or a single edge.

Let $\mathtt{asso}(G)$ and $\mathtt{diss}(G)$ denote respectively the weights of minimum association sets and minimum dissociation sets of a weighted graph G. It is routine to verify that $\mathtt{asso}(G) \leq \mathtt{diss}(G)$. Their gap can be arbitrarily large, e.g., if G is a clique on n vertices, then $\mathtt{asso}(G) = 0$ and $\mathtt{diss}(G) = n - 2$. A vertex set X is an association set or a dissociation set of a graph G if and only if $G - X$ contains no P_3 as an induced subgraph or as a subgraph, respectively. The following proposition follows from the fact that every P_3 in a C_3-free graph is induced.

---

[1] If G is a clique or an independent set, then $\widetilde{Q}(G)$ is isomorphic to G and is the largest quotient graph of G; if $\widetilde{Q}(G)$ is prime, then it is the smallest *nontrivial* quotient graph of G, both cardinality-wise and inclusion-wise (see Lemma 5). Otherwise, there can be other quotient graph larger or smaller than $\widetilde{Q}(G)$.

Proposition 2. *If a graph G is C_3-free, then* $\mathtt{asso}(G) = \mathtt{diss}(G)$.

Theorem 2 *([21,22])*. *There is an $O(mn)$-time approximation algorithm of ratio 2 for the weighted dissociation set problem.*

Note that an unweighted graph can be treated as a special weighted graph where every vertex receives a unit weight. In this case, $\mathtt{asso}(G)$ is the same as the cardinality of the minimum association set of G.

A $\{C_4, P_4\}$-free graph is called a *trivially perfect graph*. A vertex is *universal* if it is adjacent to all other vertices in this graph, i.e., has degree $n - 1$. It is easy to verify that each universal vertex is a maximal strong module of the graph.

Proposition 3 *([14,23,24])*. *Every connected trivially perfect graph has a universal vertex. One can in $O(m)$-time either decide that a graph is a trivially perfect graph, or detect an induced P_4 or C_4.*

3 The Approximation Algorithm

The association set problem admits a naive 3-approximation algorithm [18]. It finds an induced P_3 and deletes from G all the three vertices in this P_3, and repeats. Since any minimum association set has to contain some of the three vertices, the approximation ratio is at most 3. A P_3 can be found in linear time, while the process can be repeated at most $n/3$ times, and thus the algorithm can be implemented in time $O(mn)$. We present here a very simple 2.5-approximation algorithm, which runs in a high-order polynomial time, and we will show in the next section how to implement it in an efficient way to achieve the running time claimed in Theorem 1.

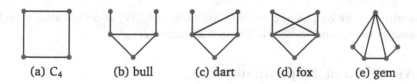

(a) C_4 (b) bull   (c) dart   (d) fox   (e) gem

Fig. 1. Small subgraphs on 4 or 5 vertices.

Let $\mathcal{F}$ denote the set of five small graphs depicted in Fig. 1, i.e., $\{C_4, \text{bull}, \text{dart}, \text{fox}, \text{gem}\}$. A quick glance of Fig. 1 convinces us that from each induced subgraph in $\mathcal{F}$, at least two vertices need to be deleted to make it P_3-free.

Proposition 4. *Let $X \subseteq V(G)$. If $G[X] \in \mathcal{F}$, then* $\mathtt{asso}(G - X) \leq \mathtt{asso}(G) - 2$.

In polynomial time we can decide whether G contains an induced subgraph in $\mathcal{F}$, and find one if it exists. We delete all its vertices if it is found. If G is not connected, then we work on its components one by one. In the rest of this section we may focus on connected $\mathcal{F}$-free graphs. In such a graph, every nontrivial module M induces a $\{C_4, P_4\}$-free subgraph: A P_4 in $G[M]$, together with any $v \in N_G(M)$ (it exists because G is connected and M is nontrivial), makes a gem.

One may use the definition of modular decomposition to derive the following combinatorial properties of $\mathcal{F}$-free graphs. Since we will present a stronger result in the next section that implies this lemma, its proof is omitted here.

Lemma 1. *Let G be an $\mathcal{F}$-free graph that is not a clique, and let $Q = \widetilde{Q}(G)$. Either G consists of a set of universal vertices and two disjoint cliques, or Q is C_3-free and the following hold for every maximal strong module M of G:*

(1) The subgraph $G[M]$ is a cluster graph. If it is not a clique, then $|N_G(M)| = 1$.
(2) If $|N_Q(M)| > 2$, then the module M is trivial (consisting of a single vertex of G).

In the first case, G has simply two intersecting cliques C_1 and C_2, and the problem is trivial: We delete either $C_1 \cap C_2$ (i.e., all universal vertices), or one of $C_1 \setminus C_2$ and $C_2 \setminus C_1$, whichever is smaller. Therefore, we focus on the other case where $\widetilde{Q}(G)$ is C_3-free. If some maximal strong module M does not induce a clique in a connected $\mathcal{F}$-free graph G, then we can delete the unique neighbor of M and consider the smaller graph $G - N_G[M]$. Now that G is not a clique but every maximal strong module M of G is, we can define a vertex-weighted graph Q isomorphic to the quotient graph $\widetilde{Q}(G)$, where the weight of each vertex in Q is the number of vertices in the corresponding module, i.e., $|M|$. We apply the algorithm of Tu and Zhou [21] to find a dissociation set of this weighted graph Q. Since Q is C_3-free, by Proposition 2 and Theorem 2, the total weight of the obtained dissociation set is at most $2\mathtt{diss}(Q) = 2\mathtt{asso}(Q) = 2\mathtt{asso}(G)$. Putting together these steps, an approximation algorithm with ratio 2.5 follows (see Fig. 2).

Theorem 3. *The output of algorithm* APPROX-ASSO *(G) is an association set of the input graph G and its size is at most $2.5\mathtt{asso}(G)$.*

4   An Efficient Implementation

We now discuss the implementation issues that lead to the claimed running time. A simpleminded implementation of the algorithm given in Fig. 2 takes $O(n^6)$ time, which is decided by the disposal of induced subgraphs in $\mathcal{F}$ (step 3). It is unclear to us how to detect them in a more efficient way than the $O(n^5)$-time enumeration. But we observe that what we need are no more than the conditions stipulated in Lemma 1, for which being $\mathcal{F}$-free is sufficient but not necessary. The following relaxation is sufficient for our algorithmic purpose: We either detect an induced subgraph in $\mathcal{F}$ or determine that G has already satisfied these conditions. Once a subgraph is found, we can delete all its vertices and repeat the process. In summary, we are after an $O(mn)$-time procedure that finds a set of subgraphs in $\mathcal{F}$ such that its deletion leaves a graph satisfying the conditions of Lemma 1.

Algorithm APPROX-ASSO(G)
INPUT: a graph G.
OUTPUT: a set X of vertices such that G − X is a cluster.

1. **if** G is a cluster graph **return** $\emptyset$;
2. **if** G is not connected **then**
        return $\bigcup$\{APPROX-ASSO(C): C is a component of G\};
3. **if** there exists X such that $G[X] \in \mathcal{F}$ **then**
        return APPROX-ASSO(G − X)∪X;
4. **if** G consists of two intersecting maximal cliques C_1 and C_2 **then**
        return the smaller of $C_1 \cap C_2$, $C_1 \setminus C_2$, and $C_2 \setminus C_1$;
5. find all maximal strong modules of G and build $\widetilde{Q}(G)$;
6. **if** a maximal strong module M is not a clique **then**
        return APPROX-ASSO(G − $N_G[M]$)∪$N_G(M)$;
7. define a weighted graph Q, and call the algorithm of [21] with it;
        return vertices of G corresponding to its output.

Fig. 2. Outline of the approximation algorithm for association set.

Toward this end a particular obstacle is the C_3-free condition in the second case of Lemma 1. Indeed, the detection of triangles in linear time is a notorious open problem that we are not able to solve. Therefore, we may have to abandon the simple "search and remove" approach. The first idea here is that we may dispose of *all* triangles of $\widetilde{Q}(G)$ in $O(mn)$ time. This is, however, still not sufficient, because after deleting a set X of some vertices, its maximal strong modules change, and more importantly, $\widetilde{Q}(G - X)$ may *not* be an (induced) subgraph of $\widetilde{Q}(G)$; see, e.g., Fig. 3. Our observation is that $\widetilde{Q}(G - X)$ is either a clique, an independent set, or an induced subgraph of $\widetilde{Q}(G[M])$ for some (not necessarily maximal) strong module M of G.

We start from recalling some simple facts about modular decomposition. For each maximal strong module M of G, we can further take the maximal strong modules and the quotient graph $\widetilde{Q}(G[M])$. This process can be recursively applied until every module consists of a single vertex. If we represent each module used in this process as a node, and add edges connecting every M with all maximal strong modules of $G[M]$, we obtain a tree rooted at $V(G)$, called the *modular decomposition tree* of G. The nodes of the modular decomposition tree are precisely all strong modules of G, where the leaves are all singleton vertex sets, and for every non-leaf node M, its children are the maximal strong modules of $G[M]$ [12]. It is known that the modular decomposition tree can be constructed in linear time [19].

Proposition 5. *If $\widetilde{Q}(G)$ is prime, then every nontrivial quotient graph of G contains $\widetilde{Q}(G)$ as an induced subgraph.*

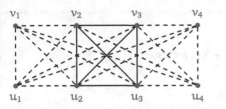
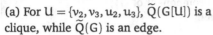
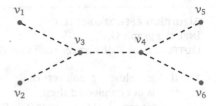

(a) For $U = \{v_2, v_3, u_2, u_3\}$, $\widetilde{Q}(G[U])$ is a clique, while $\widetilde{Q}(G)$ is an edge.

(b) For $U = \{v_1, v_2, v_5, v_6\}$, $\widetilde{Q}(G[U])$ is an independent set, while $\widetilde{Q}(G)$ is a P_4.

Fig. 3. $\widetilde{Q}(G[U])$ may not be an induced subgraph of $\widetilde{Q}(G)$.

On the one hand, since $V(G)$ itself is a strong module of G, every vertex set $U \subseteq V(G)$ is contained in some strong module. On the other hand, since two strong modules are either disjoint or one containing the other, there is a unique one that is inclusion-wise minimal of all strong modules containing U.

Theorem 4. *Let $U \subseteq V(G)$ be a subset of vertices of G, and let M be the inclusion-wise minimal strong module of G that contains U. If $\widetilde{Q}(G[U])$ is prime, then it is a subgraph of $\widetilde{Q}(G[M])$ induced by those maximal strong modules of $G[M]$ that intersect U.*

We remark that if $\widetilde{Q}(G[U])$ is a clique or independent set, then it is not necessarily an induced subgraph of $\widetilde{Q}(G[M])$; see, e.g., Fig. 3.

We are now ready to present the efficient implementation for the first phase, which would replace the first three steps of algorithm APPROX-ASSO (Fig. 2).

Lemma 2. *In $O(mn)$ time we can find a set $\mathcal{H}$ of disjoint induced subgraphs of G such that each $H \in \mathcal{H}$ is in $\mathcal{F}$ and $G - \bigcup_{H \in \mathcal{H}} V(H)$ satisfies the conditions of Lemma 1.*

Proof. We use the procedure described in Fig. 4. Step 0 is trivial. Step 1 uses the algorithm of McConnell and Spinrad [19], and step 2 uses simple enumeration, i.e., for each edge uv, we find all the common neighbors of u and v, which can be done in time $O(nm)$. This leads the disposal of triangles in step 3. During its progress, a maximal strong module M of the input graph G may not remain a maximal strong module of the current graph (i.e., $G - X$). But if M is not completely deleted (i.e., $M \not\subseteq X$), then its remnant (i.e., $M \setminus X$) is always a module of $G - X$.

Note that the three modules in each triangle must have the same parent in the modular decomposition tree. For each triangle $\{M_1, M_2, M_3\}$, we focus on their parent M (in the modular decomposition tree) and the subgraph $G[M]$ (step 3.1). All the modules mentioned in steps 3.2–3.7 are maximal strong modules of subgraph $G[M]$; they correspond to $V(Q)$. If either of the conditions of steps 3.2 and 3.3 is true, then the triangle has been disposed of and we can continue to the next one. If the deletion of vertices in previous iterations has

Procedure REDUCE(G)

INPUT: a graph G.

OUTPUT: a set of vertices specified in Lemma 2.

0. $X \leftarrow \emptyset$; if G is a clique **then return** $\emptyset$;
1. build the modular decomposition tree of G;
2. find all triangles $\mathcal{C}$ of $\tilde{Q}(G[M])$ for all strong modules M of G;
3. **while** $\mathcal{C}$ is nonempty **do**
3.0.   take a triangle $\{M_1, M_2, M_3\}$ from $\mathcal{C}$;
3.1.   let M be the parent of $\{M_1, M_2, M_3\}$ and let $Q = \tilde{Q}(G[M])$;
3.2.   if there is $1 \leqslant i \leqslant 3$ such that $M_i \subseteq X$ **then**
           delete $\{M_1, M_2, M_3\}$ from $\mathcal{C}$; **continue;**
3.3.   if there are $1 \leqslant i < j \leqslant 3$ such that M_i, M_j were merged **then**
           delete $\{M_1, M_2, M_3\}$ from $\mathcal{C}$; **continue;**
3.4.   if there are $1 \leqslant i < j \leqslant 3$ such that $N_Q[M_i] = N_Q[M_j]$ **then**
           merge M_i and M_j into a single module;
           delete $\{M_1, M_2, M_3\}$ from $\mathcal{C}$; **continue;**
3.5.   find $M' \in V(Q)$ that is adjacent to only one of M_1 and M_2;
           \\ *Assume that M' is adjacent to M_2 but not M_1.*
3.6.   if M' is adjacent to M_3 **then**
           find $M'' \in V(Q)$ adjacent to only one of M_2 and M_3;
           find gems or darts or C_4 and move their vertices to X;
3.7.   **else**   \\M' *is nonadjacent to M_3.*
           find $M'' \in V(Q)$ adjacent to only one of M_1 and M_3;
           find darts or bulls or C_4 or gem and move their vertices to X;
4. **repeat** until X is unchanged **do**
4.0. $G \leftarrow G - X$;   \\ *All modules below are maximal strong modules of G.*
4.1.   if G is not connected **then**
           return $\bigcup\{$REDUCE(C): C is a component of G$\} \cup X$;
4.2.   if G is a clique or consists of two intersecting cliques **then return** X;
4.3.   let Q denote $\tilde{Q}(G)$;   \\Q *is a clique or C_3-free.*
4.4.   if two non-clique modules are adjacent in Q **then**
           find a C_4 of G and move its vertices to X;
4.5.   **else if** Q is a clique but not an edge **then**
4.5.0.   let M be the only nontrivial module; *It exists because G is not a clique.*
4.5.1.   if G[M] has a P_4 or C_4 **then**
               find a gem or C_4 of G and move its vertices to X;
4.5.2.   **else if** G[M] has two components **then**
               find a dart of G and move its vertices to X;
4.5.3.   **else** find a fox of G and move its vertices to X;
4.6.   **else if** a module M is not a cluster **then**
4.6.1.   if G[M] has a P_4 or C_4 **then**
               find a gem or C_4 of G and move its vertices to X;
4.6.2.   **else if** G[M] is not connected **then**
               find a dart of G and move its vertices to X;
4.6.3.   **else** find a dart of G and add its vertices to X;   \\Q *is not a clique.*
4.7.   **else if** a non-clique module M has two neighbors M_1, M_2 **then**
           find a C_4 of G and move its vertices to X;
4.8.   **else if** a nontrivial module M has neighbors M_1, M_2, M_3 **then**
           find a fox of G and move its vertices to X;
5. **return** X.

Fig. 4. Procedure for the first phase.

made $N_Q[M_i] = N_Q[M_j]$ for some $1 \leq i < j \leq 3$, then $M_i \cup M_j$ is a module of $G[M \setminus X]$. Note that after they are merged, both M_i and M_j refer to the new module. Now that the procedure has passed steps 3.2–3.4, for each $1 \leq i < j \leq 3$, we can find a module adjacent to only one of M_i and M_j. This justifies step 3.5, and we may assume without loss of generality that the module M' is adjacent to M_2 but not M_1; the other case can be dealt with a symmetric way, which is omitted. In step 3.6, depending on the adjacency between M'' and M_1, M', we are in one of the following three cases: – if M'' is adjacent to neither of M_1, M', then there is a dart; – if M'' is adjacent to precisely one of M_1, M', then there is a gem; or – otherwise (M'' is adjacent to both of M_1, M'), there is a C_4;

This forbidden subgraph can be constructed by taking one vertex from each of M_1, M_2, M_3, M', and M''. We can actually find $\min\{|M_1|, |M_2|, |M_3|, |M'|, |M''|\}$ number of gems or darts, or $\min\{|M_1|, |M_3|, |M'|, |M''|\}$ number of C_4's, which we all move into X. It is similar for step 3.7.

After step 3, G might become disconnected. Then we work on its components one by one. Steps 4.1 and 4.2 are simple. The fact that the quotient graph Q built in step 4.3 is either a clique or is C_3-free can be argued using Theorem 4. Suppose for contradiction that Q is not a clique but contains a C_3; by Proposition 1, Q is prime. Then by Theorem 4, Q is a subgraph of $\widetilde{Q}(G[M])$ for some strong module M of G. Let $\{M_1, M_2, M_3\}$ be the triangle of $\widetilde{Q}(G[M])$ corresponding to a triangle in Q. But in step 3, either one of $\{M_1, M_2, M_3\}$ has been completely put into X, or two of them have been merged (then unless Q is a clique or an independent set, they will always be in the same maximal strong module). Therefore, Q must be C_3-free if it is not a clique.

Note that the algorithm enters at most one of steps 4.4–4.8. The correctness of step 4.4 is clear. If Q is a clique and it passes step 4.4, then all but one maximal strong module are trivial: Recall that each universal vertex is a maximal strong module. A P_4 of M together with a vertex in $N_G(M)$ makes a gem (4.5.1). Now that $G[M]$ is $\{P_4, C_4\}$-free, and has no universal vertex (a universal vertex of $G[M]$ is a universal vertex of G as well), according to Proposition 3, $G[M]$ is disconnected. Since step 4.2 does not apply, in step 4.5.2, at least one component is not a clique, and has a P_3, which, together with a vertex $u \in N_G(M)$ and any vertex from another component of $G[M]$, makes a dart. Otherwise (step 4.5.3), $G[M]$ has at least three components, and we can find a fox by taking three vertices from different components of $G[M]$ and two vertices from $N_G(M)$: Recall that when it enters step 4.5, G must have at least two universal vertices. In step 4.6, Q is not a clique, and assume that there is a module M that is not a cluster, then it must be $\{P_4, C_4\}$-free since the same reason as step 4.5.1. Now that if M is not connected, then a P_3 can be found in some component, which, together with some vertex in some other component of M and a vertex in $N_G(M)$, forms a dart. Therefore, M is connected and contains a P_3. It is sure that $N_G(N[M])$ is not empty since Q is not a clique, thus a dart can be found: Recall that if there are only two modules then $G[M]$ cannot have universal vertices. After step 4.6, Q is always C_3-free. Therefore, modules M_1, M_2 in step 4.7 and modules M_1, M_2, M_3 in step 4.8 are (pairwise) nonadjacent.

If G consists of two intersecting cliques, the procedure returns at step 4.2. Hence we may assume that it is not the case. The quotient graph $\tilde{Q}(G)$ is C_3-free because Theorem 4 and the algorithm has passed step 4.5. Conditions (1) and (2) of Lemma 1 follow from the correctness argument for steps 4.6–4.8. We now calculate the running time of the procedure. Note that the total number of edges of the subgraphs induced the strong modules of G is upper bounded by m. Thus, all the triangles can be listed in $O(mn)$ time in step 2. Each iteration of step 3 takes $O(m)$ time, and it decreases the order of Q by at least one, and thus step 3 takes $O(mn)$ time in total. Each iteration of step 4 takes $O(m)$ time, and it decreases the order of G by at least one, and hence step 4 takes $O(mn)$ time in total. This concludes the proof of this lemma.   □

References

1. Ailon, N., Charikar, M., Newman, A.: Aggregating inconsistent information: ranking and clustering. J. ACM **55**(5), (Article 23) 1–27 (2008). doi:10.1145/1411509. 1411513. A preliminary version appeared in STOC 2005
2. Bansal, N., Blum, A., Chawla, S.: Correlation clustering. Mach. Learn. **56**(1), 89–113 (2004). doi:10.1023/B:MACH.0000033116.57574.95. A preliminary version appeared in FOCS 2002
3. Ben-Dor, A., Shamir, R., Yakhini, Z.: Clustering gene expression patterns. J. Comput. Biol. **6**(3/4), 281–297 (1999). doi:10.1089/106652799318274
4. Boral, A., Cygan, M., Kociumaka, T., Pilipczuk, M.: A fast branching algorithm for cluster vertex deletion. Theory Comput. Syst. **58**(2), 357–376 (2016). doi:10.1007/s00224-015-9631-7
5. Bruhn, H., Chopin, M., Joos, F., Schaudt, O.: Structural parameterizations for boxicity. Algorithmica **74**(4), 1453–1472 (2016). doi:10.1007/s00453-015-0011-0. A preliminary version appeared in WG 2014
6. Cai, L.: Fixed-parameter tractability of graph modification problems for hereditary properties. Inf. Process. Lett. **58**(4), 171–176 (1996). doi:10.1016/0020-0190(96)00050-6
7. Cao, Y., Chen, J.: Cluster editing: kernelization based on edge cuts. Algorithmica **64**(1), 152–169 (2012). doi:10.1007/s00453-011-9595-1
8. Charikar, M., Guruswami, V., Wirth, A.: Clustering with qualitative information. J. Comput. Syst. Sci. **71**(3), 360–383 (2005). doi:10.1016/j.jcss.2004.10.012. A preliminary version appeared in FOCS 2003
9. Chopin, M., Nichterlein, A., Niedermeier, R., Weller, M.: Constant thresholds can make target set selection tractable. Theory Comput. Syst. **55**(1), 61–83 (2014). doi:10.1007/s00224-013-9499-3
10. Doucha, M., Kratochvíl, J.: Cluster vertex deletion: a parameterization between vertex cover and clique-width. In: Rovan, B., Sassone, V., Widmayer, P. (eds.) MFCS 2012. LNCS, vol. 7464, pp. 348–359. Springer, Heidelberg (2012)
11. Fiorini, S., Joret, G., Schaudt, O.: Improved approximation algorithms for hitting 3-vertex paths. In: Louveaux, Q., Skutella, M. (eds.) IPCO 2016. LNCS, vol. 9682, pp. 238–249. Springer, Heidelberg (2016). doi:10.1007/978-3-319-33461-5_20
12. Gallai, T.: Transitiv orientierbare graphen. Acta Mathematica Academiae Scientiarum Hungaricae **18**, 25–66 (1967). (Trans: Maffray, F., Preissmann, M.: Perfect Graphs. In: Ramírez-Alfonsín, J.L., Reed, B.A. (eds.), pp. 25–66. Wiley (2001). doi:10.1007/BF02020961

13. Habib, M., Paul, C.: A survey of the algorithmic aspects of modular decomposition. Comput. Sci. Rev. **4**(1), 41–59 (2010). doi:10.1016/j.cosrev.2010.01.001
14. Heggernes, P., Kratsch, D.: Linear-time certifying recognition algorithms and forbidden induced subgraphs. Nord. J. Comput. **14**(1–2), 87–108 (2007)
15. Hüffner, F., Komusiewicz, C., Moser, H., Niedermeier, R.: Fixed-parameter algorithms for cluster vertex deletion. Theory Comput. Syst. **47**(1), 196–217 (2010). doi:10.1007/s00224-008-9150-x
16. Lewis, J.M., Yannakakis, M.: The node-deletion problem for hereditary properties is NP-complete. J. Comput. Syst. Sci. **20**(2), 219–230 (1980). doi:10.1016/0022-0000(80)90060-4. Preliminary versions independently presented in STOC 1978
17. Liu, Y., Wang, J., You, J., Chen, J., Cao, Y.: Edge deletion problems: branching facilitated by modular decomposition. Theoret. Comput. Sci. **573**, 63–70 (2015). doi:10.1016/j.tcs.2015.01.049
18. Lund, C., Yannakakis, M.: The approximation of maximum subgraph problems. In: Lingas, A., Karlsson, R.G., Carlsson, S. (eds.) Automata, Languages, Programming (ICALP). LNCS, vol. 700, pp. 40–51. Springer, Heidelberg (1993). doi:10.1007/3-540-56939-1_60
19. McConnell, R.M., Spinrad, J.P.: Modular decomposition, transitive orientation. Discrete Math. **201**(1–3), 189–241 (1999). doi:10.1016/S0012-365X(98)00319-7. Preliminary versions appeared in SODA 1994 and SODA 1997
20. Sumner, D.P.: Graphs indecomposable with respect to the X-join. Discrete Math. **6**(3), 281–298 (1973). doi:10.1016/0012-365X(73)90100-3
21. Tu, J., Zhou, W.: A factor 2 approximation algorithm for the vertex cover P_3 problem. Inf. Process. Lett. **111**(14), 683–686 (2011). doi:10.1016/j.ipl.2011.04.009
22. Tu, J., Zhou, W.: A primal-dual approximation algorithm for the vertex cover P_3 problem. Theoret. Comput. Sci. **412**(50), 7044–7048 (2011). doi:10.1016/j.tcs.2011.09.013
23. Wolk, E.S.: The comparability graph of a tree. Proc. Am. Math. Soc. **13**, 789–795 (1962). doi:10.1090/S0002-9939-1962-0172273-0
24. Yan, J.-H., Chen, J.-J., Chang, G.J.: Quasi-threshold graphs. Discrete Appl. Math. **69**(3), 247–255 (1996). doi:10.1016/0166-218X(96)00094-7
25. Yannakakis, M.: Node-deletion problems on bipartite graphs. SIAM J. Comput. **10**(2), 310–327 (1981). doi:10.1137/0210022

Geodetic Convexity Parameters for Graphs with Few Short Induced Paths

Mitre C. Dourado[1], Lucia D. Penso[2], and Dieter Rautenbach[2(✉)]

[1] Departamento de Ciência da Computação, Instituto de Matemática,
Universidade Federal do Rio de Janeiro, Rio de Janeiro, Brazil
mitre@dcc.ufrj.br

[2] Institute of Optimization and Operations Research, Ulm University, Ulm, Germany
{lucia.penso,dieter.rautenbach}@uni-ulm.de

Abstract. We study several parameters of geodetic convexity for graph classes defined by restrictions concerning short induced paths. Partially answering a question posed by Araujo et al., we show that computing the geodetic hull number of a given P_9-free graph is NP-hard. Similarly, we show that computing the geodetic interval number of a given P_5-free graph is NP-hard. On the positive side, we identify several graph classes for which the geodetic hull number can be computed efficiently. Furthermore, following a suggestion of Campos et al., we show that the geodetic interval number, the geodetic convexity number, the geodetic Carathéodory number, and the geodetic Radon number can all be computed in polynomial time for $(q, q - 4)$-graphs.

Keywords: Geodetic convexity · Hull number · Geodetic number · Interval number · Convexity number · Carathéodory number · Radon number · P_k-free graphs · $(q, q - 4)$-graphs

1 Introduction

In the present paper we study five prominent graph parameters of geodetic convexity, the hull number, the interval number, the convexity number, the Carathéodory number, and the Radon number, for graph classes defined by restrictions concerning short induced paths. Our motivation mainly comes from two recent papers. In [7] Campos, Sampaio, Silva, and Szwarcfiter show that for the P_3-convexity, the above parameters can be determined in linear time for $(q, q - 4)$-graphs. In their conclusion they suggest to consider the geodetic versions of the parameters for these graphs. In [3] Araujo, Morel, Sampaio, Soares, and Weber study the geodetic hull number of P_5-free graphs. They show that this number can be computed in polynomial time for triangle-free P_5-free graphs, and ask about the computational complexity of the geodetic hull number of P_k-free graphs in general.

Before we discuss further related work and our own contribution, we collect some relevant definitions. All graphs will be finite, simple, and undirected, and

© Springer-Verlag GmbH Germany 2016
P. Heggernes (Ed.): WG 2016, LNCS 9941, pp. 25–37, 2016.
DOI: 10.1007/978-3-662-53536-3_3

we use standard terminology and notation. A graph G is *F-free* for some graph F if G does not contain an induced subgraph that is isomorphic to F. For a positive integer n, let K_n, P_n, $K_{1,n-1}$, and C_n be the complete graph, the path, the star, and the cycle of order n, respectively. For an integer q at least 4, a graph G is a $(q, q-4)$-*graph* [5] if every set of q vertices of G induces at most $q-4$ distinct P_4s. The *clique number* $\omega(G)$ of a graph G is the maximum order of a clique in G, which is a set of pairwise adjacent vertices. The *independence number* $\alpha(G)$ of a graph G is the maximum order of an independent set in G, which is a set of pairwise non-adjacent vertices. A vertex of a graph is *simplicial* if its neighborhood is a clique. For an integer k, let $[k]$ be the set of all positive integers at most k.

For a set X of vertices of a graph G, the *interval* $I_G(X)$ of X in G is the set of vertices of G that contains X as well as all vertices of G that lie on shortest paths between vertices from X. If $I_G(X) = X$, then X is a *convex* set. The *hull* $H_G(X)$ of X in G is the smallest convex set that contains X. If $H_G(X) = V(G)$, then X is a *hull set* of G, and if $I_G(X) = V(G)$, then X is an *interval set* of G. The *hull number* $h(G)$ of G [24] is the smallest order of a hull set of G. Similarly, the *interval number* $i(G)$ of G, also known as the *geodetic number* [28], is the smallest order of an interval set of G. The *convexity number* $cx(G)$ of G [11] is the maximum cardinality of a convex set that is a proper subset of the vertex set of G. Inspired by a classical theorem of Carathéodory [6], the *Carathéodory number* $cth(G)$ of G [19] is the minimum integer k such that for every set X of vertices of G, and every vertex x in $H_G(X)$, there is a subset Y of X of order at most k such that x belongs to $H_G(Y)$. Similarly, inspired by a classical theorem by Radon [30], the *Radon number* $r(G)$ of G [13,14] is the minimum integer k such that for every set X of at least k vertices of G, there is a subset X_1 of X such that $H_G(X_1) \cap H_G(X \setminus X_1) \neq \emptyset$. A set A of vertices of G is *anti-Radon* if A has no subset A_1 with $H_G(A_1) \cap H_G(A \setminus A_1) \neq \emptyset$. It is easy to see that the Radon number is exactly one more than the maximum cardinality of an anti-Radon set. Note that a clique is anti-Radon.

For reduction arguments useful to prove Theorem 6 below, we consider a second kind of convexity. For a set X of vertices of a graph G, the *restricted interval* $I'_G(X)$ of X in G is the set of vertices of G that contains X as well as all vertices of G that lie on an induced P_3 between vertices from X, that is, we only consider shortest paths of order 3. This leads to a convexity that has recently been studied on its own right [4], and is different from the above-mentioned P_3-convexity. If $I'_G(X) = X$, then X is *restricted convex*. The *restricted hull* $H'_G(X)$ of X in G, a *restricted hull set* of G, a *restricted interval set* of G, the *restricted hull number* $h'(G)$ of G, the *restricted interval number* $i'(G)$ of G, the *restricted convexity number* $cx'(G)$ of G, the *restricted Carathéorody number* $cth'(G)$ of G, the *restricted Radon number* $r'(G)$ of G, and *restricted anti-Radon sets* are all defined in the obvious way. Note that $I'_G(X) \subseteq I_G(X)$, which implies $H'_G(X) \subseteq H_G(X)$. Hence, every anti-Radon set is a restricted anti-Radon set.

We briefly survey some known related results. The hull number is NP-hard for bipartite graphs [2] and even for partial cubes [1], but can be computed in polynomial time for cographs [12], $(q, q - 4)$-graphs [2], $\{C_3, P_5\}$-free graphs [3], distance-hereditary graphs [29], and chordal graphs [29]. Bounds on the hull number are given in [2, 15, 24]. The interval number is NP-hard for cobipartite graphs [22] and for chordal graphs as well as for chordal bipartite graphs [16], but can be computed in polynomial time for split graphs [16], proper interval graphs [23], block-cactus graphs [22], and monopolar chordal graphs [22]. The convexity number is NP-hard for bipartite graphs [17, 27]. Finally, also the Carathéodory number [19] as well as the Radon number [14] are NP-hard. Next to the geodetic convexity and the P_3-convexity, further well-studied graphs convexities are the induced paths convexity, also known as the monophonic convexity [18, 21, 25], the all paths convexity [8], the triangle path convexity [9, 10], and the convexity based on induced paths of order at least 4 [20].

Our contributions are as follows. Partially answering the question posed by Araujo et al. [3], we show that computing the hull number of a given P_ℓ-free graph is NP-hard for every $\ell \geq 9$. Similarly, we show that computing the interval number of a given P_ℓ-free graph is NP-hard for every $\ell \geq 5$. Furthermore, we extend the result of Araujo et al. [3] that the hull number can be computed in polynomial time for $\{C_3, P_5\}$-free graphs to $\{\text{paw}, P_5\}$-free graphs, to triangle-free graphs in which every six vertices induce at most one P_5, and to $\{C_3, \ldots, C_{k-2}, P_k\}$-free graphs for every integer k. Following the suggestion of Campos et al. [7], we show that the interval number, the convexity number, the Carathéodory number, and the Radon number as well as their restricted versions can all be computed in polynomial time for $(q, q - 4)$-graphs. Section 2 contains our complexity results. In Sect. 3 we present the efficiently solvable cases, and in Sect. 4 we list some open problems.

2   Complexity Results

Theorem 1. *For a given P_9-free graph G, and a given integer k, it is NP-complete to decide whether $h(G) \leq k$.*

Proof. Since the hull of a set of vertices can be computed in polynomial time, the considered decision problem belongs to NP. In order to prove NP-completeness, we describe a polynomial reduction from a restricted version of SATISFIABILITY. Therefore, let $\mathcal{C}$ be an instance of SATISFIABILITY consisting of m clauses $C_1, \ldots, C_m$ over n boolean variables $x_1, \ldots, x_n$ such that every clause in $\mathcal{C}$ contains at most three literals, and, for every variable x_i, there are at most three clauses in $\mathcal{C}$ that contain either x_i or $\bar{x}_i$. Note that SATISFIABILITY is still NP-complete for such instances (cf. [LO1] in [26]).

Clearly, we may assume that no clause in $\mathcal{C}$ contains a variable as well as its negation, and that $n \geq 2$. If, for some variable x_i, no clause in $\mathcal{C}$ contains $\bar{x}_i$, then setting x_i to true, and removing all clauses from $\mathcal{C}$ that contain x_i, leads to an equivalent instance. Therefore, by symmetry, we may assume that, for every variable x_i, some clause in $\mathcal{C}$ contains x_i, and some clause in $\mathcal{C}$ contains $\bar{x}_i$. If, for some variable x_i, there is only one clause in $\mathcal{C}$ containing x_i, and only one clause in $\mathcal{C}$ containing $\bar{x}_i$, then introducing two new variables x_{n+1} and x_{n+2}, and adding the three clauses $x_i \vee x_{n+1} \vee x_{n+2}$, $\bar{x}_{n+1} \vee x_{n+2}$, and $x_{n+1} \vee \bar{x}_{n+2}$, leads to an equivalent instance. Therefore, by symmetry, we may assume that, for every variable x_i, there are exactly three clauses in $\mathcal{C}$ that contain either x_i or $\bar{x}_i$. If, for some variable x_i, there is one clause in $\mathcal{C}$ containing x_i, and two clauses in $\mathcal{C}$ containing $\bar{x}_i$, then exchanging x_i with $\bar{x}_i$ within $\mathcal{C}$, leads to an equivalent instance. Altogether, we may assume that, for every variable x_i, there are exactly two clauses in $\mathcal{C}$, say $C_{j_i^{(1)}}$ and $C_{j_i^{(2)}}$, that contain x_i, and exactly one clause in $\mathcal{C}$, say $C_{j_i^{(3)}}$, that contains $\bar{x}_i$. Furthermore, these three clauses are distinct.

Let the graph G be constructed as follows starting with the empty graph:

- For every $j \in [m]$, add a vertex c_j.
- For every $i \in [n]$, add a copy G_i of the graph in Fig. 1, and denote the vertices as indicated in the figure.
- Add two further vertices w_1 and w_2.
- Add further edges to turn the set

$$C = \{c_j : j \in [m]\} \cup \{y_i : i \in [n]\} \cup \{v_i : i \in [n]\}$$

into a clique.
- For every i in $[n]$, add the three edges $x_i^{(1)}c_{j_i^{(1)}}$, $x_i^{(2)}c_{j_i^{(2)}}$, and $\bar{x}_i c_{j_i^{(3)}}$.
- For every i in $[n]$, add the two edges $v_i w_1$ and $v_i w_2$.

See Fig. 2 for a partial illustration.

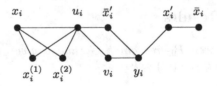

Fig. 1. The graph G_i.

Let $k = 2n + 2$. Note that the order of G is $9n + m + 2$. It remains to show that G is P_9-free, and that $\mathcal{C}$ is satisfiable if and only if $h(G) \leq k$.

Let P be an induced path in G. Since C is a clique, the subgraph $G[V(P) \cap C]$ of P induced by C is a (possibly empty) path of order at most 2. Note that all components of $G[V(G) \setminus C]$ have order at most 5, and only contain induced paths of order at most 3. This implies that P has order at most $3 + 2 + 3$, that is, G is P_9-free.

First, let $\mathcal{S}$ be a satisfying truth assignment for $\mathcal{C}$. Let

$$S = \{w_1, w_2\} \cup \bigcup_{i \in [n] : x_i \text{ true in } \mathcal{S}} \{x_i, x_i'\} \cup \bigcup_{i \in [n] : x_i \text{ false in } \mathcal{S}} \{\bar{x}_i, \bar{x}_i'\}.$$

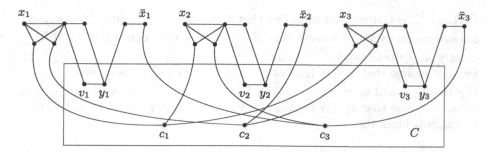

Fig. 2. Part of G, where $C_1 : x_1 \vee x_2 \vee x_3$, $C_2 : x_1 \vee \bar{x}_2 \vee x_3$, and $C_3 : \bar{x}_1 \vee x_2 \vee \bar{x}_3$. For the sake of visibility, the edges within C as well as the vertices w_1 and w_2 are not shown.

Clearly, $|S| = k$. Since $\{v_1, \ldots, v_n\} \subseteq H_G(\{w_1, w_2\})$, and $y_i \in H_G(\{x'_i, v_i\}) \cap H_G(\{\bar{x}'_i, v_i\})$ for $i \in [n]$, we obtain $\{v_1, \ldots, v_n\} \cup \{y_1, \ldots, y_n\} \subseteq H_G(S)$. If $i \in [n]$ is such that x_i is true in S, and $\ell \in [n] \setminus \{i\}$, then $u_i \in H_G(\{x_i, v_i\})$, $\bar{x}'_i \in H_G(\{y_i, u_i\})$, and $\left\{ c_{j_i^{(1)}}, c_{j_i^{(2)}}, x_i^{(1)}, x_i^{(2)} \right\} \subseteq H_G(\{x_i, v_\ell\})$. If $i \in [n]$ is such that x_i is false in S, then $u_i \in H_G(\{\bar{x}'_i, v_i\})$, $x'_i \in H_G(\{\bar{x}_i, y_i\})$, and $c_{j_i^{(3)}} \in H_G(\{\bar{x}_i, v_i\})$. Since S is a satisfying truth assignment, this implies $\{x'_1, \ldots, x'_n\} \cup \{\bar{x}'_1, \ldots, \bar{x}'_n\} \cup \{u_1, \ldots, u_n\} \cup \{c_1, \ldots, c_m\} \subseteq H_G(S)$. For $i \in [n]$, we have $\left\{ x_i^{(1)}, x_i^{(2)} \right\} \subseteq H_G \left(\left\{ u_i, c_{j_i^{(1)}}, c_{j_i^{(2)}} \right\} \right)$, $x_i \in H_G \left(\left\{ x_i^{(1)}, x_i^{(2)} \right\} \right)$, $\bar{x}_i \in H_G \left(\left\{ x'_i, c_{j_i^{(3)}} \right\} \right)$. This implies

$$\bigcup_{i \in [n]} \left\{ x_i, x_i^{(1)}, x_i^{(2)}, \bar{x}_i \right\} \subseteq H_G(S).$$

Altogether, it follows that S is a hull set, and, hence, $h(G) \leq |S| = k$.

Conversely, let S be a hull set of order at most $2n + 2$. Since w_1 and w_2 are simplicial, we have $w_1, w_2 \in S$. For $i \in [n]$, let

$$V_i^{(1)} = \left\{ x_i, x_i^{(1)}, x_i^{(2)}, u_i, \bar{x}'_i \right\}, \quad V_i^{(2)} = \{x'_i, \bar{x}_i\}, \text{ and } V_i^{(3)} = \{\bar{x}'_i, y_i, x'_i\}.$$

Since $N_G \left(V_i^{(1)} \right) \setminus V_i^{(1)} \subseteq C$, $N_G \left(V_i^{(2)} \right) \setminus V_i^{(2)} \subseteq C$, and C is a clique, the two sets $V(G) \setminus V_i^{(1)}$ and $V(G) \setminus V_i^{(2)}$ are convex, which implies that S intersects $V_i^{(1)}$ as well as $V_i^{(2)}$. Since $u_i v_i c_{j_i^{(3)}} \bar{x}_i$ is a path of order 4 between u_i and $\bar{x}_i$, no shortest path between two vertices in $V(G) \setminus V_i^{(3)}$ contains the two vertices $\bar{x}'_i$ and x'_i. Since $N_G(y_i) \setminus \{\bar{x}'_i, x'_i\} \subseteq C$, no shortest path between two vertices in $V(G) \setminus V_i^{(3)}$ intersects $V_i^{(3)}$ only in y_i. Since u_i has distance at most 2 from every vertex in C, no shortest path between two vertices in $V(G) \setminus V_i^{(3)}$ contains $\bar{x}'_i$ and y_i. Since $\bar{x}_i$ has distance at most 2 from every vertex in C, no shortest path between two vertices in $V(G) \setminus V_i^{(3)}$ contains y_i and x'_i. Altogether, it follows, that

$V(G) \setminus V_i^{(3)}$ is convex, which implies that S intersects $V_i^{(3)}$. Since $S \setminus \{w_1, w_2\}$ has order exactly $2n$, it follows that S contains exactly one vertex from $V_i^{(1)}$, exactly one vertex from $V_i^{(2)}$, and intersects $\{\bar{x}_i', x_i'\}$. Since $y_i, x_i' \in H_G(\{\bar{x}_i', \bar{x}_i\})$, we may assume that $\bar{x}_i' \in S$ implies $S \cap V(G_i) = \{\bar{x}_i, \bar{x}_i'\}$. Since $\{v_1, \ldots, v_n\} \subseteq H_G(\{w_1, w_2\})$ and $u_i, \bar{x}_i', y_i, x_i^{(1)}, x_i^{(2)} \in H_G(\{x_i, x_i', v_i, v_\ell\})$ for every $\ell \in [n] \setminus \{i\}$, we may assume that $x_i' \in S$ implies $S \cap V(G_i) = \{x_i, x_i'\}$. Altogether, for every $i \in [n]$, we obtain that

$$S \cap V(G_i) \in \{\{x_i, x_i'\}, \{\bar{x}_i, \bar{x}_i'\}\}. \tag{1}$$

Let $\mathcal{S}$ be the truth assignment, where we set x_i to be true exactly if $S \cap V(G_i) = \{x_i, x_i'\}$.

For $j \in [m]$, let

$$V_j = \{c_j\} \cup \bigcup_{i \in [n]: j = j_i^{(1)}} \left\{x_i, x_i^{(1)}\right\} \cup \bigcup_{i \in [n]: j = j_i^{(2)}} \left\{x_i, x_i^{(2)}\right\} \cup \bigcup_{i \in [n]: j = j_i^{(3)}} \{\bar{x}_i\}.$$

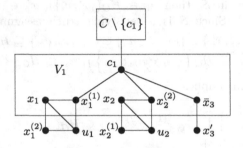

Fig. 3. The set V_1 for the clause $C_1 = x_1 \vee x_2 \vee \bar{x}_3$, where $j_1^{(1)} = j_2^{(2)} = 1$.

See Fig. 3 for an illustration. Note that $N_G(c_j) \setminus V_j = C \setminus \{c_j\}$. Furthermore, if $i \in [n]$ is such that $j = j_i^{(1)}$, then $N_G\left(\left\{x_i, x_i^{(1)}\right\}\right) \setminus V_j = \left\{u_i, x_i^{(2)}\right\}$, if $i \in [n]$ is such that $j = j_i^{(2)}$, then $N_G\left(\left\{x_i, x_i^{(2)}\right\}\right) \setminus V_j = \left\{u_i, x_i^{(1)}\right\}$, and, if $i \in [n]$ is such that $j = j_i^{(3)}$, then $N_G(\{\bar{x}_i\}) \setminus V_j = \{x_i'\}$. Since $C \setminus \{c_j\}$ is a clique, no shortest path between two vertices in $V(G) \setminus V_j$ intersects V_j only in c_j. If a shortest path P between two vertices in $V(G) \setminus V_j$ contains a vertex $x_i^{(r)}$ from V_j for some $r \in [2]$, then, possibly exchanging x_i with u_i on P, we may assume that P contains the vertex u_i. Since every two vertices in $\{u_1, \ldots, u_n\} \cup \{x_1', \ldots, x_n'\}$ have distance at most three, no shortest path between two vertices in $V(G) \setminus V_j$ contains two vertices from

$$V_j \cap \left(\left\{x_1^{(1)}, \ldots, x_n^{(1)}\right\} \cup \left\{x_1^{(2)}, \ldots, x_n^{(2)}\right\} \cup \{\bar{x}_1, \ldots, \bar{x}_n\}\right).$$

This implies that, since u_i has distance at most two from each vertex in $C \setminus \{c_j\}$ for every $i \in [n]$, no shortest path between two vertices in $V(G) \setminus V_j$ contains a vertex from

$$V_j \cap \left(\left\{ x_1^{(1)}, \ldots, x_n^{(1)} \right\} \cup \left\{ x_1^{(2)}, \ldots, x_n^{(2)} \right\} \right).$$

Similarly, since x_i' has distance at most two from each vertex in $C \setminus \{c_j\}$ for every $i \in [n]$, no shortest path between two vertices in $V(G) \setminus V_j$ contains a vertex from

$$V_j \cap \{ \bar{x}_1, \ldots, \bar{x}_n \}.$$

Altogether, it follows that $V(G) \setminus V_j$ is convex, which implies that S intersects

$$\bigcup_{i \in [n]: j = j_i^{(1)}} \left\{ x_i, x_i^{(1)} \right\} \cup \bigcup_{i \in [n]: j = j_i^{(2)}} \left\{ x_i, x_i^{(2)} \right\} \cup \bigcup_{i \in [n]: j = j_i^{(3)}} \{ \bar{x}_i \}$$

for every $j \in [m]$. By (1) and the definition of $\mathcal{S}$, this implies that $\mathcal{S}$ is a satisfying truth assignment for $\mathcal{C}$, which completes the proof. $\qquad\Box$

Theorem 2. *For a given P_5-free graph G, and a given integer k, it is NP-complete to decide whether $i(G) \leq k$.*

3 Efficiently Solvable Cases

As observed in the introduction, Araujo et al. [3] show that the hull number can be computed in polynomial time for $\{C_3, P_5\}$-free graphs. We extend their result in several ways.

The *paw* is the unique graph with degree sequence $1, 2, 2, 3$. Note that the paw arises by attaching an endvertex to one vertex of a triangle.

Theorem 3. *The hull number of a given $\{\text{paw}, P_5\}$-free graph can be computed in polynomial time.*

Proof. Let G be a $\{\text{paw}, P_5\}$-free graph. Clearly, we may assume that G is connected. If G is P_4-free, then G is a cograph, and the statement follows from [12]. Hence, we may assume that G contains an induced path $P : u_1 u_2 u_3 u_4$ of order 4. For a positive integer d, let $V_d = \{v \in V(G) : \text{dist}_G(v, V(P)) = d\}$, where $\text{dist}_G(v, V(P)) = \min\{\text{dist}_G(v, u) : u \in V(P)\}$. Let X be the union of $V(P)$ and the set of all simplicial vertices. We will show that adding at most one vertex to X yields a hull set of G, which implies that a minimum hull set can be found efficiently. In fact, every hull set contains all simplicial vertices, and considering the polynomially many extensions of the set of simplicial vertices by at most 5 vertices will yield a minimum hull set.

Let $v \in V_1$. First, we assume that v is adjacent to u_1. If v is adjacent to u_2, then, since G is paw-free, v is adjacent to all vertices of P. If v is not adjacent to u_2, then, since G is $\{\text{paw}, P_5\}$-free, $N_G(v) \cap V(P) \in \{\{u_1, u_3\}, \{u_1, u_4\}\}$.

Next, we assume that v is not adjacent to u_1 or u_4. Since G is paw-free, we obtain $N_G(v) \cap V(P) \in \{\{u_2\}, \{u_3\}\}$. Altogether, by symmetry, we obtain that $N_G(v) \cap V(P)$ is one of the sets $\{u_2\}$, $\{u_3\}$, $\{u_1, u_3\}$, $\{u_2, u_4\}$, $\{u_1, u_4\}$, and $V(P)$. Note that a vertex v in V_1 does not lie in $H_G(V(P)) \subseteq H_G(X)$ only if $N_G(v) \cap V(P) \in \{\{u_2\}, \{u_3\}\}$.

If $w \in V_2$, and v is a neighbor of w in V_1, then, since G is $\{\text{paw}, P_5\}$-free, $N_G(v) \cap V(P)$ is one of the sets $\{u_1, u_3\}$ and $\{u_2, u_4\}$. Since G is P_5-free, this implies $V_3 = \emptyset$, that is, $V(G) = V(P) \cup V_1 \cup V_2$. Suppose that w_1 and w_2 are adjacent vertices in V_2. Let $w_1 v u_i$ be a path between w_1 and $V(P)$. By symmetry, we may assume that $i < 4$. Since G is paw-free, the vertex w_2 is not adjacent to v. Now, $w_2 w_1 v u_i u_{i+1}$ is a P_5. Hence, V_2 is independent. Recall that every neighbor v in V_1 of a vertex in V_2 lies in $H_G(V(P))$. Therefore, every non-simplicial vertex in V_2 lies in $H_G(V(P)) \subseteq H_G(X)$, that is, $V(P) \cup V_2 \subseteq H_G(X)$.

Let V_1' be the set of non-simplicial vertices in V_1 that do not belong to $H_G(X)$. If $V_1' = \emptyset$, then $V_1 \subseteq H_G(X)$, and X is a hull set. Hence, we may assume that V_1' is not empty. If $A = \{v \in V_1' : N_G(v) \cap V(P) = \{u_2\}\}$, and $B = \{v \in V_1' : N_G(v) \cap V(P) = \{u_3\}\}$, then $V_1' = A \cup B$. Since G is paw-free, the sets A and B are independent. Since every vertex in V_1' is non-simplicial, it has two non-adjacent neighbors, at least one of which does not belong to $H_G(X)$. It follows that every vertex in A has a neighbor in B, and every vertex in B has a neighbor in A. Note that this implies that A and B are both not empty. Let H be the bipartite induced subgraph $G[A \cup B]$ of G with partite sets A and B. If $a_1, a_2 \in A$ and $b_1, b_2 \in B$ are such that $a_1 b_1, a_2 b_2 \in E(G)$ and $a_1 b_2, a_2 b_1 \notin E(G)$, then $b_1 a_1 u_2 a_2 b_2$ is a P_5. Hence, H is $2K_2$-free. Let a_1 be a vertex in A of maximum degree $d_H(a_1)$ in H. Suppose that a_1 is not adjacent to some vertex b_2 in B. Let a_2 be a neighbor of b_2 in A. Since $d_H(a_1) \geq d_H(a_2)$, there is a neighbor b_1 of a_1 in B that is not adjacent to a_2. Now, $a_1, a_2 \in A$ and $b_1, b_2 \in B$ are as above, which is a contradiction. Hence, $N_H(a_1) = B$, which implies that $B \subseteq H_G(\{a_1, u_3\}) \subseteq H_G(X \cup \{a_1\})$. Since $A \subseteq H_G(B \cup \{u_2\})$, it follows that $V_1' \subseteq H_G(X \cup \{a_1\})$, that is, $X \cup \{a_1\}$ is a hull set, which completes the proof.   □

We proceed to our next generalization of the result of Araujo et al. [3].

Theorem 4. *Let k be a fixed positive integer.*

For a given $\{C_i : 3 \leq i \leq k - 2\} \cup \{P_k\}$-free graph G, the hull number $h(G)$ can be computed in polynomial time.

Proof. The proof is by induction on k. For $k \leq 4$, the graph G is a cograph, and the statement follows from [12]. For $k = 5$, the statement follows from the result of Araujo et al. [3], or from Theorem 3. Now, let $k \geq 6$. The proof for $k = 6$ is similar to the proof of Theorem 3, and is given in the appendix. Hence, let $k \geq 7$.

Let G be a connected $\{C_i : 3 \leq i \leq k-2\} \cup \{P_k\}$-free graph. If G is P_{k-1}-free, then the result follows by induction. Hence, we may assume that G contains an induced path $P : u_1 \ldots u_{k-1}$ of order $k - 1$. Let X be the union of $V(P)$ and the set of all simplicial vertices. We will show that X is a hull set of G, which implies that a minimum hull set can be found efficiently. $\qquad\square$

Claim 1. *G has only cycles of orders $k - 1$ and k, and every cycle of G is induced.*

Proof. Suppose that G has a cycle of order at least $k + 1$. Let $C : x_0 \ldots x_{\ell-1} x_0$ be a shortest cycle in G of order ℓ at least $k + 1$. Since G is P_k-free, the cycle C is not induced. Let $x_i x_j$ be an edge such that $j - i = \text{dist}_C(x_i, x_j)$ is minimum. By symmetry, we may assume that $i = 0$. Since $x_0 x_1 \ldots x_j x_0$ is an induced cycle of order $j + 1$, we obtain $j \geq k - 2$. Since $x_0 x_1 \ldots x_{j-1}$ is an induced path of order j, we obtain $j \leq k - 1$, that is, $j \in \{k - 2, k - 1\}$. First, we assume that $j = k - 1$. Since the path $x_1 x_2 \ldots x_k$ is not induced, the choice of $x_i x_j$ implies that $x_1 x_k$ is an edge. Now, $x_1 x_k x_{k-1} x_0 x_1$ is a cycle of order 4, which is a contradiction. Hence $j = k - 2$. Since the path $x_1 x_2 \ldots x_k$ is not induced, the choice of $x_i x_j$ implies that there is an edge between $\{x_1, x_2\}$ and $\{x_{k-1}, x_k\}$. Since G is $\{C_4, C_5\}$-free, we obtain that $x_2 x_k$ is an edge. Since $x_0 x_1 x_2 x_k x_{k-1} x_{k-2} x_0$ is an induced cycle of order 6, we obtain $k = 7$, which implies $j = 5$ and $\ell \geq 8$. Since $x_0 x_5 x_4 x_3 x_2 x_7 x_8 \ldots x_0$ is a cycle of order $\ell - 2$, the choice of C implies $\ell \leq 9$. Now, $x_0 x_5 x_6 \ldots x_{\ell-1} x_0$ is a cycle of order $\ell - 4 \leq 5$, which is a contradiction. Hence, G has no cycle of order at least $k + 1$. In view of the forbidden induced subgraphs, this implies that G has only cycles of orders $k - 1$ or k, and that every cycle of G is induced. $\qquad\square$

For a positive integer d, let $V_d = \{v \in V(G) : \text{dist}_G(v, V(P)) = d\}$.

Claim 2. *$V_d \neq \emptyset$ implies $d < \lfloor \frac{k}{2} \rfloor - 1$.*

Proof. Suppose that V_d is non-empty for some $d \geq \lfloor \frac{k}{2} \rfloor$. Let $x_0 \ldots x_d$ be a shortest path between a vertex x_0 in V_d and some vertex x_d of P. By symmetry, we may assume that $x_d = u_i$, where $i \geq \lceil \frac{k}{2} \rceil$. If x_{d-1} has a neighbor in $\{u_j : i - \lceil \frac{k-2}{2} \rceil \leq j \leq i - 1\}$, then G has a cycle of order at most $\lceil \frac{k-2}{2} \rceil + 2 \leq k - 2$, which is a contradiction. Hence, $x_0 \ldots x_d u_{i-1} u_{i-2} \ldots u_{i-\lceil \frac{k-2}{2} \rceil}$ is an induced path of order $d + 1 + \lceil \frac{k-2}{2} \rceil \geq \lfloor \frac{k}{2} \rfloor + 1 + \lceil \frac{k-2}{2} \rceil = k$, which is a contradiction. $\qquad\square$

Claim 3. *For every d at least 2, every vertex in V_d has exactly one neighbor in V_{d-1}.*

Proof. Suppose that for some d at least 2, some vertex in V_d has two neighbors in V_{d-1}. This implies the existence of two distinct paths $Q : x_0 x_1 \ldots x_d$ and $Q' : x_0 x_1' \ldots x_d'$ between some vertex x_0 in V_d and vertices x_d and x_d' of P. Since $2d \leq k - 2$, we obtain that $V(Q) \cap V(Q') = \{x_0\}$. Let $x_d = u_i$ and $x_d' = u_j$, where $j > i$. The union of Q, Q', and the path $u_i \ldots u_j$ is a cycle C. First, we assume that C has order k. Recall that, by Claim 1, all cycles of G are induced. Since $d \geq 2$, we obtain $j - i = k - 2d \leq k - 4$. Hence, since P has order $k - 1$, we may assume that $i \geq 2$. Since the path $u_{i-1} \ldots u_j x_{d-1}' \ldots x_1' x_0 x_1 \ldots x_{d-2}$ of order k

is not induced, we obtain that u_{i-1} is adjacent to x'_{d-1}. By Claim 1, this implies that $j - i \leq 2$. Now, $u_{i-1} \ldots u_j x'_{d-1} u_{i-1}$ is a cycle of order at most 5, which is a contradiction. Hence, by Claim 1, the order of C is $k - 1$. Since $d \geq 2$, we obtain $j - i \leq k - 5$. Hence, since P has order $k - 1$, we may assume that $i \geq 3$. Similarly as above, we obtain that x'_{d-1} is adjacent to u_{i-1} or u_{i-2}. If x'_{d-1} is adjacent to u_{i-1}, then $j - i = 1$, and $u_{i-1} u_i u_j x'_{d-1} u_{i-1}$ is a cycle of order 4, which is a contradiction. Hence, x'_{d-1} is adjacent to u_{i-2}. By Claim 1, this implies that $j - i \leq 2$. Similarly as above, if $j - i = 1$, then G contains a cycle of order 5, which is a contradiction. Hence $j - i = 2$, and $u_{i-2} \ldots u_j x'_{d-1} u_{i-2}$ is a cycle of order 6, which implies that $k = 7$ and $d = 2$. If $j \leq 5$, then $u_{i-1} x_2 x_1 x_0 x'_1 x'_2 u_{j+1}$ is an induced path of order 7, which is a contradiction. Hence $j = 6$, which implies that $i = j - 2 = 4$. Now, $u_1 u_2 x'_1 x_0 x_1 u_4 u_5$ is an induced path of order 7, which is a contradiction. $\qquad \square$

Claim 4. *For every d at least 2, the set V_d is independent.*

Proof. Suppose that for some d at least 2, the set V_d is not independent. This implies the existence of two distinct paths $Q : x_0 x_1 \ldots x_d$ and $Q' : x'_0 x'_1 \ldots x'_d$ between two adjacent vertices x_0 and x'_0 in V_d and vertices x_d and x'_d of P. If $V(Q) \cap V(Q') \neq \emptyset$, then Claims 1 and 2 imply that $x_d = x'_d$, $V(Q) \cap V(Q') = \{x_d\}$, and $d = \frac{k}{2} - 1$. By symmetry, we may assume that $x_d = u_i$, where $i \geq 3$. Now, $u_{i-2} u_{i-1} x_d x_{d-1} \ldots x_1 x_0 x'_0 x'_1 \ldots x'_{d-2}$ is an induced path of order k, which is a contradiction. Hence, $V(Q) \cap V(Q') = \emptyset$. Let $x_d = u_i$ and $x'_d = u_j$, where $j > i$. The union of Q, Q', the edge $x_0 x'_0$, and the path $u_i \ldots u_j$ is a cycle C. First, we assume that C has order k. Since $d \geq 2$, we obtain $j - i = k - 2d - 1 \leq k - 5$. Hence, since P has order $k - 1$, we may assume that $i \geq 2$. Since the path $u_{i-1} \ldots u_j x'_{d-1} \ldots x'_1 x'_0 x_0 x_1 \ldots x_{d-2}$ of order k is not induced, we obtain that u_{i-1} is adjacent to x'_{d-1}. By Claim 1, this implies that $j - i \leq 2$. Now, $u_{i-1} \ldots u_j x'_{d-1} u_{i-1}$ is a cycle of order at most 5, which is a contradiction. Hence, by Claim 1, the order of C is $k - 1$. Since $d \geq 2$, we obtain $j - i = k - 1 - 2d - 1 \leq k - 6$. Hence, since P has order $k - 1$, we may assume that $i \geq 3$. Similarly as in the proof of Claim 3, we obtain that x'_{d-1} is adjacent to u_{i-2}. By Claim 1, this implies that $j - i \leq 2$. Similarly as above, if $j - i = 1$, then G contains a cycle of order 5, which is a contradiction. Hence $j - i = 2$, and $u_{i-2} \ldots u_j x'_{d-1} u_{i-2}$ is a cycle of order 6, which implies that $k = 7$. Now, $j - i \leq k - 6 = 1$, which is a contradiction. $\qquad \square$

Recall that X is the union of $V(P)$ and the set of all simplicial vertices.

Let $u \in V_d$ for some $d \geq 2$. Let $d' \geq d$ and $u' \in V_{d'}$ be such that u lies on a shortest path between u' and some vertex of P, and d' is maximum. Note that $d' = d$ and $u' = u$ is allowed. Since u' has no neighbor in $V_{d'+1}$, Claims 3 and 4 imply that u' is simplicial, and, hence, $u \in H_G(\{u'\} \cup V(P)) \subseteq H_G(X)$.

Let $u \in V_1 \setminus H_G(X)$. It follows that u does not have two neighbors on P, and also no neighbor in V_2. Since u is not simplicial, this implies that u has exactly one neighbor u_i on P as well as some neighbor u' in V_1. Let u_j be a neighbor of u' on P. Clearly, $i \neq j$. Note that $u_i u u' u_j$ is a path of order 4. Hence, since $u \notin H_G(V(P))$, the distance in G between u_i and u_j is at most 2, which implies the contradiction that G contains a cycle of order at most 5.

It follows that X is a hull set of G, which completes the proof.   □

We present another generalization of the result of Araujo et al. [3], and consider triangle-free graphs in which every six vertices induce at most one P_5. Obviously, these graphs have been inspired by the $(q, q - 4)$-graphs [5], and, consequently, we refer to them as $(6,1)$-*graphs*.

Theorem 5. *If G is a connected triangle-free $(6,1)$-graph, then either G is P_5-free or G arises from a star $K_{1,p}$ with $p \geq 2$ by subdividing two edges once and all remaining edges at most once.*

Corollary 1. *The hull number of a given triangle-free $(6,1)$-graph can be computed in polynomial time.*

Using the structural properties of $(q, q - 4)$-graphs [5,7], and establishing suitable recursions for the convexity parameters, we obtain our final result.

Theorem 6. *Let q be a fixed integer at least 4.*
For a given $(q, q-4)$-graph G, all parameters $h(G)$, $h'(G)$, $i(G)$, $i'(G)$, $cx(G)$, $cx'(G)$, $cth(G)$, $cth'(G)$, $r(G)$, and $r'(G)$ can be computed in polynomial time.

4 Conclusion

We conclude with a number of questions. Can the considered parameters be determined in linear time for $(q, q - 4)$-graphs? What is the complexity of the hull number for P_k-free graphs for $k \in \{5, 6, 7, 8\}$? What is the complexity of the other convexity parameters for $\{C_i : 3 \leq i \leq k - 2\} \cup \{P_k\}$-free graphs or (triangle-free) $(6, 1)$-graphs?

For an integer q at least 5, let the $(q, q-5)$-graphs be those graphs in which every q vertices induce at most $q-5$ distinct P_5s. Do the (triangle-free) $(q, q-5)$-graphs allow a similar decomposition as the $(q, q - 4)$-graphs [5,7]? Do these graphs have nice structural features?

References

1. Albenque, M., Knauer, K.: Convexity in partial cubes: the hull number. In: Pardo, A., Viola, A. (eds.) LATIN 2014. LNCS, vol. 8392, pp. 421–432. Springer, Heidelberg (2014)
2. Araujo, J., Campos, V., Giroire, F., Nisse, N., Sampaio, L., Soares, R.: On the hull number of some graph classes. Theoret. Comput. Sci. **475**, 1–12 (2013)
3. Araujo, J., Morel, G., Sampaio, L., Soares, R., Weber, V.: Hull number: P_5-free graphs and reduction rules. Discrete Appl. Math. **210**, 171–175 (2016)
4. Araujo, J., Sampaio, R., Santos, V., Szwarcfiter, J.L.: The convexity of induced paths of order three, applications: complexity aspects, manuscript
5. Babel, L., Olariu, S.: On the structure of graphs with few P_4's. Discrete Appl. Math. **84**, 1–13 (1998)

6. Carathéodory, C.: Über den Variabilitätsbereich der Fourierschen Konstanten von positiven harmonischen Funktionen. Rendiconti del Circolo Matematico di Palermo **32**, 193–217 (1911)
7. Campos, V., Sampaio, R.M., Silva, A., Szwarcfiter, J.L.: Graphs with few P_4's under the convexity of paths of order three. Discrete Appl. Math. **192**, 28–39 (2015)
8. Changat, M., Klavžar, S., Mulder, H.M.: The all-paths transit function of a graph. Czechoslovak Math. J. **51**, 439–448 (2001)
9. Changat, M., Mathew, J.: On triangle path convexity in graphs. Discrete Math. **206**, 91–95 (1999)
10. Changat, M., Prasanth, G.N., Mathews, J.: Triangle path transit functions, betweenness and pseudo-modular graphs. Discrete Math. **309**, 1575–1583 (2009)
11. Chartrand, G., Wall, C.E., Zhang, P.: The convexity number of a graph. Graphs Comb. **18**, 209–217 (2002)
12. Dourado, M.C., Gimbel, J.G., Kratochvíl, J., Protti, F., Szwarcfiter, J.L.: On the computation of the hull number of a graph. Discrete Math. **309**, 5668–5674 (2009)
13. Dourado, M.C., Pereira de Sá, V.G., Rautenbach, D., Szwarcfiter, J.L.: On the geodetic Radon number of grids. Discrete Math. **313**, 111–121 (2013)
14. Dourado, M.C., de Sá, V.G.P., Rautenbach, D., Szwarcfiter, J.L.: Near-linear-time algorithm for the geodetic Radon number of grids. Discrete Appl. Math. **210**, 277–283 (2016)
15. Dourado, M.C., Protti, F., Rautenbach, D., Szwarcfiter, J.L.: On the hull number of triangle-free graphs. SIAM J. Discrete Math. **23**, 2163–2172 (2010)
16. Dourado, M.C., Protti, F., Rautenbach, D., Szwarcfiter, J.L.: Some remarks on the geodetic number of a graph. Discrete Math. **310**, 832–837 (2010)
17. Dourado, M.C., Protti, F., Rautenbach, D., Szwarcfiter, J.L.: On the convexity number of graphs. Graphs Comb. **28**, 333–345 (2012)
18. Dourado, M.C., Protti, F., Szwarcfiter, J.L.: Complexity results related to monophonic convexity. Discrete Appl. Math. **158**, 1269–1274 (2010)
19. Dourado, M.C., Rautenbach, D., dos Santos, V., Schäfer, P.M., Szwarcfiter, J.L.: On the Carathéodory number of interval and graph convexities. Theoret. Comput. Sci. **510**, 127–135 (2013)
20. Dragan, F., Nicolai, F., Brandstädt, A.: Convexity and HHD-free graphs. SIAM J. Discrete Math. **12**, 119–135 (1999)
21. Duchet, P.: Convex sets in graphs II: minimal path convexity. J. Comb. Theory Ser. B **44**, 307–316 (1988)
22. Ekim, T., Erey, A.: Block decomposition approach to compute a minimum geodetic set. RAIRO Recherche Opérationnelle **48**, 497–507 (2014)
23. Ekim, T., Erey, A., Heggernes, P., van 't Hof, P., Meister, D.: Computing minimum geodetic sets of proper interval graphs. In: Fernández-Baca, D. (ed.) LATIN 2012. LNCS, vol. 7256, pp. 279–290. Springer, Heidelberg (2012)
24. Everett, M.G., Seidman, S.B.: The hull number of a graph. Discrete Math. **57**, 217–223 (1985)
25. Farber, M., Jamison, R.E.: Convexity in graphs and hypergraphs. SIAM J. Algebraic Discrete Methods **7**, 433–444 (1986)
26. Garey, M.R., Johnson, D.S., Computers, I.: A Guide to the Theory of NP-Completeness. W.H. Freeman & Co., New York (1979)
27. Gimbel, J.: Some remarks on the convexity number of a graph. Graphs Comb. **19**, 357–361 (2003)
28. Harary, F., Loukakis, E., Tsouros, C.: The geodetic number of a graph. Math. Comput. Model. **17**, 89–95 (1993)

29. Kanté, M.M., Nourine, L.: Polynomial time algorithms for computing a minimum hull set in distance-hereditary and chordal graphs. In: Emde Boas, P., Groen, F.C.A., Italiano, G.F., Nawrocki, J., Sack, H. (eds.) SOFSEM 2013. LNCS, vol. 7741, pp. 268–279. Springer, Heidelberg (2013)

30. Radon, J.: Mengen konvexer Körper, die einen gemeinsamen Punkt enthalten. Math. Ann. **83**, 113–115 (1921)

Weighted Efficient Domination for P_6-Free and for P_5-Free Graphs

Andreas Brandstädt[1]($^{(\boxtimes)}$) and Raffaele Mosca[2]

[1] Institut für Informatik, Universität Rostock, 18051 Rostock, Germany
ab@informatik.uni-rostock.de
[2] Dipartimento di Economia,
Universitá degli Studi "G. D'Annunzio", 65121 Pescara, Italy
r.mosca@unich.it

Abstract. In a finite undirected graph $G = (V, E)$, a vertex $v \in V$ *dominates* itself and its neighbors in G. A vertex set $D \subseteq V$ is an *efficient dominating set* (*e.d.s.* for short) of G if every $v \in V$ is dominated in G by exactly one vertex of D. The *Efficient Domination* (ED) problem, which asks for the existence of an e.d.s. in G, is known to be NP-complete for P_7-free graphs and solvable in polynomial time for P_5-free graphs. The P_6-free case was the last open question for the complexity of ED on F-free graphs.

Recently, Lokshtanov, Pilipczuk and van Leeuwen showed that weighted ED is solvable in polynomial time for P_6-free graphs, based on their quasi-polynomial algorithm for the Maximum Weight Independent Set problem for P_6-free graphs. Independently, by a direct approach which is simpler and faster, we found an $\mathcal{O}(n^5 m)$ time solution for weighted ED on P_6-free graphs. Moreover, we showed that weighted ED is solvable in linear time for P_5-free graphs which solves another open question for the complexity of (weighted) ED.

1 Introduction

Let $G = (V, E)$ be a finite undirected graph without loops and multiple edges; let $|V| = n$ and $|E| = m$. A vertex $v \in V$ *dominates* itself and its neighbors. A vertex subset $D \subseteq V$ is an *efficient dominating set* (*e.d.s.* for short) of G if every vertex of G is dominated by exactly one vertex in D.

The notion of efficient domination was introduced by Biggs [1] under the name *perfect code*. Note that not every graph has an e.d.s.; the EFFICIENT DOMINATING SET (ED) problem asks for the existence of an e.d.s. in a given graph G. We can assume that G is connected; if not then the ED problem for G can be splitted into ED for each of its connected components. If a graph G with vertex weight function $w : V \to \mathbb{Z} \cup \{\infty, -\infty\}$ and an integer k is given, the MINIMUM WEIGHT EFFICIENT DOMINATING SET (WED) problem asks whether G has an e.d.s. D of total weight $w(D) := \Sigma_{x \in D} w(x) \leq k$.

As mentioned in [4], the maximization version of WED can be defined analogously, replacing the condition $w(D) \leq k$ with $w(D) \geq k$. Since negative weights

© Springer-Verlag GmbH Germany 2016
P. Heggernes (Ed.): WG 2016, LNCS 9941, pp. 38–49, 2016.
DOI: 10.1007/978-3-662-53536-3_4

are allowed, the maximization version of WED is equivalent to its minimization version; subsequently we restrict the problem to the minimization version WED. The vertex weight ∞ plays a special role; vertices which are definitely not in any e.d.s. get weight ∞, and thus, in the WED problem we are asking for an e.d.s. of finite minimum weight. Let $\gamma_{ed}(G, w)$ denote the minimum weight of an e.d.s. in G. We call a vertex $x \in V$ *forced* if x is contained in every e.d.s. of finite weight in G.

The importance of the ED problem for graphs mostly results from the fact that it is a special case of the EXACT COVER problem for hypergraphs (problem [SP2] of [9]); ED is the Exact Cover problem for the closed neighborhood hypergraph of G.

For a subset $U \subseteq V$, let $G[U]$ denote the *induced subgraph* of G with vertex set U. For a graph F, a graph G is F-*free* if G does not contain any induced subgraph isomorphic to F. Let P_k denote a chordless path with k vertices. $F + F'$ denotes the disjoint union of graphs F and F'; for example, $2P_3$ denotes $P_3 + P_3$.

Many papers have studied the complexity of ED on special graph classes - see e.g. [3–6,12] for references. In particular, a standard reduction from the Exact Cover problem shows that ED remains NP-complete for $2P_3$-free (and thus, for P_7-free) chordal graphs. For P_6-free graphs, the question whether ED can be solved in polynomial time was the last open case for F-free graphs [5]; it was the main open question in [6]. As a first step towards a dichotomy, it was shown in [3] that for P_6-free chordal graphs, WED is solvable in polynomial time.

Recently, it has been shown by Lokshtanov et al. [10] that WED is solvable in polynomial time for P_6-free graphs (the time bound is more than $\mathcal{O}(n^{500})$). Their result for WED is based on their quasi-polynomial algorithm for the Maximum Weight Independent Set problem for P_6-free graphs. Independently, in [7] we found a polynomial time solution for WED on P_6-free graphs using a direct approach which is simpler than the one in [10] and leads to the much better time bound $\mathcal{O}(n^5 m)$. According to [5], the results of [7,10] finally lead to a dichotomy for the WED problem on P_k-free graphs and moreover on F-free graphs.

In our approach, we need the following notion: A graph $G = (V, E)$ is *unipolar* if there is a partition of V into sets A and B such that $G[A]$ is P_3-free (i.e., the disjoint union of cliques) and $G[B]$ is a complete graph. See e.g. [8,11] for recent work on unipolar graphs. Note that ED remains NP-complete for unipolar graphs [8] (which can also be seen by the standard reduction from Exact Cover; there, every clique in $G[A]$ has only two vertices). Clearly, every unipolar graph is $2P_3$-free and thus P_7-free. It follows that for each $k \geq 7$, WED is NP-complete for P_k-free unipolar graphs.

The main results of this paper are the following:

1. In Sect. 2, we give a polynomial time reduction of the WED problem for P_6-free graphs to WED for P_6-free unipolar graphs.
2. In Sect. 3, we solve WED for P_6-free unipolar graphs in polynomial time.
3. In Sect. 4, we describe the polynomial time algorithm for the WED problem on P_6-free graphs. Thus, we obtain a dichotomy for the WED problem on F-free graphs, and in particular on P_k-free graphs and on P_k-free unipolar graphs.

In the full version of this paper, we describe a linear time algorithm for WED on P_5-free graphs based on modular decomposition (see [2] for details); this answers another open question in [6].

Due to space limitation, most of the proofs and the linear time algorithm for WED on P_5-free graphs cannot be given here.

2 Reducing WED on P_6-Free Graphs to WED on P_6-Free Unipolar Graphs

Throughout this section, let $G = (V, E)$ be a connected P_6-free graph. Subsequently we consider the distance levels of $v \in V$ according to the usual approach which is used already in various papers such as [6]. For $v \in V$, let $N_i(v)$ denote the i-th distance level of v, that is $N_i(v) = \{u \in V \mid d_G(u, v) = i\}$. Then, since G is P_6-free, we have $N_i(v) = \emptyset$ for each $i \geq 5$. If $v \in D$ for an e.d.s. D of G then, by the e.d.s. property, we have

$$(N_1(v) \cup N_2(v)) \cap D = \emptyset. \tag{1}$$

Let $G_v := (N_2(v) \cup N_3(v) \cup N_4(v), E_v)$ such that $N_2(v)$ is turned into a clique by correspondingly adding edges, i.e., $E_v = E' \cup F$ where E' is the set of the original edges in $G[N_2(v) \cup N_3(v) \cup N_4(v)]$ and F is the set of new edges turning $N_2(v)$ into a clique, and let $w(x) := \infty$ for every $x \in N_2(v)$. We first claim:

Proposition 1. G_v is P_6-free.

Proof. Suppose to the contrary that there is a P_6 P in G_v, say with vertices a, b, c, d, e, f and edges ab, bc, cd, de, ef. If $\{ab, bc, cd, de, ef\} \cap F = \emptyset$ then P would be a P_6 in G which is a contradiction. Thus, $\{ab, bc, cd, de, ef\} \cap F \neq \emptyset$. Then clearly, $|\{ab, bc, cd, de, ef\} \cap F| = 1$ since $N_2(v)$ is a clique in G_v. Now in any case, we get a P_6 in G by adding $N[v]$ and the corresponding edges in $N[v]$ and between $N_1(v)$ and $N_2(v)$ which is a contradiction. $\qquad\square$

Obviously, the following holds:

Proposition 2.

(i) For vertex $v \in V$ with $w(v) < \infty$, D is a finite weight e.d.s. in G with $v \in D$ if and only if $D \setminus \{v\}$ is a finite weight e.d.s. in G_v.

(ii) Thus, if for every $v \in V$, WED is solvable in time T for G_v then WED is solvable in time $n \cdot T$ for G.

From now on, let $D(v)$ denote an e.d.s. of finite weight of G_v. We call a vertex x v-*forced* if $x \in D(v)$ for every e.d.s. $D(v)$ of finite weight of G_v and v-*excluded* if $x \notin D(v)$ for every such e.d.s. $D(v)$ of G_v. Clearly, if x is v-excluded then we can set $w(x) = \infty$, e.g., for all $x \in N_2(v)$, $w(x) = \infty$ as above.

Let $Q_1, \ldots, Q_r$ denote the connected components of $G_v[N_3(v) \cup N_4(v)]$ (i.e., of $G[N_3(v) \cup N_4(v)]$). By (1), we have:

$$\text{For each } i \in \{1, \ldots, r\}, \text{ we have } |Q_i \cap D(v)| \geq 1. \tag{2}$$

Clearly, the $D(v)$-candidates in Q_i must have finite weight.

A component Q_i is *trivial* if $|Q_i| = 1$. Obviously, by (2), the vertices of the trivial components are v-forced.

Clearly, since $D(v)$ is an e.d.s. of finite weight, every $x \in N_2(v)$ must contact a component Q_i for some $i \in \{1, \ldots, r\}$.

2.1   Join-Reduction

Now we consider a graph $G = (A \cup B, E)$ such that $A_1, \ldots, A_k$ are the components of $G[A]$, and a vertex weight function w with $w(b) = \infty$ for all $b \in B$. Assume that G has an e.d.s. D of finite weight. As above, we can assume that every component A_i is nontrivial since any trivial component A_i consists of a forced D-vertex.

By the e.d.s. property of D, we have (analogously to condition (2)):

$$\text{For every } x \in B, x \textcircled{1} A_i \text{ for at most one } i \in \{1, \ldots, k\}. \tag{3}$$

Thus, from now on, we can assume that every vertex $x \in B$ has a join to at most one component A_i. Moreover, if $x \textcircled{1} A_i$ for some $i \in \{1, \ldots, k\}$ then for every neighbor $y \in A_j$ of x, $j \neq i$, $y \notin D$, i.e., we can set $w(y) = \infty$, and thus, $y \notin D$ for any e.d.s. D of finite weight of G.

For any vertex $x \in B$ with $x \textcircled{1} A_i$ for exactly one $i \in \{1, \ldots, k\}$, by the e.d.s. property of D, $|D \cap A_i| \geq 2$ is impossible. Thus, x is correctly dominated if $|D \cap A_i| = 1$, that is, the D-candidates in A_i are universal for A_i; let U_i denote the set of universal vertices in A_i (note that U_i is a clique). Clearly, for $x \textcircled{1} A_i$ we have:

$$\text{If } U_i = \emptyset \text{ then } G \text{ has no finite weight e.d.s.} \tag{4}$$

Thus, for every A_i such that there is a vertex $x \in B$ with $x \textcircled{1} A_i$, we can reduce A_i to the clique U_i, we can omit x in B, and for every neighbor $y \in A_j$ of x, $j \neq i$, we set $w(y) = \infty$. The following algorithm is needed twice in this manuscript:

Join-Reduction Algorithm

Given: A graph $G = (A \cup B, E)$ such that $A_1, \ldots, A_k$ are the components of $G[A]$, and a vertex weight function w with $w(b) = \infty$ for all $b \in B$.

Task: Reduce G in time $\mathcal{O}(n^3)$ to an induced subgraph $G' = (A' \cup B', E')$ with weight function w' and components $A'_1, \ldots, A'_k$ of $G[A']$ such that we have:

(*i*) For every $b \in B'$ and every $i \in \{1, \ldots, k\}$, if b contacts (nontrivial) component A'_i then b distinguishes A'_i.

(*ii*) $\gamma_{ed}(G, w) = \gamma_{ed}(G', w') < \infty$ or state that G has no such e.d.s.

begin

(a) Determine the sets

$B_{join} := \{b \in B \mid \text{there is an } i \in \{1, \ldots, k\} \text{ with } b①A_i\}$ and

$A_{join} := \{A_i \mid i \in \{1, \ldots, k\} \text{ and there is a } b \in B \text{ with } b①A_i\}$.

(b) **If** there is a vertex $b \in B_{join}$ and there are $i \neq j$ with $b①A_i$ and $b①A_j$ **then** STOP − G does not have an e.d.s. of finite weight.

{From now on, every $b \in B_{join}$ has a join to exactly one $A_i \in A_{join}$.}

(c) **For all** $b \in B_{join}$ and $A_i \in A_{join}$ such that $b①A_i$ **do**

begin

(c.1) Determine the set U_i of universal vertices in A_i. **If** $U_i = \emptyset$ **then** STOP − G does not have an e.d.s. of finite weight **else** set $A'_i := U_i$.

(c.2) For every neighbor $y \in A \setminus A_i$ of b, set $w'(y) := \infty$.

end

(d) For every $A_i \notin A_{join}$, set $A'_i := A_i$, and finally set $A' := A'_1 \cup \ldots \cup A'_k$, $B' := B \setminus B_{join}$ and $G' := G[A' \cup B']$.

end

Lemma 1. *The Join-Reduction Algorithm is correct and can be done in time* $\mathcal{O}(n^3)$.

For applying the Join-Reduction Algorithm to G_v, we set $B := N_2(v)$ and $A := N_3(v) \cup N_4(v)$. For reducing WED on G to WED on a unipolar graph G', this is a first step which, by condition (i) of the Task, leads to the fact that finally, for every nontrivial component Q_j of $G[N_3(v) \cup N_4(v)]$, every vertex in $N_2(v)$ which contacts Q_j also distinguishes Q_j.

2.2   Component-Reduction

Let $G'_v = (A' \cup B', E')$ be the result of applying the Join-Reduction algorithm to G_v; let B' be the corresponding subset of $N_2(v)$ and let A' be the corresponding subset of $N_3(v) \cup N_4(v)$. Recall that in G'_v, B' is a clique. In the next step, we reduce WED for G_v to WED for unipolar graphs.

We consider the components Q_i of $G'_v[A']$ which are not yet a clique; as already mentioned, we can assume that if $x \in B'$ has a neighbor in Q_i then it has a neighbor and a non-neighbor in Q_i. For $1 \leq i \leq r$, let $Q_i^+(x) := Q_i \cap N(x)$ and $Q_i^-(x) := Q_i \setminus N(x)$. Since Q_i is connected, we have: If x distinguishes Q_i then it distinguishes an edge in Q_i.

For $x, x' \in B'$ and edges $y_1 z_1$ in Q_i, $y_2 z_2$ in Q_j, $i \neq j$, let $xy_1 \in E, xz_1 \notin E$ and $x'y_2 \in E, x'z_2 \notin E$. Then, since G and G_v are P_6-free, we have:

$$xy_2 \in E \text{ or } xz_2 \in E \text{ or } x'y_1 \in E \text{ or } x'z_1 \in E. \tag{5}$$

Another useful P_6-freeness argument is the following:

$$\text{For } x \in B' \text{ and } y \in Q_i^+(x), y \text{ does not distinguish any edge in } Q_i^-(x). \tag{6}$$

We claim:

There is a vertex $b^* \in B'$ which contacts Q_i for every $i \in \{1, \ldots, r\}$.   (7)

Let $q^* \in D(v)$ be the vertex dominating b^*; without loss of generality assume that $q^* \in Q_1$, and let $D(v, q^*)$ denote a finite weight e.d.s. of G_v containing q^*. Q_1 is partitioned into

$Z := N[q^*] \cap Q_1,$
$W := Q_1 \cap N(b^*) \setminus Z,$ and
$Y := Q_1 \setminus (Z \cup W).$

Then clearly, the following properties hold:

Lemma 2.

(i) $Z \cap D(v, q^*) = \{q^*\}$.
(ii) $W \cap D(v, q^*) = \emptyset$.
(iii) $Z \oslash Y$.
(iv) For every component K of $G[Y]$, the set of $D(v, q^*)$-candidate vertices in K is a clique.

For the algorithmic approach, we set $w(y) = \infty$ for every $y \in W$ and for every non-universal vertex $y \in K$ in any component K of $G_v[Y]$.

For $i \geq 2$, let $Q_i^+ := Q_i \cap N(b^*)$ and $Q_i^- := Q_i \setminus N(b^*)$. Clearly, by the e.d.s. property, for every $i \geq 2$, $Q_i^+ \cap D(v, q^*) = \emptyset$; set $w(y) = \infty$ for every $y \in Q_i^+$. Thus, the components of $G[Q_i^-]$ must contain the corresponding $D(v, q^*)$-vertices.

Again, as in Lemma 2 (iv), for each such component K, the $D(v, q^*)$-candidates must be universal vertices for K since by (6), two such $D(v, q^*)$-candidates in K would have a common neighbor in Q_i^+, i.e., only the universal vertices of component K are the $D(v, q^*)$-candidate vertices; set $w(y) = \infty$ for every non-universal vertex $y \in K$.

Let $I := \{a \in A' : w(a) = \infty\}$. Then I admits a partition $\{I_1, I_2, I_3\}$ as defined below:

- I_1 is formed by those vertices of I which are either in Z, or in Y, or in Q_i^- for $i \geq 2$.
- I_2 is formed by those vertices of I which are either in W and contact exactly one component of $G[Y]$ or in Q_i^+ and contact exactly one component of $G[Q_i^-]$ for $i \geq 2$.
- I_3 is formed by those vertices of I which are either in $W \setminus I_2$ or in $Q_i^+ \setminus I_2$ for $i \geq 2$.

Note that we have:

(a) By construction and by the e.d.s. property, if $I_3 \neq \emptyset$ then $D(v, q^*)$ does not exist (in fact each vertex of I_3 either would not be dominated by any $D(v, q^*)$-candidate or would be dominated by more than one $D(v, q^*)$-candidate).

(b) By construction, if $D(v, q^*)$ exists then $D(v, q^*)$ is an e.d.s. of $G'_v[(A' \cup B') \setminus (I_1 \cup I_2)]$ as well; in particular, by construction and by (6), each vertex of $I_1 \cup I_2$ is dominated in G'_v by exactly one vertex of $D(v, q^*)$; then vertices of $I_1 \cup I_2$ can be removed.

(c) $G'_v[(A' \cup B') \setminus (I_1 \cup I_2)]$ is unipolar (once assuming that $I_3 = \emptyset$).

Then for every potential $D(v)$-neighbor q^* of b^*, we can reduce the WED problem for G'_v to the WED problem for $G'_v(q^*)$ consisting of B' and the P_3-free subgraph induced by $\{q^*\}$ and by the corresponding cliques of universal vertices in components K as described above with respect to $D(v, q^*)$. Clearly, the $D(v, q^*)$-candidates in the cliques of the P_3-free subgraph can be chosen corresponding to optimal weights.

Summarizing, we can do the following:

Component-Reduction Algorithm

Given: The result $H = G'_v = (A' \cup B', E')$ with vertex weight function w of applying the Join-Reduction algorithm to G_v such that $K_1, \ldots, K_s$ denote the clique components and $Q_1, \ldots, Q_r$ denote the non-clique components of $G'_v[A']$.

Task: Reduce H in time $\mathcal{O}(n^3)$ to (less than n) unipolar graphs $H_\ell = G'(q^*)$ with weight function w_ℓ, $1 \le \ell < n$, such that $\gamma_{ed}(H, w) = \min_\ell \gamma_{ed}(H_\ell, w_\ell)$ or state that H has no e.d.s. of finite weight.

begin

(a) Determine a vertex $b^* \in B'$ contacting every Q_i, $i \in \{1, \ldots, r\}$.

(b) For every $q^* \in N(b^*) \cap A'$ with $w(q^*) < \infty$, say $q^* \in Q_i$, reduce Q_i according to Lemma 2 and for all j, $j \ne i$, reduce Q_j according to the paragraph after the proof of Lemma 2 such that finally, the resulting subgraph $G'(q^*)$ is unipolar.

end

Lemma 3. *The Component-Reduction Algorithm is correct and can be done in time* $\mathcal{O}(n^3)$.

Corollary 1. *If WED is solvable in polynomial time on P_6-free unipolar graphs then it is solvable in polynomial time on P_6-free graphs.*

3   Solving WED on P_6-Free Unipolar Graphs in Polynomial Time

Throughout this section, let $G = (V, E)$ be a connected P_6-free unipolar graph with partition $V = A \cup B$ such that $G[A]$ is the disjoint union of cliques $A_1, \ldots, A_k$, and $G[B]$ is a complete subgraph. Clearly, if $k \le 3$ then every e.d.s. of G contains at most four vertices. Thus, from now on, we can assume that $k \ge 4$. In particular, for any e.d.s. D of G, $|D \cap B| \le 1$. Thus, WED for such graphs is solvable in polynomial time if and only if WED is solvable in polynomial time for e.d.s. D with $B \cap D = \emptyset$.

Clearly, for a P_6-free unipolar graph, the following holds (recall (5)):

Claim 1. *If for distinct $b_1, b_2 \in B$, b_1 distinguishes an edge $x_1 x_2$ in A_i and b_2 distinguishes an edge $y_1 y_2$ in A_j, $i \neq j$, then either b_2 contacts $\{x_1, x_2\}$ or b_1 contacts $\{y_1, y_2\}$.*

The key result of this section is the following:

Lemma 4. *For connected unipolar graphs fulfilling Claim 1, it can be checked in polynomial time whether G has a finite weight e.d.s. D with $B \cap D = \emptyset$.*

Lemma 4 is based on various propositions described subsequently. As a first step, we again reduce G corresponding to the Join-Reduction Algorithm of Sect. 2: Since $B \cap D = \emptyset$, clearly, $|D \cap A_i| = 1$ for every $i \in \{1, \ldots, k\}$. Thus, if $A_i = \{a_i\}$ then a_i is a forced D-vertex; from now on, we can assume that every A_i is nontrivial.

Moreover, every $b \in B$ must contact at least one A_i, and if b has a join to two components A_i, A_j, $i \neq j$, then G does not have an e.d.s. Thus, by (3) and the subsequent paragraph in Sect. 2, from now on, we can assume that no vertex $b \in B$ has a join to any A_i, i.e., if b contacts A_i then it distinguishes A_i.

Again, as by (7), there is a vertex $b^* \in B$ which contacts every A_i. However, we need a stronger property. For this, we define the following notions:

Definition 1. *For vertices $b_1, b_2 \in B$ and a nontrivial component $K = A_i$ of A, we say:*

 (i) *b_2 overtakes b_1 for K if b_2 distinguishes an edge in $K \setminus N(b_1)$.*
 (ii) *b_2 includes b_1 for K if $N(b_2) \cap K \supseteq N(b_1) \cap K$.*
(iii) *b_2 strictly includes b_1 for K if $N(b_2) \cap K \supset N(b_1) \cap K$.*
 (iv) *b_1 and b_2 cover K if $N(b_1) \cup N(b_2) = K$.*
 (v) *$b_1 \to b_2$ if b_2 overtakes b_1 for at least three distinct nontrivial components of A.*
 (vi) *$b^* \in B$ is a good vertex of B if for none of the vertices $b \in B \setminus \{b^*\}$, $b^* \to b$ holds.*

Assume that G has an e.d.s. D of finite weight.

Claim 2. *For vertices $b_1, b_2 \in B$, we have:*

 (i) *b_1 and b_2 cover at most two A_i, A_j, $i, j \in \{1, \ldots, k\}$, $i \neq j$.*
 (ii) *If b_2 overtakes b_1 for A_i then for any A_j, $j \neq i$, b_1 does not overtake b_2.*
(iii) *If b_2 overtakes b_1 for some A_i, A_j, $i \neq j$, then b_2 strictly includes b_1 for A_i, A_j.*
 (iv) *If b_2 overtakes b_1 for some A_i, A_j, $i \neq j$, then b_2 includes b_1 for all but at most two A_ℓ, $1 \leq \ell \leq k$.*
 (v) *If b_2 strictly includes b_1 for some A_i then b_2 includes b_1 for all but at most two A_ℓ, $1 \leq \ell \leq k$.*

Let $H = (B, F)$ denote the directed graph with vertex set B and edges $b \to b' \in F$ as in Definition 1 (v). Thus, a good vertex of B is one with outdegree 0 with respect to H. As usual, H is a *directed acyclic graph* (*dag* for short) if there is no directed cycle in H.

Claim 3. *H is a dag.*

It is well known that any dag has a vertex with outdegree 0. Thus, Claim 3 implies:

Claim 4. *There is a good vertex $b^* \in B$.*

Let b^* be such a good vertex. Then, since by the condition in Lemma 4, $B \cap D = \emptyset$, b^* must have a D-neighbor $a^* \in A \cap N(b^*) \cap D$; the algorithm tries all possible vertices in $A \cap N(b^*)$. Let $D(a^*)$ denote an e.d.s. with $a^* \in D(a^*)$; without loss of generality, assume that $a^* \in A_1$. Clearly, $(A_1 \setminus \{a^*\}) \cap D(a^*) = \emptyset$. Without loss of generality, let us assume that $A_1 = \{a^*\}$. Since a^* dominates b^*, each neighbor of b^* in A_i, $i \geq 2$, is not in $D(a^*)$. For $i \in \{2, \ldots, k\}$, let $A'_i := A_i \setminus N(b^*)$, and let $A' = \{a^*\} \cup A'_2 \cup \ldots \cup A'_k$. Obviously, we have:

(a) For each A'_i, $|A'_i \cap D(a^*)| = 1$.

Moreover, as before, we can assume:

(b) For each vertex $b \in B$, b does not have a join to two distinct A'_i, A'_j, $i \neq j$.

(c) If vertex $b \in B$ has a join to exactly one A'_i then it does not contact the remaining components A'_j, $j \neq i$.

Thus, again by (3) and the subsequent paragraph in Sect. 2, from now on, we can assume that no vertex $b \in B$ has a join to any A_i, i.e., if b contacts A_i then it distinguishes A_i. Next we claim:

(d) At most two distinct components A'_i, A'_j are distinguished by some vertex of $B \setminus \{b^*\}$.

Summarizing, by the above, $D(a^*)$ exists if and only if

(i) the above properties hold and

(ii) $G[A' \cup B]$ has a (weighted) e.d.s. $D(a^*)$ with $B \cap D(a^*) = \emptyset$.

Checking (i) can be done in polynomial time (actually one should just check if some of the above properties hold). Checking (ii) can be done in polynomial time as shown below: For the components of $G[A']$, let

– $C_1(A')$ be the set of those components of $G[A']$ which are not distinguished by any vertex of B, and

– $C_2(A')$ be the set of those components of $G[A']$ which are distinguished by some vertex of B.

For each member K of $C_1(A')$, any vertex of K (of minimum weight, for WED) can be assumed to be the only vertex in $D(a^*) \cap K$, without loss of generality since such vertices form a clique and have respectively the same neighbors in $G[(A' \cup B) \setminus K]$ (for WED, one can select a vertex of minimum weight).

Concerning $C_2(A')$, we have $|C_2(A')| \leq 2$ by property (d). Then the set $\{(a^*, a_2, \ldots, a_k) \mid a_i \in A'_i, i \in \{2, \ldots, k\}\}$ of k-tuples of candidate vertices in $D(a^*)$ contains $\mathcal{O}(n^2)$ members by property (d). Thus one can check in polynomial time if $D(a^*)$ exists.

Algorithm WED for P_6-free unipolar graphs

Given: A connected P_6-free unipolar graph $G = (A \cup B, E)$ such that B is a clique and $G[A]$ is the disjoint union of cliques $A_1, \ldots, A_k$.

Task: Determine an e.d.s. of G with minimum finite weight if there is one or state that G does not have such an e.d.s.

(a) Reduce G to G' by the Join-Reduction Algorithm. {From now on, we can assume that for every $b \in B$ and every $i \in \{1, \ldots, k\}$, b distinguishes A_i if b contacts A_i.}

(b) Construct the dag H according to Definition 1 (v), and determine a good vertex $b^* \in B$ in H.

(c) For every neighbor $a^* \in A'$ of b^*, determine the $\mathcal{O}(n^2)$ possible tuples of $D(a^*)$-candidates and check whether they are an e.d.s. of finite weight.

(d) Finally, choose an e.d.s. of minimum finite weight or state that G' does not have such an e.d.s.

Theorem 1. *Algorithm WED for P_6-free unipolar graphs is correct and can be done in time $\mathcal{O}(n^3 m)$.*

4   The Algorithm for WED on P_6-Free Graphs

By combining the principles described above (and in particular by Corollary 1, Lemma 4, and Theorem 1) we obtain:

Algorithm WED for P_6-free graphs

Given: A P_6-free graph $G = (V, E)$.

Task: Determine an e.d.s. of G with minimum finite weight if there is one or state that G does not have such an e.d.s.

For every $v \in V$ do
begin

(a) Determine the distance levels $N_i(v)$, $1 \le i \le 4$.

(b) For G_v as defined in Sect. 2, with $B = N_2(v)$ and $A = N_3(v) \cup N_4(v)$, reduce G_v to G'_v by the Join-Reduction Algorithm. {From now on, we can assume that for every $b \in B$ and every $i \in \{1, \ldots, k\}$, b distinguishes A_i if b contacts A_i.}

(c) According to the Component-Reduction Algorithm, determine a vertex $b^* \in B$ contacting every component in $G[A]$ which is not a clique, and for every neighbor $q^* \in N(b^*) \cap A$, do:

(c.1) Reduce G'_v to $G'(v, q^*)$ by the Component-Reduction Algorithm. {Now, $G'(v, q^*)$ is P_6-free unipolar.}

(c.2) Carry out the Algorithm WED for P_6-free unipolar graphs for input $G'(v, q^*)$ with its weight function.

(d) Finally, for every resulting candidate e.d.s., check whether it is indeed a finite weight e.d.s. of G, choose an e.d.s. of minimum finite weight of G or state that G does not have such an e.d.s.

end

Theorem 2. *Algorithm WED for P_6-free graphs is correct and can be done in time $\mathcal{O}(n^5 m)$.*

5   Conclusion

As mentioned, the direct approach for solving WED on P_6-free graphs gives a dichotomy result for the complexity of WED on F-free graphs. In [3], using an approach via G^2, it was shown that WED can be solved in polynomial time for P_6-free chordal graphs, and a conjecture in [3] says that for P_6-free graphs with e.d.s., the square is perfect which would also lead to a polynomial time algorithm for WED on P_6-free graphs but anyway, the time bound of our direct approach is better than in the case when the conjecture would be true.

Acknowledgments. The first author thanks Martin Milanič for discussions and comments about the WED problem for P_5-free graphs and for some subclasses of P_6-free graphs.

References

1. Biggs, N.: Perfect codes in graphs. J. Comb. Theory (B) **15**, 289–296 (1973)
2. Brandstädt, A.: Weighted efficient domination for P_5-free graphs in linear time. CoRR arXiv:1507.06765v1 (2015)
3. Brandstädt, A., Eschen, E.M., Friese, E.: Efficient domination for some subclasses of P_6-free graphs in polynomial time. In: Extended Abstract to Appear in the Conference Proceedings of WG 2015. Full version: CoRR arXiv:1503.00091v1 (2015)
4. Brandstädt, A., Fičur, P., Leitert, A., Milanič, M.: Polynomial-time algorithms for weighted efficient domination problems in AT-free graphs and dually chordal graphs. Inf. Process. Lett. **115**, 256–262 (2015)
5. Brandstädt, A., Giakoumakis, V.: Weighted efficient domination for $(P_5 + kP_2)$-free graphs in polynomial time. CoRR arXiv:1407.4593v1 (2014)
6. Brandstädt, A., Milanič, M., Nevries, R.: New polynomial cases of the weighted efficient domination problem. In: Chatterjee, K., Sgall, J. (eds.) MFCS 2013. LNCS, vol. 8087, pp. 195–206. Springer, Heidelberg (2013)
7. Brandstädt, A., Mosca, R.: Weighted efficient domination for P_6-free graphs in polynomial time. CoRR arXiv:1508.07733v1 (2015). (based on a manuscript by R. Mosca, Weighted Efficient Domination for P_6-Free Graphs, July 2015)
8. Eschen, E., Wang, X.: Algorithms for unipolar and generalized split graphs. Discrete Appl. Math. **162**, 195–201 (2014)
9. Garey, M.R., Johnson, D.S.: Computers and Intractability - A Guide to the Theory of NP-Completeness. Freeman, San Francisco (1979)

10. Lokshtanov, D., Pilipczuk, M., van Leeuwen, E.J.: Independence and efficient domination on P_6-free graphs. In: Conference Proceedings SODA 2016, pp. 1784–1803. CoRR arXiv:1507.02163v2 (2015)
11. McDiarmid, C., Yolov, N.: Recognition of unipolar and generalized split graphs. Algorithms **8**, 46–59 (2015)
12. Milanič, M.: Hereditary efficiently dominatable graphs. J. Graph Theory **73**, 400–424 (2013)

Saving Colors and Max Coloring:
Some Fixed-Parameter Tractability Results

Bruno Escoffier[✉]

Sorbonne Universités, UPMC Univ Paris 06, CNRS, LIP6 UMR 7606,
4 Place Jussieu, 75005 Paris, France
bruno.escoffier@upmc.fr

Abstract. Max coloring is a well known generalization of the usual
Min Coloring problem, widely studied from (standard) complexity and
approximation viewpoints. Here, we tackle this problem under the framework of parameterized complexity. In particular, we first show to what
extend the result of [3] - saving colors from the trivial bound of n on the
chromatic number - extends to Max Coloring. Then we consider possible improvements of these results by considering the problem of saving
colors/weight with respect to a better bound on the chromatic number.
Finally, we consider the fixed parameterized tractability of Max Coloring
in restricted graph classes under standard parameterization.

1 Introduction

As it is well known for decades, deciding whether a given simple graph is
3-colorable is an NP-complete problem. As an immediate consequence, dealing with parameterized complexity, the coloring problem parameterized by the
value of the solution (number of colors) is not in XP unless $P = NP$. This result
stimulates the study of this fundamental problem under other parameterizations.
For instance, in [2,10] is considered the coloring problem parameterized by the
distance (adding/removing vertices or edges) of the input graph to a polynomial
time solvable class of graphs. Chor et al. [3] considered the problem of "saving
k colors", by addressing the following question: given a graph $G = (V, E)$ with
n vertices and an integer k, is it possible to color G with $n - k$ colors? In other
words, can we save k colors with respect to the trivial coloring (with n colors)?
They show that the problem is FPT, and admits a linear kernel.

On the other hand, among the many variants or generalizations of the coloring problem, Max Coloring (sometimes called Weighted Coloring) has been
deeply investigated in the last ten years, see for instance [1,4,5,11,12] and references therein. In this problem, coming from batch scheduling, each vertex v has
a nonnegative integer weight $w(v)$; the weight of a color is the *maximum* weight
of its vertices, and the weight of a coloring is the sum of the weights of its colors.
The goal is to find a coloring of minimum weight.[1] This is a generalization of

---

[1] In batch scheduling, a color is a set of tasks (batch) processed together; then the
weight of the color is the time to process the batch, and the total weight is the time
to process every batch, i.e., every task.

© Springer-Verlag GmbH Germany 2016
P. Heggernes (Ed.): WG 2016, LNCS 9941, pp. 50–61, 2016.
DOI: 10.1007/978-3-662-53536-3_5

the usual coloring problem, which corresponds to the particular case where all the vertices have weight 1. Max Coloring is much harder than coloring; it is for instance NP-hard in bipartite graphs [5], in interval graphs [6], and has been recently shown even not solvable in polynomial time in trees under ETH [1].

The goal of this work is threefold, by addressing the following questions:

1. Is it possible to save weights for Max Coloring problem as it is possible to save colors in the usual coloring problem?
2. Is it possible to strengthen these results by starting with a less trivial upper bound on $\chi(G)$? This question holds for the usual coloring problem: can we save k colors not with respect to n but with respect to another bound $b < n$? If so, how does this extend to saving weights for Max Coloring?
3. Under the standard parameterization (number of colors), could we say something on Max Coloring in classes of graphs where the usual coloring problem is polynomial?

Let us first deal with the first point. The bound of n for min coloring consists of giving one color per vertex; this coloring has value $W = \sum_{i=1}^{n} w(v_i)$ for Max Coloring. Assume that weights are in non-increasing order. Then, there are two questions that naturally generalizes the question of coloring a graph with $n - k$ colors: given k, can we save a *total weight* of k, i.e., does there exist a coloring of weight at most $W - k$? Or, can we save k *vertex weights*, i.e., does there exist a coloring of weight at most $\sum_{i=1}^{n-k} w(v_i)$? We tackle these questions in Sect. 2 and show that both problems are FPT with respect to k.

Let us now consider the second point. For the usual coloring problem, saving k colors in FPT-time can be reached as follows: starting from the initial coloring with n colors we save colors by merging together pairs of non adjacent vertices. If it is possible to do this k times, then we have saved k colors. Otherwise, the graph is composed by a clique plus $O(k)$ vertices, and an optimal coloring is computable by matching techniques in FPT-time. Let us call a coloring *non trivial* if it is not possible to merge two colors of this coloring. It is very easy to build a non trivial coloring (in polynomial time), or to transform a coloring into a non trivial one without increasing the number of colors. Let c be the number of colors used by a non trivial coloring, and m be the number of edges in the graph. Since we have an edge between any pair of colors, we have $m \geq c(c-1)/2$, meaning $\chi(G) \leq c \leq f(m) = \left\lfloor \frac{1}{2} + \sqrt{2m + \frac{1}{4}} \right\rfloor$. This bound $f(m)$ is generally much smaller than n, and it is worth considering the possibility to save colors with respect to this bound $f(m)$. In other words, the following question holds: *Given a graph G and an integer k, is it possible to color G with $f(m) - k$ colors?*

We will focus on this question in Sect. 3. Interestingly, we show that while the problem of saving k colors wrt $f(m)$ is still FPT, it *does not* admit a linear kernel under ETH, but (only) a quadratic one. Then we tackle the possible extension of this result to Max Coloring. More precisely, we consider the question of determining whether a coloring of weight at most $\sum_{i=1}^{f(m)} w(v_i) - k$ exists or not. We show that the problem is in XP. The core of the proof is a *polynomial time* reduction to a very specific class of graphs, namely graphs made of two cliques plus $O(k)$ vertices. This reduction should be also useful while determining if the problem is FPT, that we leave as an open question.

Then, in Sect. 4, we derive from previous works some results on Max Coloring under standard parameterization, in classes of graphs where min coloring is polynomial but Max Coloring remains hard: Max Coloring is FPT in chordal graphs (hence in interval graphs and trees), while it is not in XP for bipartite graphs (hence for perfect graphs). We give in Sect. 5 some concluding remarks.

Due to space constraints, some proofs are omitted or sketched.

2   Saving Weight for Max Coloring

In the Max Coloring problem, the input is a vertex-weighted graph (G, w), where $w(v_i) \in \mathbb{N}^*$ denotes the weight of the vertex v_i. Given a coloring $\mathcal{C} = (S_1, S_2, \ldots, S_c)$ of G, the weight of a color S_j is $w(S_j) = \max\{w(v_i), v_i \in S_j\}$, and the weight of the coloring - to be minimized - is $w(\mathcal{C}) = \sum_{j=1}^{c} w(S_j)$. We assume that the vertices are ordered in non increasing weight, i.e. $w(v_1) \geq w(v_2) \geq \cdots \geq w(v_n)$. We also denote $W = \sum_{i=1}^{n} w(v_i)$.

As a generalization of the unweighted case, we consider the following problem.

Problem 1.

- Input: a vertex-weighted graph (G, w), and an integer k.
- Parameter: k
- Question: does there exist a coloring of G of weight at most $W - k$?

Let us also define **Problem 1'** which is the same as Problem 1 up to the fact that the question is now to determine if there exists a coloring of weight at most $\sum_{i=1}^{n-k} w_i$.

We show that these problems are FPT. The technique generalizes the one for the unweighted case, by using matchings in a properly weighted graph. More precisely, we first show how matching techniques allow to solve Max Coloring in complement of bipartite graphs in polynomial time (this result will be also useful later in Sect. 3.2). Then we reduce the problem to the case where the graph has a very large clique, and use brute force to produce an FPT number of instances of Max Coloring in complement of bipartite graphs.

Lemma 1. *Max Coloring is polynomially solvable in the class of complement of bipartite graphs.*

Proof. Let G be a graph, whose vertices are partitioned into 2 cliques C_1 and C_2. We consider the complement graph $\overline{G}$ of G (same vertex set, and an edge is in $\overline{G}$ iff it is not in G), and we assign to edge (v_i, v_j) in $\overline{G}$ the weight $\min\{w(v_i), w(v_j)\}$. This corresponds to the amount of weight saved by using the same color for v_i and v_j.

Consider a matching in $\overline{G}$. This defines a coloring of G: an unmatched vertex constitutes one color, two matched vertices constitute one color. Reciprocally, any coloring of G defines a matching in $\overline{G}$. The weight of a matching M in $\overline{G}$ is $w(M) = \sum_{(v_i,v_j) \in M} \min\{w(v_i), w(v_j)\}$, and the weight of the associated coloring $\mathcal{C}$ is:

$$w(\mathcal{C}) = \sum_{v \in \overline{M}} w(v) + \sum_{(v_i,v_j) \in M} \max\{w(v_i), w(v_j)\}$$

$$= \sum_{v \in V} w(v) - \sum_{(v_i,v_j) \in M} \min\{w(v_i), w(v_j)\} = \sum_{v \in V} w(v) - w(M)$$

where $\overline{M}$ is the set of vertices unmatched in M. So we can solve Max Coloring in G by computing a maximum weight matching in $\overline{G}$. □

Theorem 1. *Problems 1 and 1' are FPT, solvable in time $2^{O(k \log k)}p(n)$ for some polynomial p.*

Proof. As for the unweighted case, consider first a maximal anti-matching (matching in the complement graph). If it contains at least k anti-edges, it means that we can save at least k weights, so a global weight at least k, and we answer YES (for both problems). Otherwise, we have $t \leq k$ anti-edges in the matching, meaning that the other $n - 2t$ vertices form a clique. Let C be this clique, and R be the endpoints of the anti-matching.

We consider all possible colorings of the vertices in R. Consider one of these colorings, with colors $\mathcal{C}_R = (R_1, \cdots, R_s)$. We contract in G all colors R_i, meaning that we replace vertices in R_i by one vertex r_i, adjacent to all neighbors of the contracted vertices, and assign the weight $w(r_i) = max\{w(v), v \in r_i\}$ to this new vertex. We also make the set of r_i's a clique. Clearly, there exists in G a coloring of weight at most B where vertices in R are colored as in $\mathcal{C}_R$ if and only if there exists a coloring of weight at most B in the contracted graph. This contracted graph is made of 2 cliques, so we can solve Max Coloring in polynomial time due to Lemma 1.

Then it is sufficient to produce all the possible colorings of R and to see if one allows to answer YES. There are at most $2^{O(t \log t)} = 2^{O(k \log k)}$ such colorings, so the result follows. □

Speaking about kernels, it does not seem obvious to generalize the approach of [3] leading to a linear kernel (vertex size). Indeed, the crown reduction does not seem to directly lead to such a result. So the following question occurs: *what size of kernel can be achieved for Problems 1 and 1'?*

3 Saving Colors from Non Trivial Colorings

3.1 Coloring with $f(m) - k$ Colors

We now concentrate on saving colors with respect to an (improved) upper bound on the chromatic number. As explained in the introduction, we consider here the problem of saving k colors with respect to the 'non trivial coloring bound', formalized as follows - where $f(m) = \left\lfloor \frac{1}{2} + \sqrt{2m + \frac{1}{4}} \right\rfloor$.

Problem 2.

- Input: A graph $G = (V, E)$ on n edges and m vertices, an integer k.
- Parameter: k
- Question: is $\chi(G) \le f(m) - k$?

We first show that the problem is FPT and provide a quadratic kernel for it. The idea of the FPT algorithm is to reduce the problem to the one of saving k colors from the trivial coloring with n colors. Note that the bounds n and $f(m)$ become equal (only) when the graph is complete. The distance between the bounds gets bigger while the density of the graph decreases. Even for graphs of density $1/2$, we have roughly $n^2/4$ edges, so $f(m)$ is roughly $n/\sqrt{2}$, much lower than n. So for the reduction we need to reach a graph which is almost complete; this is achievable by iteratively removing vertices of 'low' degree.

Theorem 2. *Problem 2 is FPT. It admits a kernel of vertex size - and edge size - $O(k^2)$.*

Proof. Let (G, k) be an instance of the problem. Let $b = f(m) - k$.

It is easy to see that if a vertex v of a graph H has degree $d(v) < c$, then H is c-colorable if and only if $H \setminus \{v\}$ is c-colorable. Then starting from (G, k) we apply the following reduction rule:

Rule 1: while there exists a vertex of degree at most $b - 1$, remove it from G.

After the application of Rule 1, we get a graph $G' = (V', E')$ with n' vertices and m' edges, and every vertex of degree at least $b = f(m) - k \ge f(m') - k$. Clearly, G' is b-colorable iff G is b-colorable. Note that if $f(m') \le 2k$ then the problem is trivially solvable in FPT-time, so assume now that $f(m') \ge 2k + 1$. Summing up the degrees, we get $2m' = \sum_{v \in V'} d_{G'}(v) \ge n'(f(m') - k)$. Then, since $\sqrt{2m'} \le f(m') + 1/2$:

$$n' - f(m') \le \frac{(f(m') + 1/2)^2 - f(m')(f(m') - k)}{f(m') - k}$$
$$= \frac{(k + 1)(f(m') - k) + (k + 1/2)^2}{f(m') - k} \le k + 1 + \frac{(k + 1)^2}{f(m') - k} \le 2(k + 1)$$

where the last inequality uses $f(m') \ge 2k + 1$.

So, $b = f(m) - k \ge f(m') - k \ge n' - 3k - 2$. If $b \ge n'$ the answer is trivially yes (G' is b-colorable, and so is G), otherwise we solve the problem in FPT-time using the FPT algorithm of [3] for deciding whether G' is $b = n' - k'$ colorable or not, where $k' = n' - b \le 3k + 2$ (and $k' \ge 0$).

The previous arguments can be easily turned into a kernelization algorithm:

- We first apply Rule 1, and get the graph G'. (G, k) is a yes-instance iff (G', k_1) with $k_1 = k + f(m') - f(m) \le k$ is a yes-instance.
- If $f(m') \le 2k$, we have $m' = O(k^2)$ and $n' \le 2m' = O(k^2)$ (since isolated vertices can be removed), so we have the claimed quadratic kernel.

– Otherwise we have $f(m') - k_1 = n' - k'$ for $k' = O(k)$. Applying the kernelization by Chor et al. [3], either we find an answer, or we have a graph (G'', k'') where G'' has $O(k') = O(k)$ vertices - hence $O(k^2)$ edges - such that G'' has an $n'' - k''$ coloring iff G' has a $n' - k'$ coloring, hence iff G has a $f(m) - k$ coloring. $n'' - k'' = f(m'') - k_2$ for some $k_2 \leq k''$, and the result follows. $\square$

The kernelization algorithm produces a graph with $O(k^2)$ vertices and edges. While being of the same order in term of size of the graph $O(m + n)$, the size of the kernel in term of number of vertices is significantly bigger than the one obtained by [3] with the problem of coloring a graph with $n - k$ colors. However, we show that finding a smaller kernel for Problem 2 is quite improbable, since this would contradict ETH.

Theorem 3. *Problem 2 does not admit a kernel of vertex size $o(k^2)$ under ETH.*

Proof. Min coloring is not solvable in subexponential time under ETH even in graphs with a linear number of edges[2]. More precisely, under ETH there exist two constant $\epsilon > 0$ and d such that min coloring is not solvable in time $O(2^{\epsilon n})$ in graphs where $m \leq dn$.

Suppose that there is for Problem 2 a kernelization algorithm that, given an instance (G, k), produces an equivalent instance (G', k') where the number of vertices in G' is $n' = o(k^2)$. Take a graph G with $m \leq dn$ edges. Let $c \leq n$, we want to determine whether G is c-colorable or not. Let $k = f(m) - c$. If $k \leq 0$ we can easily find a c-coloring of G in polynomial time. Otherwise, apply the kernelization algorithm to get in polynomial time (G', k') where G' is $(f(m') - k')$-colorable iff G is c-colorable, with $n' = o(k^2)$. $k \leq f(m) = O(\sqrt{m}) = O(\sqrt{n})$ (since $m \leq dn$), so $n' = o(k^2) = o(n)$. We can optimally color G' in time $2^{O(n')} = 2^{o(n)}$, thus determining in time $2^{o(n)}$ if G is c-colorable or not. Then we can determine the chromatic number of G in subexpontential time, thus contradicting ETH. $\square$

To conclude this section, let us briefly mention other bounds on the chromatic number. A well known upper bound is $\chi(G) \leq \Delta + 1$ where Δ is the maximum degree of vertices in the graph. However, 3-coloring is NP-complete in graphs of maximum degree 4 [7], meaning that saving k colors with respect to the bound $\Delta + 1$ is not in XP if $P \neq NP$. The same occurs for a refinement of this bound, namely $\chi(G) \leq \max\{\min\{d(v_i) + 1, i\}, i = 1, \cdots, n\}$. Dealing with the lower bound $\omega(G)$ (size of the maximum clique), the NP-completeness of 3-coloring [7] also shows that determining whether a graph is $\omega(G) + k$ colorable is not in XP if $P \neq NP$ (even if a maximum clique is given).

3.2   Coloring with Weight $\sum_{i=1}^{f(m)} w_i - k$

We now generalize Problem 2 to Max Coloring. Since any non trivial coloring uses at most $f(m) = \left\lfloor \frac{1}{2} + \sqrt{2m + \frac{1}{4}} \right\rfloor$ colors, there always exists a coloring of

---

[2] A classical reduction from 3-SAT produces a graph with $\Theta(var + clauses)$ vertices and edges, where var and $clauses$ are the number of variables and clauses, see for instance http://cgi.csc.liv.ac.uk/~igor/COMP309/3CP.pdf.

weight at most $\sum_{i=1}^{f(m)} w(v_i)$ (recall that v_i are in non-increasing order of weight). So we consider the following problem.

Problem 3.

- Input: a vertex-weighted graph (G, w), and an integer k.
- Parameter: k
- Question: does there exist a coloring of G of weight at most $\sum_{i=1}^{f(m)} w(v_i) - k$?

In this problem, we can save weight either by reducing the number of colors (wrt $f(m)$), or by producing colors with "small" weights, for instance by grouping together vertices of large weights (note that an optimal solution for max coloring may use a number of colors much larger than the chromatic number).

A first idea would be to use Rule 1 devised for the unweighted case (removing vertices of low degree), but this is no longer sound, because of this possibility to save on color weights. Even if there is an isolated vertex of large weight, removing it does not produce an equivalent instance (putting it with another vertex of large weight allows to save something). This case of isolated vertices can be handled (by properly modifying some weight in the remaining instance) but vertices of large weights cannot be removed even if they have small degree.

Then, vertices of large weight behave differently than vertices of small weight. Thus we decompose the vertex set according to the weights of the vertices, and this distinction is crucial in the solution of the problem. To make this distinction, we define $w^* = w_{f(m)}$: this is the smallest weight considered in the bound $\sum_{i=1}^{f(m)} w(v_i)$. We define $V_>$ (resp. $V_\leq$) as the set of vertices of weight greater than w^* (resp. at most w^*), and $n_> = |V_>|$, $n_\leq = |V_\leq|$.

The idea is to show that, unless a particular case occurs where we can safely answer YES, we can reduce the instance to a very particular case where the input graph is made of two cliques, plus $O(k)$ vertices. Then, in XP-time we can further reduce the graph to two cliques, instance on which we can solve the Max Coloring problem in polynomial time using Lemma 1. It is worth noticing that the only 'XP-time' step consists of dealing with the $O(k)$ extra-vertices.

We first consider two reduction rules.

Rule 2. While there exists in $V_\leq$ a vertex of degree at most $f(m)-k-1$, remove it.
The following Lemma states that this rule is sound.

Lemma 2. *Let (G, w, k) be an instance of Problem 3, v_ℓ a vertex in $V_\leq$ of degree at most $f(m) - k - 1$, and G' the graph obtained from G by removing v_ℓ[3]. Let $k' = \sum_{i=1}^{f(m')} w(v_i') + k - \sum_{i=1}^{f(m)} w(v_i)$. Then (G, w, k) is a YES-instance iff (G', w, k') is a YES-instance.*

Note that after this step if $k' \leq 0$ we can safely answer YES.

Rule 3. Compute a maximal matching M in the complement of $G[V_>]$. If M has size at least k, answer YES.

---

[3] We use v_i' to denote vertices of G' (so $v_i' = v_i$ for $i < \ell$ and $v_i' = v_{i+1}$ for $i \geq \ell$).

Lemma 3. *Rule 3 is sound.*

As a consequence, once Rules 2 and 3 have been applied, we get an equivalent instance where:

- Condition 1: every vertex in $V_\le$ has degree at least $f(m) - k$;
- Condition 2: $V_>$ is made of a clique $C_>$ on at least $n_> - 2k$ vertices, plus at most $2k$ vertices.

Now, we consider Algorithm 1.

Algorithm 1
STEP 0: Compute a coloring $\mathcal{C} = (S_1, S_2, \dots)$ of G;
STEP 1: While there exists a vertex v in S_j which has no neighbor in some color S_i with $i < j$, color v with S_i;
STEP 2: If it is possible to empty a color S_i, do it and go to STEP 1
STEP 3: Take any pair of colors of size 2, any pair of colors of size 1. If it is possible to re-color these 6 vertices with 3 colors, do it and go to STEP 1.

Lemma 4. *Algorithm 1 runs in polynomial time. Moreover, let (G, w, k) be an instance satisfying conditions 1 and 2. If G has more than $(k+1)^4$ edges, then Algorithm 1:*

- *Either produces a coloring with at most $f(m) - k$ colors;*
- *Or allows to find a decomposition of V into (C_1, C_2, T) where C_1 and C_2 are two cliques, and $|T| = O(k)$.*

Proof. Only the general idea of the proof is given. We first explain why Algorithm 1 takes polynomial time:

- Step 1 is clearly polynomial, since each vertex changes color at most n times.
- For the second step, S_i being an independent set, emptying S_i is *not* possible if and only if there is a vertex in S_i adjacent to all other colors, so this step is also polynomial.
- Step 3 is clearly polynomial, since it involves 4 colors and 6 vertices.
- Since the number of colors reduces in steps 2 and 3, the steps are executed $O(n)$ times.

Now we deal with the claimed property. Let $\mathcal{C} = (S_1, \dots, S_p)$ be the coloring output by the algorithm.

Thanks to Step 1, $\mathcal{C}$ is non trivial, hence $p \le f(m)$. If $p \le f(m) - k$, then the property of the Lemma holds (case 1). So we consider that $p > f(m) - k$, meaning $f(m) \le p + k - 1$. Since $f(m) + 1/2 \ge \sqrt{2m}$, we get that $m < (p+k)^2/2$. This bound on the number of edges allows to show that, thanks to Step 1, almost all the colors in $\mathcal{C}$ have size one. More precisely, let s be the number of colors of size at least 2.

Fact 1. The number of colors of size at least 2 is $s = o(p)$.

Assume[4] that these are the first s colors in $\mathcal{C}$. Now, thanks to Step 2, in each color S_i there is a vertex which is adjacent to all other colors, let z_i be such a vertex. Note that for $i > s$ z_i is the unique vertex of S_i. Denote $Z_1 = \{z_i, i \leq s\}$, $Z_2 = \{z_i, i > s\}$ and $Z = Z_1 \cup Z_2$. The second step of the proof is to show that almost all vertices of 'small' weights are in Z. More precisely, let Q be the set of vertices outside Z of 'small' weight: $Q = V_{\leq} \setminus Z$.

Fact 2. The number of vertices in Q is at most $4k + 10$

Now, thanks to condition 2, the set of vertices of large weight $V_>$ is made of a clique $C_>$ plus $O(k)$ vertices. Then, besides Z and $C_>$, G contains $O(k)$ vertices: $O(k)$ vertices from $V_{\leq}$ (the set Q), and the remaining $O(k)$ vertices from $V_>$. The third step of the proof is to show that Z is almost a clique, and this is where Step 3 of the algorithm intervenes: we can show that Z is almost a clique otherwise we could recolor 2 colors of size 2 and 2 colors of size 1 with 3 colors in total.

Fact 3. Z is made of a clique plus $O(k)$ vertices.

Then, from the previous facts we immediately get that our graph is made of:

- A clique C_Z which is a subset of Z
- A clique $C_>$ which is a subset of $V_>$
- $O(k)$ other vertices.   □

Now, we are ready to show that the Problem 3 is in XP.

Theorem 4. *Problem 3 is in XP.*

Proof. We apply rules 2 and 3. Either we answer YES, or we have an equivalent instance satisfying conditions 1 and 2.

First consider the case where $m < (k + 1)^4$. In this case we can solve the problem in FPT time. Indeed, if there are at least two isolated vertices y_1, y_2 in the graph with $w(y_1) \geq w(y_2)$, they can be replaced by one vertex y with weight $w(y_1)$: optimal colorings have obviously the same weight before and after this replacement, so we just have to adjust (reduce) the parameter if $w(y_2) \geq w^*$, so that the value $\sum_{i=1}^{f(m)} w(v_i) - k$ remains the same.

Hence, we can assume that there is at most one isolated vertex in G, and $n \leq 2m + 1 \leq 2(k + 1)^4 + 1$, so the graph is already a kernel.

Now, suppose that $m \geq (k + 1)^4$. Then we apply Algorithm 1. If this gives a coloring using at most $f(m) - k$ colors, then this coloring has a weight at most $\sum_{i=1}^{f(m)-k} w(v_i) \leq \sum_{i=1}^{f(m)} w(v_i) - k$ so we answer YES.

Otherwise, using Lemma 4 we get a partition of the graph into 2 cliques C_1 and C_2 and a set T of size $|T| = O(k)$.

Vertices of $T \cup C_2$ will be colored with between $|C_2|$ and $|T| + |C_2|$ colors: for vertices of T, we consider the $O((|C_2| + |T|)^{|T|}) = n^{O(k)}$ possibilities for vertices in T, by putting each vertex in one of the possible colors. In each of

---

[4] Thanks to Step 2, each color of size 1 is adjacent to all other colors, so we can reorder colors if needed to put colors of size 1 at the end while preserving the fact that each vertex in S_j is adjacent to each color S_i, $i < j$.

these cases, we contract the vertices in the same color into one vertex (with weight the maximum weight of merged vertices), and produce in each case a clique C_2'. Then we only have to solve the problem on the obtained graph made of 2 cliques, in polynomial time, using Lemma 1. The global running time is $n^{O(k)}$, so the result follows. □

Theorem 4 leaves the question of fixed parameterized tractability open: *is problem 3 FPT?* In the light of our proof, it suffices to show that the problem is (or is not) FPT on the class of graphs that are the union of two cliques and $O(k)$ vertices.

4 Max Coloring Under Standard Parameterization

As mentioned in introduction, Max Coloring is hard to solve in classes of graphs where the usual coloring problem is polynomial. In particular:

- It is NP-hard in bipartite and interval graphs (hence in chordal and perfect graphs) [5,6];
- It is not solvable in polynomial time in trees under ETH [1].

Note that Max Coloring is in APX in perfect graphs [12], and admits an approximation scheme in interval graphs [11].

The question of the parameterized complexity of this problem under standard parameterization occurs for Max Coloring in these graph classes.

Problem 4.

- Input: a vertex-weighted graph (G, w), an integer B.
- Parameter: B.
- Question: does there exist a coloring with weight at most B?

Dealing with bipartite graphs, the following result settles the question.

Theorem 5. [5] *In bipartite graphs, it is NP-hard to distinguish between instances where there exists a coloring of weight at most 7 from instances where each coloring has weight at least 8.*

As an immediate corollary, we get that *Problem 4 is not in XP in bipartite graphs* assuming $P \neq NP$.

Let us now concentrate on trees/interval graphs/chordal graphs. We show that Problem 4 is FPT in these classes of graphs, using dynamic programming on tree decompositions, based on the following result.

Theorem 6. [8] *In graphs of maximum treewidth k, the problem of finding the lightest coloring having at most r colors is solvable in $O(n^{r+1})$.*

Here, the O-notation hides the dependency in k, which is $r^{O(k)}$. Actually, it can be easily seen from the proof that this result can be restated as follows: *given any sets of r integers $a_1 \geq a_2 \geq \cdots \geq a_r \geq 0$, determining whether there exists a coloring with r colors $S_1, \ldots, S_r$ such that $w(S_i) \leq a_i$ can be done in time $r^{O(k)} n^{O(1)}$ in graphs of treewidth at most k.*

Restated like this, we can derive that Problem 4 is FPT in chordal graphs.

Theorem 7. *Problem 4 is FPT in chordal graphs.*

Proof. Given a chordal graph G, let c be the size of a maximum clique (computable in polynomial time). If $c > B$ then we can safely answer NO. Otherwise, $c \leq B$. G being chordal, its treewidth is $c - 1 \leq B - 1$.

On the other hand, there exists at most $(B+1)^B$ ways to choose B weights $a_1 \geq a_2 \geq \cdots \geq a_B \geq 0$ whose sum is at most B, hence $(B+1)^B$ tuples of color weights to check in order to determine if there exists a coloring of weight at most B. Applying the result of [8] restated as above, the problem is solved FPT-time $2^{O(B \log B)} n^{O(1)}$. □

To conclude this section, let us mention that another parameterization is very natural: the maximum weight of vertices. When parameterized by the maximum weight of vertices w_{max}, Max Coloring is:

- not in XP in bipartite graphs (from the fact that in Theorem 5 weights at most 4 are used):
- FPT in trees, from Theorem 7 since in trees the optimum value is at most twice the maximum weight (take a coloring with two colors).

However, what happens in interval (or chordal) graphs? As far as we know, the (standard) complexity of Max Coloring in interval graphs with bounded weights is still an open question. Dealing with parameterized complexity: *is Max Coloring FPT in interval graphs when parameterized by the maximum weight? Is it in XP?*

5   Conclusion

Besides the open questions mentioned above, and besides other coloring problems where a similar approach could be considered, analogous questions should be also interesting for other kinds of problems. Let us mention that the same arguments as for coloring show that the problem of *saving k bins* in bin packing is FPT, as well as other partitioning problems. Indeed, consider a problem where we have to partition a set X of size n into a minimum number of subsets, where only some subsets are admissible - stable sets in coloring, elements with sum of lengths at most 1 in bin packing - (see [9] for similar questions dealing with approximation). If any subset of an admissible subset is admissible, as it is the case in the two previous problems, then the question of determining whether there exists a partition with at most $n - k$ sets or not is FPT. Indeed, one computes a maximal set of admissible pairs. If there are at least k pairs then we are done (singletons are admissible, otherwise the problem has no feasible solution), otherwise the remaining $n - O(k)$ elements must be each in a separate subset. Then using brute force on the $O(k)$ elements in the pairs and solving maximum matchings, we get an FPT algorithm. Improving upon better bounds could be also interesting for these problems.

References

1. Araújo, J., Nisse, N., Pérennes, S.: Weighted coloring in trees. SIAM J. Discrete Math. **28**(4), 2029–2041 (2014)
2. Cai, L.: Parameterized complexity of vertex colouring. Discrete Appl. Math. **127**(3), 415–429 (2003)
3. Chor, B., Fellows, M., Juedes, D.W.: Linear kernels in linear time, or how to save k colors in $O(n^2)$ steps. In: Hromkovič, J., Nagl, M., Westfechtel, B. (eds.) WG 2004. LNCS, vol. 3353, pp. 257–269. Springer, Heidelberg (2004)
4. de Werra, D., Demange, M., Escoffier, B., Monnot, J., Paschos, V.T.: Weighted coloring on planar, bipartite and split graphs: complexity and approximation. Discrete Appl. Math. **157**(4), 819–832 (2009)
5. Demange, M., de Werra, D., Monnot, J., Paschos, V.T.: Time slot scheduling of compatible jobs. J. Sched. **10**(2), 111–127 (2007)
6. Escoffier, B., Monnot, J., Paschos, V.T.: Weighted coloring: further complexity and approximability results. Inf. Process. Lett. **97**(3), 98–103 (2006)
7. Garey, M.R., Johnson, D.S., Stockmeyer, L.J.: Some simplified NP-complete graph problems. Theor. Comput. Sci. **1**(3), 237–267 (1976)
8. Guan, D.J., Zhu, X.: A coloring problem for weighted graphs. Inf. Process. Lett. **61**(2), 77–81 (1997)
9. Hassin, R., Monnot, J.: The maximum saving partition problem. Oper. Res. Lett. **33**(3), 242–248 (2005)
10. Jansen, B.M.P., Kratsch, S.: Data reduction for graph coloring problems. Inf. Comput. **231**, 70–88 (2013)
11. Nonner, T.: Clique clustering yields a PTAS for max-coloring interval graphs. In: Aceto, L., Henzinger, M., Sgall, J. (eds.) ICALP 2011, Part I. LNCS, vol. 6755, pp. 183–194. Springer, Heidelberg (2011)
12. Pemmaraju, S.V., Raman, R.: Approximation algorithms for the max-coloring problem. In: Caires, L., Italiano, G.F., Monteiro, L., Palamidessi, C., Yung, M. (eds.) ICALP 2005. LNCS, vol. 3580, pp. 1064–1075. Springer, Heidelberg (2005)

Finding Two Edge-Disjoint Paths
with Length Constraints

Leizhen Cai and Junjie Ye[(✉)]

Department of Computer Science and Engineering,
The Chinese University of Hong Kong, Shatin, Hong Kong SAR, China
{lcai,jjye}@cse.cuhk.edu.hk

Abstract. We consider the problem of finding, for two pairs (s_1, t_1) and (s_2, t_2) of vertices in an undirected graph, an (s_1, t_1)-path P_1 and an (s_2, t_2)-path P_2 such that P_1 and P_2 share no edges and the length of each P_i satisfies L_i, where $L_i \in \{\leq k_i, = k_i, \geq k_i, \leq \infty\}$. We regard k_1 and k_2 as parameters and investigate the parameterized complexity of the above problem when at least one of P_1 and P_2 has a length constraint (note that $L_i = $ " $\leq \infty$" indicates that P_i has no length constraint). For the nine different cases of (L_1, L_2), we obtain FPT algorithms for seven of them. Our algorithms uses random partition backed by some structural results. On the other hand, we prove that the problem admits no polynomial kernel for all nine cases unless $NP \subseteq coNP/poly$.

Keywords: Edge-disjoint paths · Random partition · Parameterized complexity · Kernelization

1 Introduction

Disjoint paths in graphs are fundamental and have been studied extensively in the literature. Given k pairs of *terminal vertices* (s_i, t_i) for $1 \leq i \leq k$ in an undirected graph G, the classical EDGE-DISJOINT PATHS problem asks whether G contains k pairwise edge-disjoint paths P_i between s_i and t_i for all $1 \leq i \leq k$. The problem is NP-complete as shown by Even et al. [8], but is solvable in time $O(mn)$ by network flow [16] if all vertices s_i (resp., t_i) are the same vertex s (resp., t). When we regard k as a parameter, a celebrated result of Robertson and Seymour [17] on vertex-disjoint paths can be used to obtain an FPT algorithm for EDGE-DISJOINT PATHS. On the other hand, Bodlaender et al. [4] have shown that EDGE-DISJOINT PATHS admits no polynomial kernel unless $NP \subseteq coNP/poly$.

In this paper, we study EDGE-DISJOINT PATHS with length constraints L_i on (s_i, t_i)-paths P_i and focus on the problem for two pairs of terminal vertices. The length constraints $L_i \in \{\leq k_i, = k_i, \geq k_i, \leq \infty\}$ indicate that the length of P_i need to satisfy L_i. We regard k_1 and k_2 as parameters, and study the parameterized complexity of the following problem.

Partially supported by GRF grant CUHK410212 of the Research Grants Council of Hong Kong.

© Springer-Verlag GmbH Germany 2016
P. Heggernes (Ed.): WG 2016, LNCS 9941, pp. 62–73, 2016.
DOI: 10.1007/978-3-662-53536-3_6

EDGE-DISJOINT (L_1, L_2)-PATHS
INSTANCE: Graph $G = (V, E)$, two pairs (s_1, t_1) and (s_2, t_2) of vertices.
QUESTION: Does G contain (s_i, t_i)-paths P_i for $i = 1, 2$ such that P_1 and P_2 share no edge and the length of P_i satisfies L_i?

There are nine different length constraints on two paths (note that EDGE-DISJOINT $(\leq \infty, \leq \infty)$-PATHS puts no length constraint on two paths). For instance, EDGE-DISJOINT $(= k_1, \leq \infty)$-PATHS requires that $|P_1| = k_1$ but P_2 has no length constraint, and EDGE-DISJOINT $(= k_1, \geq k_2)$-PATHS requires that $|P_1| = k_1$ and $|P_2| \geq k_2$.

Related Work. EDGE-DISJOINT (L_1, L_2)-PATHS has been studied under the framework of classical complexity. Ohtsuki [15], Seymour [18], Shiloah [20], and Thomasssen [21] independently gave polynomial-time algorithms for EDGE-DISJOINT $(\leq \infty, \leq \infty)$-PATHS. Tragoudas and Varol [22] proved the NP-completeness of EDGE-DISJOINT $(\leq k_1, \leq k_2)$-PATHS, and Eilam-Tzoreff [7] showed the NP-completeness of EDGE-DISJOINT $(\leq k_1, \leq \infty)$-PATHS even when k_1 equals the (s_1, t_1)-distance. For EDGE-DISJOINT (L_1, L_2)-PATHS with $L_1 = k_1$ or $\geq k_1$ (same for $L_2 = k_2$ or $\geq k_2$), we can easily establish its NP-completeness by reductions from the classical HAMILTONIAN PATH problem.

As for the parameterized complexity, there are a few results in connection with our EDGE-DISJOINT (L_1, L_2)-PATHS. Golovach and Thilikos [13] obtained an $2^{O(kl)}m \log n$-time algorithm for EDGE-DISJOINT PATHS when every path has length at most l. For a single pair (s, t) of vertices, an (s, t)-path of length exactly l can be found in time $O(2.6181^l m \log^2 n)$ [9,19] and $O^*(2.5961^l)$ [23]. (Note that l-PATH that finds a path of length l can be solved in time $O(2.6181^l n \log^2 n)$ [9,19].) For the problem of finding an (s, t)-path of length at least l, Bodlaender [1] derived an $O(2^{2l}(2l)!n + m)$-time algorithm, Gabow and Nie [12] designed an $l^l 2^{O(l)} mn \log n$-time algorithm, and FPT algorithms of Fomin et al. [10] for cycles and paths can be adapted to yield an $8^{l+o(l)} m \log^2 n$-time algorithm.

Our Contributions. In this paper, we investigate the parameterized complexity of EDGE-DISJOINT (L_1, L_2)-PATHS for the nine different length constraints and have obtained FPT algorithms for seven of them (see Table 1 for a summary).

In particular, we use random partition in an interesting way to obtain FPT algorithms for EDGE-DISJOINT $(= k_1, \leq \infty)$-PATHS and EDGE-DISJOINT $(= k_1, \geq k_2)$-PATHS. This is achieved by bounding the number of some special edges, called "nearby-edges", in the two paths P_1 and P_2 by a function of k_1 and k_2 alone. We also consider polynomial kernels and prove that all nines cases admit no polynomial kernel unless $NP \subseteq coNP/poly$.

Notation and Definitions. All graphs in the paper are simple undirected connected graphs. For a graph G, we use $V(G)$ and $E(G)$ to denote its vertex set and edge set respectively, and n and m, respectively, are numbers of vertices and edges of G. For two vertices s and t, the distance between s and t is denoted by $d(s, t)$.

Table 1. Running times of FPT algorithms for EDGE-DISJOINT (L_1, L_2)-PATHS with length constraints $L_i \in \{\le k_i, = k_i, \ge k_i, \le \infty\}$ for $i = 1, 2$. Note that $r_1 = k_1 + k_2, r_2 = k_1^2 + 5k_2$, and $r_3 = k_2^2 + 5k_1$.

Constraints	$\|P_2\| \le k_2$	$\|P_2\| = k_2$	$\|P_2\| \ge k_2$	$\le \infty$
$\|P_1\| \le k_1$	$O(2.01^{r_1} m \log n)$		$O(2.01^{r_2} m \log^3 n)$	$O(2.01^{k_1^2} m \log n)$
$\|P_1\| = k_1$	$O(5.24^{r_1} m \log^3 n)$			$O(2.01^{k_1^2} m \log^3 n)$
$\|P_1\| \ge k_1$	$O(2.01^{r_3} m \log^3 n)$		Open	

An instance (I, k) of a parameterized problem Π consists of two parts: an input I and a parameter k. We say that a parameterized problem Π is fixed-parameter tractable (FPT) if there is an algorithm solving every instance (I, k) in time $f(k)|I|^{O(1)}$ for some computable function f. A kernelization algorithm for a parameterized problem Π maps an instance (I, k) in time polynomial in $|I| + k$ into a smaller instance (I', k') such that (I, k) is a yes-instance iff (I', k') is a yes-instance and $|I'| + k' \le g(k)$ for some computable function g. Problem Π has a polynomial kernel if $g(k)$ is a polynomial function.

For simplicity, we write $O(2.01^{f(k)})$ for $2^{f(k)+o(f(k))}$ as the latter is $O((2 + \epsilon)^{f(k)})$ for any constant $\epsilon > 0$ and we choose $\epsilon = 0.01$. In particular, $2^k k^{O(\log k)} = 2^{k+O(\log^2 k)} = O(2.01^k)$.

In the rest of the paper, we present FPT algorithms for seven cases in Sect. 2, and show the nonexistence of polynomial kernels in Sect. 3. We conclude with some open problems in Sect. 4.

2 FPT Algorithms

Random partition provides a natural tool for finding edge-disjoint (L_1, L_2)-paths in a graph G: We randomly partition edges of G to form two graphs G_1 and G_2, and then independently find paths P_1 in G_1 (resp., P_2 in G_2) whose lengths satisfy L_1 (resp., L_2).

When our problem satisfies the following two conditions, the above approach yields a randomized FPT algorithm and can typically be derandomized by universal sets.

1. Whenever G has a solution, the probability of "G_1 contains required P_1 and G_2 contains required P_2" is bounded below by a function of k_1 and k_2 alone.
2. It takes FPT time to find required paths P_1 in G_1 and P_2 in G_2.

Indeed, straightforward applications of the above method yield FPT algorithms for EDGE-DISJOINT (L_1, L_2)-PATHS when $L_i \in \{\le k_i, = k_i\}$ for $i = 1, 2$.

Theorem 1. EDGE-DISJOINT (L_1, L_2)-PATHS *can be solved in* $O(2.01^{k_1+k_2} m \log n)$ *time for* $(L_1, L_2) = (\le k_1, \le k_2)$, *and* $O(5.24^{k_1+k_2} m \log^3 n)$ *time for* $(L_1, L_2) = (\le k_1, = k_2)$ *or* $(= k_1, = k_2)$.

Proof. Let $r = k_1 + k_2$. We randomly color each edge by color 1 or 2 with probability $1/2$ to define a random partition of edges. Denote by $G_i, i = 1, 2$, the graph consisting of edges of color i. Then for all three cases of (L_1, L_2), the probability that both G_1 and G_2 contain required paths is at least $1/2^r$ when EDGE-DISJOINT (L_1, L_2)-PATHS has a solution.

We can use BFS starting from s_i to determine whether G_i contains an (s_i, t_i)-path of length at most k_i in time $O(m)$, and an algorithm of Fomin et al. [10] to determine whether G_i contains an (s_i, t_i)-path of length exactly l in time $O(2.6181^l m \log^2 n)$. Furthermore, we use a family of (m, r)-universal sets of size $2^r r^{O(\log r)} \log m$ [14] for derandomization. Therefore EDGE-DISJOINT (L_1, L_2)-PATHS can be solved in time

$$2^r r^{O(\log r)} \log m * m = 2^r r^{O(\log r)} m \log n = O(2.01^r m \log n)$$

for $(L_1, L_2) = (\leq k_1, \leq k_2)$, and time

$$2^r r^{O(\log r)} \log m * (2.6181^{k_1} + 2.6181^{k_2}) m \log^2 n = O(5.24^r m \log^3 n)$$

for $(L_1, L_2) = (\leq k_1, = k_2)$ or $(= k_1, = k_2)$. $\square$

For other cases of (L_1, L_2), a random edge partition of G does not, unfortunately, gurantee condition (1) because of the possible existence of a long path in a solution. To handle such cases, we will compute some special edges and then use random partition on such edges to ensure condition (1). For this purpose, we call a vertex v a *nearby-vertex* if $d(s_1, v) + d(v, t_1) \leq k_1$, and call an edge a *nearby-edge* if its two endpoints are both nearby-vertices. We will show that there exists a solution where the number of nearby-edges is bounded above by a polynomial in k_1 and k_2 alone, which enables us to apply random partition to nearby-edges to ensure condition (1) and hence to obtain FPT algorithms. We note that such a clever way of applying random partition has been used by Cygan et al. [6] in obtaining an Eulerian graph by deleting at most k edges.

In the next two subsections, we rely on random partition of nearby-edges to obtain FPT algorithms to solve EDGE-DISJOINT (L_1, L_2)-PATHS for the following four cases of (L_1, L_2): $(\leq k_1, \leq \infty), (= k_1, \leq \infty), (\leq k_1, \geq k_2)$ and $(= k_1, \geq k_2)$.

2.1 One Short and One Unconstrained

In this subsection, we use random partition on nearby-edges to obtain FPT algorithms for EDGE-DISJOINT (L_1, L_2)-PATHS when (L_1, L_2) is $(\leq k_1, \leq \infty)$ or $(= k_1, \leq \infty)$. To lay the foundation of our FPT algorithms, we first present the following crucial property on the number of nearby-edges in a special solution. Recall that a nearby-vertex v satisfies $d(s_1, v) + d(v, t_1) \leq k_1$ and both endpoints of a nearby-edge are nearby-vertices.

Lemma 1. *Let (s_1, t_1) and (s_2, t_2) be two pairs of vertices in a graph $G = (V, E), P_1$ an (s_1, t_1)-path of length at most k_1, and P_2 a minimum-length (s_2, t_2)-path edge-disjoint from P_1. Then*

1. *all edges in P_1 are nearby-edges, and*
2. *P_2 contains at most $(k_1 + 1)^2$ nearby-edges.*

Proof. Statement 1 is obvious and we focus on Statement 2.

For a vertex v in P_2, we say that v is a P_1-*near vertex* if there is a vertex u in P_1 such that G contains a (u, v)-path of length at most $k_1/2$ that is edge-disjoint from P_1. We call v a u-*near vertex* when we want to emphasize the endpoint u, and refer to such a (u, v)-path as a P_1-*near (u, v)-path*.

Let v^* be a nearby-vertex in P_2. Since $d(s_1, v^*) + d(v^*, t_1) \leq k_1$, there is an (s_1, v^*)-path or a (t_1, v^*)-path of length at most $k_1/2$. As s_1 and t_1 are vertices of P_1, v^* must be a P_1-near vertex. Therefore each nearby-vertex in P_2 is a P_1-near vertex, and we bound the number of P_1-near vertices to prove this lemma.

Suppose to the contrary that P_2 contains at least $(k_1 + 1)^2 + 1$ P_1-near vertices. Then by pigeonhole principle, there exists a vertex u in P_1 that has at least $k_1 + 2$ u-near vertices. Sort these vertices along P_2 from s_2 to t_2. Let v_1 and v_2 be the first and last vertex respectively. Then the (v_1, v_2)-section of P_2 has length at least $k_1 + 1$. Let W be the (v_1, v_2)-walk concatenating the P_1-near (u, v_1)-path and the P_1-near (u, v_2)-path. Then W contains at most k_1 edges and is edge-disjoint from P_1 by the definition of P_1-near path. So we can replace the (v_1, v_2)-section by W to obtain an (s_2, t_2)-walk that contains an (s_2, t_2)-path shorter than P_2, contradicting to the minimality of P_2. Therefore P_2 contains at most $(k_1 + 1)^2$ P_1-near vertices and thus nearby-vertices, which implies that P_2 contains at most $(k_1 + 1)^2$ nearby-edges.   □

The above lemma lays the ground for an FPT algorithm based on random partition. Let $\{E_1, E_2\}$ be a random partition of nearby-edges, and construct $G_1 = G[E_1]$ and $G_2 = G - E(G_1)$. Note that whenever G admits a solution, it has a solution (P_1, P_2) such that P_2 is a minimum-length (s_2, t_2)-path edge disjoint from P_1. Lemma 1 implies that P_1 is inside G_1 with probability $\geq 1/2^{k_1}$, and P_2 is inside G_2 with probability $\geq 1/2^{(k_1+1)^2}$. This ensures that, with probability $\geq 1/2^{k_1}$, G_1 contains an (s_1, t_1)-path of length at most k_1 and, with probability at least $1/2^{(k_1+1)^2}$, G_2 contains an (s_2, t_2)-path. Therefore with probability $\geq 1/2^{k_1+(k_1+1)^2}$, we will be able to find a solution for G by finding an (s_1, t_1)-path of length at most k_1 in G_1 and an (s_2, t_2)-path in G_2. This paves the way for the following randomized FPT algorithm for EDGE-DISJOINT $(\leq k_1, \leq \infty)$-PATHS. Note that the algorithm also works for EDGE-DISJOINT $(= k_1, \leq \infty)$-PATHS once we change "length $\leq k_1$" to "length k_1" in Step 3.

Algorithm 1.

1. Find all nearby-edges in $O(m)$ time by two rounds of BFS, one from s_1 and the other from t_1.
2. Randomly color each nearby-edge by color 1 or 2 with probability $1/2$, and color all remaining edges of G by color 2. Let G_i $(i = 1, 2)$ be the graph consisting of edges of color i.
3. Find an (s_1, t_1)-path P_1 of length $\leq k_1$ in G_1, and an (s_2, t_2)-path P_2 in G_2. Return (P_1, P_2) as a solution if both P_1 and P_2 exist, and return "No" otherwise.

Algorithm 1 solves EDGE-DISJOINT ($\leq k_1$, $\leq \infty$)-PATHS with probability $\geq 1/2^{k_1+(k_1+1)^2}$ and runs in $O(m)$ time, as the two tasks in Step 3 for G_1 and G_2 also take $O(m)$ time. Let m' be the number of nearby-edges and $r = k_1 + (k_1+1)^2$. We can use (m', r)-universal sets to derandomize our algorithm, and obtain a deterministic FPT algorithm running in time

$$2^r r^{O(\log r)} \log n * m' = O(2.01^{k_1^2} m \log n).$$

For EDGE-DISJOINT ($= k_1$, $\leq \infty$)-PATHS, Step 3 takes more time as it takes $O(2.6181^{k_1} m \log^2 n)$ time to find an (s_1, t_1)-path P_1 of length k_1. Therefore our deterministic FPT algorithm for the problem takes time

$$2^r r^{O(\log r)} \log m' * 2.6181^{k_1} m \log^2 n = O(2.01^{k_1^2} m \log^3 n).$$

Theorem 2. EDGE-DISJOINT ($\leq k_1, \leq \infty$)-PATHS *and* EDGE-DISJOINT ($= k_1, \leq \infty$)-PATHS *can be solved in time* $O(2.01^{k_1^2} m \log n)$ *and* $O(2.01^{k_1^2} m \log^3 n)$ *respectively.*

2.2   One Short and One Long

Now we consider EDGE-DISJOINT (L_1, L_2)-PATHS when (L_1, L_2) is ($\leq k_1, \geq k_2$) or ($= k_1, \geq k_2$). The main difficulty lies in the possibility that one path may be long, and we overcome this obstacle by the following lemma similar to Lemma 1 to upper bound the number of nearby-edges in a special solution. Again, the lemma enables us to use random partition on nearby-edges to obtain FPT algorithms for both cases.

For an (s_1, t_1)-path P, a *P-valid* (s_2, t_2)-*path* is an (s_2, t_2)-path that is edge-disjoint from P and has length at least k_2.

Lemma 2. *Let* (s_1, t_1) *and* (s_2, t_2) *be two pairs of vertices in a graph* $G = (V, E)$, P *an* (s_1, t_1)-*path of length at most* k_1, *and* Q *a P-valid* (s_2, t_2)-*path of minimum length. Then*

1. *all edges in* P *are nearby-edges, and*
2. *at most* $k_1^2 + 3k_1 + 2k_2$ *edges of* Q *are nearby-edges.*

The proof is omitted due to the space limit, and will appear in the full version of this paper.

The above lemma enables us to obtain a randomized FPT for EDGE-DISJOINT ($\leq k_1$, $\geq k_2$) by replacing Step 3 of Algorithm 1 as follows:

Step 3: Find an (s_1, t_1)-path P_1 of length $\leq k_1$ in G_1, and an (s_2, t_2)-path P_2 of length $\geq k_2$ in G_2. Return (P_1, P_2) as a solution if both P_1 and P_2 exist, and return "No" otherwise.

By Lemma 2, the randomized algorithm solves EDGE-DISJOINT ($\leq k_1$, $\geq k_2$)-PATHS with probability $\geq 1/2^{k_1^2+4k_1+2k_2}$. Since an (s_2, t_2)-path P_2 of length $\geq k_2$ can be found in time $8^{k_2+o(k_2)} m \log^2 n$ [10] as mentioned earlier in the introduction, the two tasks in Step 3 takes $8^{k_2+o(k_2)}m\log^2 n$ time and thus the randomized algorithm runs in the same time. Let m' be the number of nearby-edges and $r = k_1^2 + 4k_1 + 2k_2$. We can use (m', r)-universal sets to derandomize our algorithm, and obtain a deterministic FPT algorithm for EDGE-DISJOINT $(\leq k_1, \geq k_2)$-PATHS running in time

$$2^r r^{O(\log r)} \log m' * 8^{k_2+o(k_2)}m\log^2 n = O(2.01^{k_1^2+5k_2}m\log^3 n).$$

For EDGE-DISJOINT $(= k_1, \geq k_2)$-PATHS, Step 3 needs to find an (s_1, t_1)-path P_1 of length k_1 which takes $O(2.6181^{k_1}m\log^2 n)$ time. Therefore our deterministic FPT algorithm for the problem takes time

$$2^r r^{O(\log r)} \log m' * O(2.6181^{k_1}m\log^2 n + 8^{k_2+o(k_2)}m\log^2 n) = O(2.01^{k_1^2+5k_2}m\log^3 n).$$

Theorem 3. *Both* EDGE-DISJOINT $(\leq k_1, \geq k_2)$-PATHS *and* EDGE-DISJOINT $(= k_1, \geq k_2)$-PATHS *can be solved in time* $O(2.01^{k_1^2+5k_2}m\log^3 n)$.

3   Incompressibility

Having obtained FPT algorithms, we are impelled to investigate the existence of polynomial kernels for EDGE-DISJOINT (L_1, L_2)-PATHS. Our findings are negative as we will show that, unless $NP \subseteq coNP/poly$, the problem admits no polynomial kernel for all nine different cases of length constraints (L_1, L_2).

We start with relaxed-composition algorithms defined by Cai and Cai [5], which is a relaxation of composition algorithms introduced by Bodlaender et al. [2] in their pioneer work on the nonexistence of polynomial kernels, and a clipped version of cross-composition [3] without polynomial equivalence relations.

Definition 1 (relaxed-composition [5]). A relaxed-composition algorithm for a parameterized problem Π takes w instances $(I_1, k), \ldots, (I_w, k) \in \Pi$ as input and, in time polynomial in $\sum_{i=1}^{w} |I_i| + k$, outputs an instance $(I, k) \in \Pi$ such that

1. (I, k) is a yes-instance of Π iff some (I_i, k) is a yes-instance of Π, and
2. k' is polynomial in $\max_{i=1}^{w} |I_i| + \log w$.

Note that relaxed-composition algorithms relax the requirement in composition algorithms [2] for parameter k' from polynomial in k to polynomial in $\max_{i=1}^{w} |I_i| + \log w$. As observed by Cai and Cai [5], the following important result is implicitly established in Bodlaender et al. [2].

Theorem 4 [2,3,11]. *If an NP-complete parameterized problem admits a relaxed-composition algorithm, then it has no polynomial kernel, unless* $NP \subseteq coNP/poly$.

We also need the following polynomial parameter transformation (ppt-reduction in short).

Definition 2 (ppt-reduction [4,5]). A ppt-reduction from a parameterized problem Π to another parameterized problem Π' is an algorithm that, for input $(I, k) \in \Pi$, takes time polynomial in $|I| + k$ and outputs an instance $(I', k) \in \Pi'$ such that

1. (I, k) is a yes-instance of Π iff (I', k') is a yes-instance of Π', and
2. parameter k' is bounded above by a polynomial of k.

Theorem 5 [4]. *If there is a ppt-reduction from a parameterized problem Π to another parameterized problem Π', then Π' admits no polynomial kernel whenever Π admits no polynomial kernel.*

Now we show the nonexistence of polynomial kernels for seven easy cases. We first use relaxed-compositions to show the nonexistence of polynomial kernels of (s,t)-k-PATH (resp., LONG (s,t)-PATH) that are NP-complete problems of finding an (s,t)-path of length k (resp., $\geq k$). Then we present ppt-reductions from these two problems to EDGE-DISJOINT (L_1, L_2)-PATHS problems.

Lemma 3. *Both (s,t)-k-PATH and LONG (s,t)-PATH admit no polynomial kernel unless $NP \subseteq coNP/poly$.*

Proof. Given w instances of (s,t)-k-PATH with s_i and t_i being the two terminal vertices of the i-th instance for $1 \leq i \leq w$, we can relaxed-composite these w instances into one instance by identifying s_i (resp., t_i) as one vertex for all $1 \leq i \leq w$. Then, by Theorem 4, (s,t)-k-PATH admits no polynomial kernel unless $NP \subseteq coNP/poly$. By the same relaxed-composition, we can deduce that LONG (s,t)-PATH admits no polynomial kernel unless $NP \subseteq coNP/poly$.   □

Theorem 6. EDGE-DISJOINT (L_1, L_2)-PATHS *for* (L_1, L_2) *being* $(\leq k_1, = k_2), (\leq k_1, \geq k_2), (= k_1, = k_2), (= k_1, \leq \infty), (= k_1, \geq k_2), (\geq k_1, \leq \infty)$ *or* $(\geq k_1, \geq k_2)$, *admits no polynomial kernel unless* $NP \subseteq coNP/poly$.

Proof. Given an instance of (s,t)-k-PATH, we construct an instance of EDGE-DISJOINT $(= k_1, \leq \infty)$-PATHS as following:

1. Set $s_1 = s$ and $t_1 = t$, and $k_1 = k$,
2. add new vertices s_2 and t_2, and edge $s_2 t_2$.

The above reduction is clearly a ppt-reduction, and thus EDGE-DISJOINT $(= k_1, \leq \infty)$-PATHS admits no polynomial kernel unless $NP \subseteq coNP/poly$. For the other six cases, similar ppt-reductions from (s,t)-k-PATH or LONG (s,t)-PATH will work.   □

Now we consider the remaining two cases of length constraints $(\leq k_1, \leq k_2)$ and $(\leq k_1, \leq \infty)$. Following our argument for the other cases, we can easily construct ppt-reductions from the problem of determining whether G contains an (s, t)-path of length at most k. Unfortunately, this short path problem is solvable in polynomial time and thus admits a polynomial kernel, which makes such ppt-reductions meaningless for the purpose of proving the nonexistence of polynomial kernels. In fact, these two cases are difficult to deal with, and we will design delicate relaxed-composition algorithms to establish the nonexistence of their polynomial kernels.

Theorem 7. *Both* EDGE-DISJOINT $(\leq k_1, \leq k_2)$-PATHS *and* EDGE-DISJOINT $(\leq k_1, \leq \infty)$-PATHS *admit no polynomial kernel unless* $NP \subseteq coNP/poly$.

Proof. Let $(G^1, \leq k_1, \leq k_2), \ldots, (G^w, \leq k_1, \leq k_2)$ be w instances of EDGE-DISJOINT $(\leq k_1, \leq k_2)$-PATHS, and $n = \max_{i=1}^{w} |V(G_i)|$. Let (s_1^i, t_1^i) and (s_2^i, t_2^i) be the two pairs of vertices of the i-th instance for $1 \leq i \leq w$. Assume that w is a power of two, say $w = 2^d$. Otherwise we can add some redundant no-instances to make w a power of two.

We first show how to composite two instances into one instance, which is the crucial step of our relaxed-composition. Given the i-th instance and j-th instance, we construct a new instance $(G', \leq k_1', \leq k_2')$ as following (See Fig. 1 for an illustration.):

1. Create two pairs of vertices (s_1', t_1') and (s_2', t_2'), and four vertices u_1, u_2, v_1 and v_2.
2. Connect these new vertices with graph G^i and G^j as showed in Fig. 1, where each dashed/dotted edge is a *short-path* of length one, and each normal edge is a *long-path* of length $k_1 + 4$.
3. Denote by G' the new graph and set $k_1' = k_1 + 4, k_2' = k_2 + 3(k_1 + 4) + 1$.

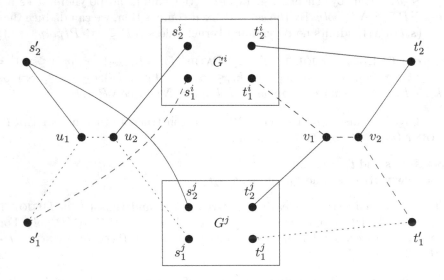

Fig. 1. The relaxed-composition for two instances. Here a dashed/dotted edge is a short-path of length one, and a normal edge is a long-path of length $k_1' = k_1 + 4$.

We claim that $(G', \leq k_1', \leq k_2')$ is a yes-instance iff one of these two instances is a yes-instance.

Suppose that one of these two instances has a solution. Without loss of generality, assume that $(G^i, \leq k_1, \leq k_2)$ has a solution (P_1, P_2). Let P_1' be the (s_1', t_1')-path concatenated by P_1 and the four dashed short-paths, and P_2' be the (s_2', t_2')-path going through u_1, u_2, s_2^i and t_2^i, whose (s_2^i, t_2^i)-section is P_2. By the edge-disjointness between P_1 and P_2, P_1' and P_2' are edge-disjoint. Furthermore, we have $|P_1'| \leq k_1'$ and $|P_2'| \leq k_2'$ as $|P_1'| \leq k_1$ and $|P_2'| \leq k_2$. Then (P_1', P_2') is a solution of $(G', \leq k_1', \leq k_2')$.

Conversely, suppose that (P_1', P_2') is a solution of $(G', \leq k_1', \leq k_2')$. Since P_1' has length at most $k_1' = k_1 + 4$, and each long-path has length $k_1 + 4$, P_1' contains either all dotted short-paths or dashed short-paths. Assume that P_1' contains all dotted short-paths. (The argument is similar when P_1' contains all dashed short-paths.) Then the (s_1^j, t_1^j)-section P_1 of P_1' is an (s_1^j, t_1^j)-path in G^j of length at most k_1. Moreover, P_2' must be an (s_2', t_2')-path going through the (s_2', s_2^j)-long-path P_s and the (t_2^j, v_1)-long-path P_t. Since $d(v_1, t_2') = k_1 + 5$, the (s_2^j, t_2^j)-section $P_2 \in G^j$ of P_2' has length at most

$$|P_2'| - |P_s| - |P_t| - d(v_1, t_2) \leq (k_2 + 3k_1 + 13) - 2(k_1 + 4) - (k_1 + 5) \leq k_2.$$

Then (P_1, P_2) is a solution of $(G^j, \leq k_1, \leq k_2)$.

Now we are ready to present our relaxed-composition that contains $d = \log w$ iterations. In the i-th iteration, there are 2^{d-i+1} instances and we group these instances into 2^{d-i} pairs for $1 \leq i \leq d$. For each pair, we composite them into one instance as presented above. Finally, there remains only one instance which completes the relaxed-composition. Let $(\leq k_1^i, \leq k_2^i)$ be the length constraints after the i-th iteration for $0 \leq i \leq d$. Note that $k_1^0 = k_1$ and $k_2^0 = k_2$. The recursion relation for k_1^i and k_2^i is

$$k_1^{i+1} = k_1^i + 4 \text{ and } k_2^{i+1} = k_2^i + 3k_1^{i+1} + 1,$$

as short-path and long-path respectively have length 1 and k_1^{i+1} in the i-th iteration. We have $k_1^i = k_1 + 4i$ and $k_2^i = k_2 + (3k_1 + 1)i + 6i(i + 1)$ for $0 \leq i \leq d$.

Let $(G'', \leq k_1'', \leq k_2'')$ be the final instance, where $k_1'' = k_1^d = k_1 + 4d$ and $k_2'' = k_2^d = k_2 + (3k_1 + 1)d + 6d(d + 1)$. By above proof for the composition of two instances, we can deduce that $(G'', \leq k_1'', \leq k_2'')$ has a solution iff one of these w instances has a solution. Both k_1'' and k_2'' are polynomially bounded in $n + \log w$ as $d = \log w$. This composition is a valid relaxed-composition. Since EDGE-DISJOINT $(\leq k_1, \leq k_2)$-PATHS is NP-complete, by Theorem 4, it admits no polynomial kernel unless $NP \subseteq coNP/poly$.

The relaxed-composition also holds if we discard the length constraint for the second path, i.e. discard the length constraints " $\leq k_2$ " and " $\leq k_2'$ ", which yields that EDGE-DISJOINT $(\leq k_1, \leq \infty)$-PATHS admits no polynomial kernel unless $NP \subseteq coNP/poly$. $\qquad\square$

4   Concluding Remarks

We have obtained FPT algorithms to solve EDGE-DISJOINT (L_1, L_2)-PATHS for seven of the nine different cases of length constraints (L_1, L_2), and also established the nonexistence of polynomial kernels for all nine cases, assuming $NP \nsubseteq coNP/poly$. However parameterized complexities of the remaining two cases are open.

Problem 1. Determine the parameterized complexities of EDGE-DISJOINT $(\geq k_1, \leq \infty)$-PATHS and EDGE-DISJOINT $(\geq k_1, \geq k_2)$-PATHS.

Note that an FPT algorithm for EDGE-DISJOINT $(\geq k_1, \geq k_2)$-PATHS will yield a new polynomial-time algorithm to solve EDGE-DISJOINT PATHS for two pairs of terminal vertices (i.e., EDGE-DISJOINT $(\leq \infty, \leq \infty)$-PATHS).

We can consider vertex-disjoint paths, instead of edge-disjoint paths, and form VERTEX-DISJOINT (L_1, L_2)-PATHS problems for nine different length constraints (L_1, L_2). It is straightforward to obtain FPT algorithms by color-coding or random partition for the three cases of (L_1, L_2) being $(\leq k_1, \leq k_2), (= k_1, \leq k_2)$ or $(= k_1, = k_2)$. Interestingly, both VERTEX-DISJOINT $(\geq k_1, \leq \infty)$-PATHS and VERTEX-DISJOINT $(\geq k_1, \geq k_2)$-PATHS can be solved by finding a minor that is a disjoint union of two paths, and thus are FPT by the graph minor theorem. (Note that we can not use this approach to solve Problem 1 by transforming edge-disjoint paths into vertex-disjoint paths through line graphs, because paths in line graphs may not correspond to paths in original graphs.) The remaining four cases seem much harder than their corresponding edge-disjoint counterparts. We note that structural properties similar to Lemmas 1 and 2 seem not hold for vertex-disjoint paths with length constraints.

On the other hand, our proofs for the nonexistence of polynomial kernels also work for VERTEX-DISJOINT (L_1, L_2)-PATHS, and hence VERTEX-DISJOINT (L_1, L_2)-PATHS admits no polynomial kernel unless $NP \subseteq coNP/poly$ for all nine different cases of length constraints (L_1, L_2).

Finally, we can consider both edge-disjoint and vertex-disjoint paths with length constraints for digraphs, which appear to be much harder than these problems on undirected graphs.

Problem 2. For digraphs, determine the parameterized complexity of EDGE-DISJOINT (L_1, L_2)-PATHS and VERTEX-DISJOINT (L_1, L_2)-PATHS for various length constraints (L_1, L_2).

References

1. Bodlaender, H.L.: On linear time minor tests with depth-first search. J. Algorithms **14**(1), 1–23 (1993)
2. Bodlaender, H.L., Downey, R.G., Fellows, M.R., Hermelin, D.: On problems without polynomial kernels. J. Comput. Syst. Sci. **75**(8), 423–434 (2009)
3. Bodlaender, H.L., Jansen, B.M., Kratsch, S.: Kernelization lower bounds by cross-composition. SIAM J. Discrete Math. **28**(1), 277–305 (2014)

4. Bodlaender, H.L., Thomassé, S., Yeo, A.: Kernel bounds for disjoint cycles and disjoint paths. Theoret. Comput. Sci. **412**, 4570–4578 (2011)
5. Cai, L., Cai, Y.: Incompressibility of H-free edge modification problems. Algorithmica **71**(3), 731–757 (2014)
6. Cygan, M., Marx, D., Pilipczuk, M., Pilipczuk, M., Schlotter, I.: Parameterized complexity of Eulerian deletion problems. Algorithmica **68**(1), 41–61 (2014)
7. Eilam-Tzoreff, T.: The disjoint shortest paths problem. Discrete Appl. Math. **85**(2), 113–138 (1998)
8. Even, S., Itai, A., Shamir, A.: On the complexity of timetable and multicommodity flow problems. SIAM J. Comput. **5**(4), 691–703 (1976)
9. Fomin, F.V., Lokshtanov, D., Panolan, F., Saurabh, S.: Representative sets of product families. In: Schulz, A.S., Wagner, D. (eds.) ESA 2014. LNCS, vol. 8737, pp. 443–454. Springer, Heidelberg (2014)
10. Fomin, F.V., Lokshtanov, D., Saurabh, S.: Efficient computation of representative sets with applications in parameterized and exact algorithms. In: Proceedings of the 25th Annual ACM-SIAM Symposium on Discrete Algorithms, pp. 142–151. SIAM (2014)
11. Fortnow, L., Santhanam, R.: Infeasibility of instance compression and succinct PCPs for NP. J. Comput. Syst. Sci. **77**(1), 91–106 (2011)
12. Gabow, H.N., Nie, S.: Finding long paths, cycles and circuits. In: Hong, S.-H., Nagamochi, H., Fukunaga, T. (eds.) ISAAC 2008. LNCS, vol. 5369, pp. 752–763. Springer, Heidelberg (2008)
13. Golovach, P.A., Thilikos, D.M.: Paths of bounded length and their cuts: parameterized complexity and algorithms. Discrete Optim. **8**(1), 72–86 (2011)
14. Naor, M., Schulman, L.J., Srinivasan, A.: Splitters and near-optimal derandomization. In: Proceedings of the 36th Annual Symposium on Foundations of Computer Science, pp. 182–191. IEEE (1995)
15. Ohtsuki, T.: The two disjoint path problem and wire routing design. In: Saito, N., Nishizeki, T. (eds.) Graph Theory and Algorithms. LNCS, vol. 108, pp. 207–216. Springer, Heidelberg (1980)
16. Orlin, J.B.: Max flows in O(nm) time, or better. In: Proceedings of the Forty-fifth Annual ACM Symposium on Theory of Computing, pp. 765–774. ACM (2013)
17. Robertson, N., Seymour, P.D.: Graph minors. XIII. The disjoint paths problem. J. Comb. Theory Ser. B **63**(1), 65–110 (1995)
18. Seymour, P.D.: Disjoint paths in graphs. Discrete Math. **29**(3), 293–309 (1980)
19. Shachnai, H., Zehavi, M.: Representative families: a unified tradeoff-based approach. J. Comput. Syst. Sci. **82**(3), 488–502 (2016)
20. Shiloach, Y.: A polynomial solution to the undirected two paths problem. J. ACM **27**(3), 445–456 (1980)
21. Thomassen, C.: 2-linked graphs. Eur. J. Comb. **1**(4), 371–378 (1980)
22. Tragoudas, S., Varol, Y.L.: Computing disjoint paths with length constraints. In: d'Amore, F., Franciosa, P.G., Marchetti-Spaccamela, A. (eds.) Graph-Theoretic Concepts in Computer Science. LNCS, vol. 1197, pp. 375–389. Springer, Heidelberg (1997)
23. Zehavi, M.: Mixing color coding-related techniques. In: Bansal, N., Finocchi, I. (eds.) ESA 2015. LNCS, vol. 9294, pp. 1037–1049. Springer, Heidelberg (2015)

Packing and Covering Immersion Models of Planar Subcubic Graphs

Archontia C. Giannopoulou[1([⊠])], O-joung Kwon[2], Jean-Florent Raymond[3,5], and Dimitrios M. Thilikos[3,4]

[1] Technische Universität Berlin, Berlin, Germany
archontia.giannopoulou@tu-berlin.de
[2] Institute for Computer Science and Control, Hungarian Academy of Sciences, Budapest, Hungary
ojoungkwon@gmail.com
[3] AlGCo Project-Team, CNRS, LIRMM, Montpellier, France
sedthilk@thilikos.info
[4] Department of Mathematics, National and Kapodistrian University of Athens, Athens, Greece
[5] University of Warsaw, Warsaw, Poland
jean-florent.raymond@mimuw.edu.pl

Abstract. A graph H is an immersion of a graph G if H can be obtained by some subgraph G after lifting incident edges. We prove that there is a polynomial function $f : \mathbb{N} \times \mathbb{N} \to \mathbb{N}$, such that if H is a connected planar subcubic graph on $h > 0$ edges, G is a graph, and k is a non-negative integer, then either G contains k vertex/edge-disjoint subgraphs, each containing H as an immersion, or G contains a set F of $f(k, h)$ vertices/edges such that $G \setminus F$ does not contain H as an immersion.

Keywords: Erdö–Pósa properties · Graph immersions · Packings and coverings in graphs

1 Introduction

All graphs is this paper are finite, undirected, loopless, and may have multiedges. Let $\mathcal{C}$ be a class of graphs. A $\mathcal{C}$-vertex/edge cover of G is a set S of vertices/edges such that each subgraph of G that is isomorphic to a graph in $\mathcal{C}$ contains some element of S. A $\mathcal{C}$-vertex/edge packing of G is a collection of vertex/edge-disjoint subgraphs of G, each isomorphic to some graph in $\mathcal{C}$.

The original version of this chapter was revised: The affiliation of the author has been corrected. The erratum to this chapter is available at 10.1007/978-3-662-53536-3_26

A.C. Giannopoulou—The research of this author has been supported by the European Research Council (ERC) under the European Union's Horizon 2020 research and innovation programme (ERC consolidator grant DISTRUCT, agreement No 648527) and by the Warsaw Center of Mathematics and Computer Science.

O.-j. Kwon—Supported by ERC Starting Grant PARAMTIGHT (No. 280152)

J.-F. Raymond—Supported by the (Polish) National Science Centre grant PRE-LUDIUM 2013/11/N/ST6/02706.

© Springer-Verlag GmbH Germany 2016
P. Heggernes (Ed.): WG 2016, LNCS 9941, pp. 74–84, 2016.
DOI: 10.1007/978-3-662-53536-3_7

We say that a graph class C has the *vertex/edge Erdö–Pósa property* (shortly v/e-*E&P property*) for some graph class G if there is a function $f : \mathbb{N} \to \mathbb{N}$, called a *gap function*, such that, for every graph G in G and every non-negative integer k, either G has a vertex/edge C-packing of size k or G has a vertex/edge C-cover of size $f(k)$. In the case where G is the class of all graphs we simply say that C has the v/e-E&P property. An interesting topic in Graph Theory, related to the notion of duality between graph parameters, is to detect instantiations of C and G such that C has the v/e-E&P property for G and, if yes, optimize the corresponding gap. Certainly, the first result of this type was the celebrated result of Erdős and Pósa in [11] who proved that the class of all cycles has the v-E&P property with gap function $O(k \cdot \log k)$. This result have triggered a lot of research on its possible extensions. One of the most general ones was given in [22] where its was proven that the class of graphs that are contractible to some graph H have the v-E&P property iff H is planar (see also [4,5,8] for improvements on the gap function).

Other instantiations of C for which the v-E&P property has been proved concern odd cycles [16,19], long cycles [2], and graphs containing cliques as minors [9] (see also [13,15,21] for results on more general combinatorial structures).

As noticed in [8], cycles have the e-E&P property as well. Interestingly, only few more results exist for the cases where the e-E&P property is satisfied. It is known for instance that graphs contractible to θ_r (i.e. the graph consisting of two vertices and an edge of multiplicity r between them) have the e-E&P property [3]. Moreover it was proven that odd cycles have the e-E&P property for planar graphs [17] and for 4-edge-connected graphs [16].

Given two graphs G and H, we say that H is an *immersion* of G if H can be obtained from some subgraph of G by lifting incident edges (see Sect. 2 for the definition of the lift operation). Given a graph H, we denote by $\mathcal{I}(H)$ the set of all graphs that contain H as an immersion. Using this terminology, the edge variant of the original result of Erdős and Pósa in [11] implies that the class $\mathcal{I}(\theta_2)$ has the v-E&P property (and, according to [8], the e-E&P property as well). A natural question is whether this can be extended for $\mathcal{I}(H)$, for other H's, different than θ_2. This is the question that we consider in this paper. A distinct line of research is to identify the graph classes G such that for every graph H, $\mathcal{I}(H)$ has the e-E&P property for G. In this direction, it was recently proved in [18] that for every graph H, $\mathcal{I}(H)$ has the e-E&P property for 4-edge-connected graphs.

In this paper we show that if H is non-trivial (i.e., has at least one edge), connected, planar, and subcubic, i.e., each vertex is incident to at most 3 edges, then $\mathcal{I}(H)$ has the v/e-E&P property (with polynomial gap in both cases). More concretely, our main result is the following.

Theorem 1. *Let H be a connected planar subcubic graph of $h > 0$ edges, let $k \in \mathbb{N}$, and let G be a graph without any $\mathcal{I}(H)$-vertex/edge packing of size greater than k. Then G has a $\mathcal{I}(H)$-vertex/edge cover of size bounded by a polynomial function of h and k.*

The main tools of our proof are the graph invariants of tree-cut width and tree-partititon width, defined in [24] and [10] respectively (see Sect. 2 for the formal definitions). Our proof uses the fact that every graph of polynomially (on k) big tree-cut width contains a wall of height k as an immersion (as proved in [24]). This permits us to consider only graphs of bounded tree-cut width and, by applying suitable reductions, we finally reduce the problem to graphs of bounded tree partition width (Theorem 2). The result follows as we next prove that for every H, the class $\mathcal{I}(H)$ has the e-E&P property for graphs of bounded tree-partition width (Theorem 3).

One might conjecture that the result in Theorem 1 is tight in the sense that both being planar and subcubic are necessary for H in order $\mathcal{I}(H)$ to have the e-E&P property. In this direction, in Sect. 7, we give counterexamples for the cases where H is planar but not subcubic and is subcubic but not planar.

2   Definitions and Preliminary Results

We use $\mathbb{N}^+$ for the set of all positive integers and we set $\mathbb{N} = \mathbb{N}^+ \cup \{0\}$.

Graphs. As already mentioned, we deal with loopless graphs where multiedges are allowed. Given a graph G, we denote by $V(G)$ its set of vertices and by $E(G)$ its multiset of edges. The notation $|E(G)|$ stands for the total number of edges, that is, counting multiplicities. We use the term *multiedge* to refer to a 2-element set of adjacent vertices and the term *edge* to deal with one particular instanciation of the multiedge connecting two vertices. The function mult_G maps a set of two vertices of G to the multiplicity of the edge connecting them, or zero if they are not adjacent. If $\mathsf{mult}_G(\{u, v\}) = k$ for some $k \in \mathbb{N}^+$, we denote by $\{u, v\}_1, \ldots, \{u, v\}_k$ the distinct edges connecting u and v. For the sake of clarity, we identify a multiedge of multiplicity one and its edge and write $\{u, v\}$ instead of $\{u, v\}_1$ when $\mathsf{mult}_G(\{u, v\}) = 1$.

We denote by $\deg_G(v)$ the degree of a vertex v in a graph G, that is, the number of vertices that are adjacent to v. The multidegree of v, that we write $\mathrm{mdeg}_G(v)$, is the number of edges (i.e. counting multiplicities) incident with v. We drop the subscript when it is clear from the context.

Immersions. Let H and G be graphs. We say that G contains H as an *immersion* if there is a pair of functions (ϕ, ψ), called an *H-immersion model*, such that ϕ is an injection of $V(H) \rightarrow V(G)$ and ψ sends $\{u, v\}_i$ to a path of G between $\phi(u)$ and $\phi(v)$, for every $\{u, v\} \in E(H)$ and every $i \in \{1, \ldots, \mathsf{mult}_H(\{u, v\})\}$, in a way such that distinct edges are sent to edge-disjoint paths. Every vertex in the image of ϕ is called a *branch vertex*. An *H-immersion expansion* M in a graph G is a subgraph of G defined as follows: $V(M) = \phi(V(H)) \cup \bigcup_{e \in H} V(\psi(e))$ and $E(M) = \bigcup_{e \in E(H)} E(\psi(e))$ for some H-immersion model (ϕ, ψ) of G. We call the paths in $\psi(E(H))$ *certifying paths* of the H-immersion expansion M.

We say that two edges are *incident* if they share some endpoint. A *lift* of two incident edges $e_1 = \{x, y\}$ and $e_2 = \{y, z\}$ of G is the operation that removes the edges e_1 and e_2 from the graph and then, if $x \neq z$, adds the edge $\{x, z\}$ (or increases the multiplicity of $\{x, z\}$ by 1 if this edge already exists). Notice that H is an immersion of G if and only if a graph isomorphic to H can be obtained from some subgraph of G after applying lifts of incident edges[1].

Packings and Coverings. An H-cover of G is a set $C \subseteq E(G)$ such that $G \backslash C$ does not contain H as an immersion. An H-packing in G is a collection of edge-disjoint H-immersion expansions in G. We denote by $\mathbf{pack}_H(G)$ the maximum size of an H-packing and by $\mathbf{cover}_H(G)$ the minimum size of an H-cover in G.

Rooted Trees. A *rooted tree* is a pair (T, s) where T is a tree and $s \in V(T)$ is a vertex referred to as the *root*. Given a vertex $x \in V(T)$, the *descendants* of x in (T, s), denoted by $\mathrm{desc}_{(T,s)}(x)$, is the set containing each vertex w such that the unique path from w to s in T contains x. If y is a descendant of x and is adjacent to x, then it is a *child* of x. Two vertices of T are *siblings* if they are children of the same vertex. Given a rooted tree (T, s) and a vertex $x \in V(G)$, the *height* of x in (T, s) is the maximum distance between x and a vertex in $\mathrm{desc}_{(T,s)}(x)$.

We now define two types of decompositions of graphs: tree-partitions (cf. [14,23]) and tree-cut decompositions (cf. [24]).

Tree-Partitions. We introduce, especially for the needs of our proof, a multi-graph extension of the parameter of tree-partition width defined in [14,23] where we could consider the number of edges between the bags and the number of vertices in the bags. A *tree-partition* of a graph G is a pair $\mathcal{D} = (T, \mathcal{X})$ where T is a tree and $\mathcal{X} = \{X_t\}_{t \in V(T)}$ is a partition of $V(G)$ such that either $|V(T)| = 1$ or for every $\{x, y\} \in E(G)$, there exists an edge $\{t, t'\} \in E(T)$ where $\{x, y\} \subseteq X_t \cup X_{t'}$. We call the elements of $\mathcal{X}$ *bags* of $\mathcal{D}$. Given an edge $f = \{t, t'\} \in E(T)$, we define E_f as the set of edges with one endpoint in X_t and the other in $X_{t'}$. The *width* of $\mathcal{D}$ is defined as $\max\{|X_t|\}_{t \in V(T)} \cup \{|E_f|\}_{f \in E(T)}$. The *tree-partition width* of G is the minimum width over all tree-partitions of G and will be denoted by $\mathbf{tpw}(G)$. A *rooted tree-partition* of a graph G is a triple $\mathcal{D} = ((T, s), \mathcal{X})$ where (T, s) is a rooted tree and $(\mathcal{X}, T)$ is a tree-partition of G.

Tree-Cut Decompositions. A *near-partition* of a set S is a collection of pairwise disjoint subsets $S_1, \ldots, S_k \subseteq S$ (for some $k \in \mathbb{N}$) such that $\bigcup_{i=1}^{k} S_i = S$. Observe that this definition allows a set of the family to be empty. A *tree-cut decomposition* of a graph G is a pair $\mathcal{D} = (T, \mathcal{X})$ where T is a tree and $\mathcal{X} = \{X_t\}_{t \in V(T)}$ is a near-partition of $V(G)$. As in the case of tree-partitions, we call the elements of $\mathcal{X}$ *bags* of $\mathcal{D}$. A *rooted tree-cut decomposition* of a graph

---

[1] While we mentioned this definition in the introduction, we now adopt the more technical definition of immersion in terms of immersion models as this will facilitate the presentation of the proofs.

G is a triple $\mathcal{D} = ((T,s), \mathcal{X})$ where (T,s) is a rooted tree and $(T, \mathcal{X})$ is a tree-cut decomposition of G. Given that $\mathcal{D} = ((T,s), \mathcal{X})$ is a rooted tree partition or a rooted tree-cut decomposition of G and given $t \in V(T)$, we set
$$G_t = G \left[\bigcup_{t \in \mathrm{desc}_{(T,s)}(t)} X_t \right].$$
The *torso* of a tree-cut decomposition $(T, \mathcal{X})$ at a node t is the graph obtained from G as follows. If $V(T) = \{t\}$, then the torso at t is G. Otherwise let $T_1, \ldots, T_\ell$ be the connected components of $T \setminus t$. The torso H_t at t is obtained from G by *consolidating* each vertex set $\bigcup_{b \in V(T_i)} X_b$ into a single vertex z_i. The operation of consolidating a vertex set Z into z is to replace Z with z in G, and for each edge e between Z and $v \in V(G) \setminus Z$, adding an edge zv in the new graph. Given a graph G and $X \subseteq V(G)$, let the *3-center* of (G, X) be the unique graph obtained from G by suppressing vertices in $V(G) \setminus X$ of degree two and deleting vertices of degree at most 1. For each node t of T, we denote by $\widetilde{H_t}$ the *3-center* of (H_t, X_t), where H_t is the torso of $(T, \mathcal{X})$ at t.

Let $\mathcal{D} = ((T,s), \mathcal{X})$ be a rooted tree-cut decomposition of G. The adhesion of a vertex t of T, that we write $\mathrm{adh}_\mathcal{D}(t)$, is the number of edges with exactly one endpoint in G_t. The *width* of a tree-cut decomposition $(\mathcal{X}, T)$ of G is $\max_{t \in V(T)}\{\mathrm{adh}_\mathcal{D}(t), |\widetilde{H_t}|\}$. The *tree-cut width* of G, denoted by $\mathbf{tcw}(G)$, is the minimum width over all tree-cut decompositions of G.

A vertex $t \in V(T)$ is *thin* if $\mathrm{adh}_\mathcal{D}(t) \leq 2$, and *bold* otherwise. We also say that $\mathcal{D}$ is *nice* if for every thin vertex $t \in V(T)$ we have $N(V(G_t)) \cap \bigcup_{b \text{ is a sibling of } t} V(G_b) = \emptyset$. In other words, there is no edge from a vertex of G_t to a vertex of G_b, for any sibling b of t, whenever t is thin. The notion of nice tree-cut decompositions has been introduced by Ganian et al. in [12]. Furthermore, they proved the following result.

Proposition 1 [12]. *Every rooted tree-cut decomposition can be transformed into a nice one without increasing the width.*

We say than an edge of G *crosses* the bag X_t, for some $t \in V(T)$ if its endpoints belongs to bags X_{t_1} and X_{t_2}, for some $t_1, t_2 \in V(T)$ such that t belongs to the interior of the (unique) path of T connecting t_1 to t_2.

3   From Tree-Cut Decompositions to Tree-Partitions

The purpose of this section is to prove the following theorem.

Theorem 2. *For every connected graph G, and every connected graph H with at least one edge, there is a graph G' such that*

- $\mathbf{tpw}(G') \leq (\mathbf{tcw}(G) + 1)^2/2$,
- $\mathbf{pack}_{H+}(G') \leq \mathbf{pack}_H(G)$, *and*
- $\mathbf{cover}_H(G) \leq \mathbf{cover}_{H+}(G')$.

Theorem 2 will allow us in Sect. 4 to consider graphs of bounded tree-partition width instead of graphs of bounded tree-cut width.

For every graph G, we define G^+ as the graph obtained if, for every vertex v, we add two new vertices v' and v'' and the edges $\{v', v''\}$ (of multiplicity 2), $\{v, v'\}$ and $\{v, v''\}$ (both of multiplicity 1). Observe that for every G, every vertex of G^+ has degree at least 3. We also define G^* as the graph obtained by adding, for every vertex v, the new vertices $v'_1, \ldots, v'_{\mathrm{mdeg}(v)}$ and $v''_1, \ldots, v''_{\mathrm{mdeg}(v)}$ and the edges $\{v'_i, v''_i\}$ (of multiplicity 2), $\{v, v'_i\}$, and $\{v, v''_i\}$ (both of multiplicity 1), for every $i \in \{1, \ldots, \deg(v)\}$. If v is a vertex of G, then we denote by $Z_{v,i}$ the subgraph $G^*[\{v, v'_i, v''_i\}]$ for every $i \in \{1, \ldots, \mathrm{mdeg}_G(v)\}$. Our first aim is to prove the following three lemmata.

Lemma 1 $(\star^2)$. *Let G be a graph, let H be a connected graph with at least one edge and let G' be a subdivision of G^*. Then we have*

– $\mathbf{pack}_{H^+}(G^*) = \mathbf{pack}_{H^+}(G')$ *and*
– $\mathbf{cover}_{H^+}(G^*) = \mathbf{cover}_{H^+}(G')$.

Lemma 2 $(\star)$. *For every two graphs H and G such that H is connected and has at least one edge, we have $\mathbf{pack}_{H^+}(G^*) \leq \mathbf{pack}_H(G)$.*

Lemma 3 $(\star)$. *For every two graphs H and G such that H is connected and has at least one edge, we have $\mathbf{cover}_H(G) \leq \mathbf{cover}_{H^+}(G^*)$.*

We are now ready to prove Theorem 2.

Proof (of Theorem 2). Let $k = \mathbf{tcw}(G)$. We examine the nontrivial case where G is not a tree, i.e., $\mathbf{tcw}(G) \geq 2$. Let us consider the graph G^*. We claim that $\mathbf{tcw}(G^*) = \mathbf{tcw}(G)$. Indeed, starting from an optimal tree-cut decomposition of G, we can, for every vertex v of G and for every $i \in \{1, \ldots, \mathrm{mdeg}_G(v)\}$, create a bag that is a children of the one of v and contains $\{v'_i, v''_i\}$. According to the definition of G^*, this creates a tree-cut decomposition $\mathcal{D} = ((T, s), \{X_t\}_{t \in V(T)})$ of G^*. Observe that for every vertex x that we introduced to the tree of the decomposition during this process, $\mathrm{adh}_{\mathcal{D}}(x) = 2$ and the corresponding bag has size two. This proves that $\mathbf{tcw}(G^*) \leq \max(\mathbf{tcw}(G), 2) = \mathbf{tcw}(G)$. As G is a subgraph of G^*, we obtain $\mathbf{tcw}(G) \leq \mathbf{tcw}(G^*)$ and the proof of the claim is complete.

According to Proposition 1, we can assume that G^* has a nice rooted tree-cut decomposition of width $\leq k$. For notational simplicity we again denote it by $\mathcal{D} = ((T, s), \{X_t\}_{t \in V(T)})$ and, obviously, we can also assume that all leaves of T correspond to non-empty bags.

Our next step is to transform the rooted tree-cut decomposition $\mathcal{D}$ into a rooted tree-partition $\mathcal{D}' = ((T, s), \{X'_t\}_{t \in V(T)})$ of a subdivision G' of G^*. Notice that the only differences between two decompositions are that, in a tree-cut decomposition, empty bags are allowed as well as edges connecting vertices of bags corresponding to non-adjacent vertices of T.

---

[2] All proofs with a $(\star)$ have been omitted from this extended abstract.

We proceed as follows: if X is a bag crossed by edges, we subdivide every edge crossing X and add the obtained subdivision vertex to X. By repeating this process we decrease at each step the number of bags crossed by edges, that eventually reaches zero. Let G' be the obtained graph and observe that G' is a subdivision of G. As G is connected, the obtained rooted tree-cut decomposition $\mathcal{D}' = ((T, s), \{X'_t\}_{t \in V(T)})$ is a rooted tree partition of G'.

Notice that the adhesion of any bag of T in $\mathcal{D}$ is the same as in $\mathcal{D}'$. However, the bags of $\mathcal{D}'$ may grow during the construction of G'. Let t be a vertex of T and let $\{t_1, \ldots, t_m\}$ be the set of children of t. We claim that $|X'_t| \le (k+1)^2/2$.

Let E_t be the set of edges crossing X_t in G. Let H_t be the torso of $\mathcal{D}$ at t, and let $H'_t = H_t \setminus X_t$. Observe that $|E_t|$ is the same as the number of edges in H'_t. Let z_p be the vertex of H'_t corresponding to the parent of t, and similarly for each $i \in \{1, \ldots, m\}$ let z_i be the vertex of H'_t corresponding to the child t_i of t. Notice that if t_i is a thin child of t, then z_i can be adjacent to only z_p as $\mathcal{D}$ is a nice rooted tree-cut decomposition. Thus the sum of the number of incident edges with z_i in H'_t for all thin children t_i of t is at most $\mathrm{adh}_{\mathcal{D}}(t) \le k$. On the other hand, if t_i is a bold child of t, then z_i has at least 3 neighbors in H_t, and thus it is contained in the 3-center of (H_t, X_t). Thus, the number of all bold children of t is bounded by $k - |X_t|$. Since each vertex in H'_t is incident with at most k edges, the total number of edges in H'_t is at most $(k - |X_t| + 1)k/2 + k$. As $|E(H'_t)| = |E_t| = |X'_t \setminus X_t|$, it implies that $|X'_t| \le |X_t| + k \cdot (k - |X_t| + 2)/2 \le \max\{2k, k(k+2)/2\} \le (k+1)^2/2$. We conclude that G' has a rooted tree partition of width at most $(\mathbf{tcw}(G)+1)^2/2$.

Recall that G' is a subdivision of G^*. By the virtue of Lemmas 3, 2, and 1, we obtain that $\mathbf{pack}_{H^+}(G') \le \mathbf{pack}_H(G)$ and $\mathbf{cover}_H(G) \le \mathbf{cover}_{H^+}(G')$. Hence G' satisfies the desired properties. $\qquad\square$

4 Erdős-Pósa for Bounded Tree-Partition Width

Before we proceed, we require the following lemma and an easy observation.

Lemma 4 ($\star$). *Let G and H be two graphs and let $X \subseteq V(G)$. Let $\mathcal{C}$ be the collection of connected components of $G \setminus X$. If M is an H-immersion expansion of G then M contains vertices from at most $(|X| + 1) \cdot |E(H)|$ graphs of $\mathcal{C}$.*

Observation 1. *Let G and H be graphs and let $F \subseteq E(G)$. Then it holds that $\mathbf{cover}_H(G) \le \mathbf{cover}_H(G \setminus F) + |F|$.*

For a graph H, we define $\omega_H : \mathbb{N} \to \mathbb{N}$ so that $\omega_H(r) = \left\lceil r \cdot \frac{3r+1}{2} \cdot |E(H)| \right\rceil$. The next Theorem is an important ingredient of our results.

Theorem 3. *Let H be a connected graph with at least one edge. Then for every graph G it holds that $\mathbf{cover}_H(G) \le \omega_H(\mathbf{tpw}(G)) \cdot \mathbf{pack}_H(G)$*

Proof. Let us show by induction on k that if $\mathbf{pack}_H(G) \le k$ and $\mathbf{tpw}(G) \le r$ then $\mathbf{cover}_H(G) \le \omega_H(r) \cdot k$.

The case $k = 0$ is trivial. Let us now assume that $k \ge 1$ and that for every graph G of tree-partition width at most r and such that $\mathbf{pack}_H(G) = k - 1$, we have $\mathbf{cover}_H(G) \leq \omega_H(r)(k-1)$. Let G be a graph such that $\mathbf{pack}_H(G) = k$ and $\mathbf{tpw}(G) \leq r$. Let also $\mathcal{D} = ((T, s), \{X_t\}_{t \in V(T)})$ be an optimal rooted tree partition of G. We say that a vertex $t \in V(T)$ is *infected* if G_t contains an H-immersion expansion. Let t be an infected vertex of T of minimum height.

Claim. If some of the H-immersion expansions of G shares an edge with $G_{t'}$ for some child t' of t, then it also shares an edge with $E_{\{t,t'\}}$.

Proof of Claim: Let M be some H-immersion expansions of G. Notice that, by the choice of t, M cannot be entirely inside in $G_{t'}$. This fact, together with the connectivity of M, implies that $E(M) \cap E_{\{t,t'\}} \neq \emptyset$. $\square$

Suppose that M is an H-immersion expansion of G_t and let U be the set of children of t corresponding to bags which share vertices with M. We define the multisets $A = E(G[X_t]) \cap E(M)$, $B = \bigcup_{t' \in U} E_{\{t,t'\}}$ and $C = \bigcup_{t' \in U} E(G'_t)$. We also set $D = A \cup B$. By the definition of U, it follows that $E(M) \subseteq C \cup D$ **(1)**.

Let us upper-bound the size of $|D|$. Applying Lemma 4 for G_t, H, and X_t, we have $|U| \leq (r+1) \cdot |E(H)|$, hence $|B| \leq r(r+1) \cdot |E(H)|$. Besides, every path of M connecting two branch vertices meets every vertex of X_t at most once (as it is a path), thus $E(M)$ does not contain an edge of $G[X_t]$ with a multiplicity larger than $|E(H)|$. It follows that $|A| \leq \frac{r(r-1)}{2} \cdot |E(H)|$ and finally we obtain that $|D| = |A| \cup |B| \leq r \cdot \frac{3r+1}{2} \cdot |E(H)| = \omega_H(r)$.

Let $G' = G \setminus D$. We now show that $\mathbf{pack}_H(G') \leq k-1$. Let us consider an H-immersion expansion M' in G'. As $E(M') \subseteq E(G) \setminus D$, if follows that $E(M') \cap D = \emptyset$. **(2)**.

Recall that $B \subseteq D$, which together with **(2)** implies that $E(M') \cap B = \emptyset$. This fact, combined with the claim above, implies that $E(M') \cap C = \emptyset$. **(3)** From **(2)** and **(3)**, we obtain that $E(M') \cap (C \cup D) = \emptyset$, which, combined with **(1)**, implies that $E(M) \cap E(M)' \neq \emptyset$. Consequently, every maximum packing of H-immersion expansions in G' is edge-disjoint from M. If such a packing had size $\geq k$, it would form together with M a packing of size $k+1$ in G, a contradiction. Thus $\mathbf{pack}_H(G') \leq k-1$, as desired. By the induction hypothesis applied on G', $\mathbf{cover}_H(G') \leq \omega_H(r) \cdot (k-1)$ edges. Therefore, from Observation 1, $\mathbf{cover}_H(G) \leq |D| + \mathbf{cover}_H(G') \leq |D| + \omega_H(r) \cdot (k-1) \leq \omega_H(r) \cdot k$ edges as required. $\square$

Theorem 4. *Let H be a connected graph with at least one edge, $r \in \mathbb{N}$, and G be a graph where $\mathbf{tcw}(G) \leq r$. Then $\mathbf{cover}_H(G) \leq \sigma(r) \cdot (4 \cdot |V(H)| + |E(H)|) \cdot \mathbf{pack}_H(G)$, where $\sigma(r) = \lceil \frac{1}{8}(3(r+1)^4 + 2(r+1)^2) \rceil$.*

Proof. Clearly, we can assume that G is connected, otherwise we work on each of its connected components separately. By Theorem 2, there is a graph G' where $\mathbf{tpw}(G') \leq (r+1)^2/2$, $\mathbf{pack}_{H^+}(G') \leq \mathbf{pack}_H(G)$ and $\mathbf{cover}_H(G) \leq \mathbf{cover}_{H^+}(G')$. The result follows as, from Theorem 3, $\mathbf{cover}_{H^+}(G') \leq \omega_{H^+}((r+1)^2/2) \cdot \mathbf{pack}_{H^+}(G')$ and $\omega_{H^+}((r+1)^2/2) = \sigma(r) \cdot |E(H^+)| \leq \sigma(r) \cdot (4 \cdot |V(H)| + |E(H)|)$. $\square$

5   Erdős-Pósa for Immersions of Subcubic Planar Graphs

Grids and Walls. Let k and r be positive integers where $k, r \geq 2$. The $(k \times r)$-grid $\Gamma_{k,r}$ is the Cartesian product of two paths of lengths $k - 1$ and $r - 1$ respectively. We denote by Γ_k the $(k \times k)$-grid. The k-wall W_k is the graph obtained from a $((k+1) \times (2 \cdot k + 2))$-grid with vertices (x, y), $x \in \{1, \ldots, k+1\}$, $y \in \{1, \ldots, 2k + 2\}$, after the removal of the "vertical" edges $\{(x, y), (x + 1, y)\}$ for odd $x + y$, and then the removal of all vertices of degree 1. We say that k is the *height* of the wall.

Lemma 5 ($\star$). *Every connected planar subcubic graph H is an immersion of the wall $W_{|V(H)|}$.*

By combining [24, Theorem 17] with the main result of [7] (see also [6]) we can readily obtain the following.

Theorem 5. *There is a function $\tau : \mathbb{N}^+ \to \mathbb{N}$ such that the following holds: for every graph G and $r \in \mathbb{N}^+$, if $\mathbf{tcw}(G) \geq f(r)$ then W_r is an immersion of G. Moreover, $f(r) = O(r^{29}\mathrm{polylog}(r))$.*

Lemma 6. *Let G be a graph and let H be a connected planar subcubic graph on h vertices. Then $\mathbf{tcw}(G) = O(h^{29} \cdot (\mathbf{pack}_{\mathcal{I}(H)}(G))^{14.5} \cdot (\mathrm{polylog}(h) + \mathrm{polylog}(\mathbf{pack}_{\mathcal{I}(H)}(G)))$.*

Proof. Let $\mathbf{pack}_H(G) \leq k$. Let $g(h, k) = f((h+1) \cdot \lceil (k+1)^{1/2} \rceil)$, where f is the function of Theorem 5. Suppose that $\mathbf{tcw}(G) \geq g(h, k)$. Then, from Theorem 5, we obtain that G contains the wall W of height $(h + 1) \cdot \lceil (k + 1)^{1/2} \rceil$ as an immersion. Notice that W contains $k+1$ vertex-disjoint walls $W_1, W_2, \ldots, W_{k+1}$ of height h. From Lemma 5, each one of these walls contains H as an immersion and thus an H-immersion expansion. Since, these walls are vertex-disjoint they are also edge-disjoint. Hence, we have found a packing of H of size $k + 1 > k$, a contradiction. Therefore, $\mathbf{tcw}(G) \leq g(h, k)$. Notice now that, from Theorem 5, $g(h, k) = O(h^{29}k^{14.5}(\mathrm{polylog}(h) + \mathrm{polylog}(k)))$ as required. $\square$

The edge version of Theorem 1 follows as a corollary of Theorem 4 and Lemma 6.

6   The Vertex Case

To prove the vertex version of Theorem 1, is a much easier task. For this, we follow the same methodology by using the graph parameter of treewidth instead of tree-cut width, and topological minors instead of immersions. Due to lack of space the necessary definitions and the proof have been omitted.

7   Discussion

Notice that in Theorem 1 we demand that H is a connected graph. It is easy to extend this result if instead of H we consider some collection $\mathcal{H}$ of finite connected graphs containing one that is planar subcubic, and where $\mathcal{I}(\mathcal{H})$ contains all graphs containing some graph in $\mathcal{H}$ as an immersion. Moreover, it is easy to drop the connectivity condition for the vertex variant using arguments from [22]. However it remains open whether this can be done for the edge variant as well.

Naturally, the most challenging problem on the Erdö–Pósa properties of immersions is to characterize the graph classes:

$$\mathcal{H}^{\mathsf{v/e}} = \{H \mid \mathcal{I}(H) \text{ has the v/e-E\&P property}\}$$

In this paper we prove that both $\mathcal{H}^{\mathsf{v}}$ and $\mathcal{H}^{\mathsf{e}}$ contain all planar subcubic graphs. It is an interesting question whether $\mathcal{H}^{\mathsf{v/e}}$ are wider than this. Using arguments similar to [20,22] it is possible to prove the following.

Lemma 7. *None of $\mathcal{H}^{\mathsf{v}}$ and $\mathcal{H}^{\mathsf{e}}$ contains a non-planar subcubic graph.*

For the non-subcubic case, we can first observe that $K_{1,4}$, which is planar and non-subcubic belongs in both $\mathcal{H}^{\mathsf{v}}$ and $\mathcal{H}^{\mathsf{e}}$. However, this is not the case for all planar and non-subcubic graphs as is indicated in the following observation.

Observation 2 ($\star$). *There exists a 3-connected non-subcubic planar graph H that does belong neither to $\mathcal{H}^{\mathsf{v}}$ nor to $\mathcal{H}^{\mathsf{e}}$.*

Providing an exact characterization of $\mathcal{H}^{\mathsf{v}}$ and $\mathcal{H}^{\mathsf{e}}$ is a challenging open problem. A first step to deal with this problem could be the cases of $\theta_4 = $ and the 4-wheel . Especially for the 4-wheel, the structural results in [1] might be useful in this direction.

References

1. Belmonte, R., Giannopoulou, A., Lokshtanov, D., Thilikos, D.M.: The Structure of W_4-Immersion-Free Graphs. CoRR, abs/1602.02002 (2016)
2. Birmelé, E., Bondy, J.A., Reed, B.A.: The Erdős-Pósa property for long circuits. Combinatorica **27**(2), 135–145 (2007)
3. Chatzidimitriou, D., Raymond, J.-F., Sau, I., Thilikos, D.M.: Minors in graphs of large θ_r-girth. CoRR, abs/1510.03041 (2015)
4. Chekuri, C., Chuzhoy, J.: Large-treewidth graph decompositions and applications. In: 45st Annual ACM Symposium on Theory of Computing (STOC), pp. 291–300 (2013)
5. Chekuri, C., Chuzhoy, J.: Polynomial bounds for the grid-minor theorem. CoRR, abs/1305.6577 (2013)
6. Chuzhoy, J.: Excluded grid theorem: improved and simplified. In: Proceedings of the Forty-Seventh Annual ACM on Symposium on Theory of Computing, STOC 2015, Portland, OR, USA, 14–17 June 2015, pp. 645–654 (2015)

7. Chuzhoy, J.: Improved bounds for the excluded grid theorem. CoRR, abs/1602.02629 (2015)
8. Diestel, R.: Graph Theory. Graduate Texts in Mathematics, vol. 173, 3rd edn. Springer, Heidelberg (2005)
9. Diestel, R., Kawarabayashi, K., Wollan, P.: The Erdős-Pósa property for clique minors in highly connected graphs. J. Comb. Theor. Ser. B **102**(2), 454–469 (2012)
10. Ding, G., Oporowski, B.: On tree-partitions of graphs. Discrete Math. **149**(1–3), 45–58 (1996)
11. Erdős, P., Pósa, L.: On independent circuits contained in a graph. Can. J. Math. **17**, 347–352 (1965)
12. Ganian, R., Kim, E.J., Szeider, S.: Algorithmic applications of tree-cut width. In: Italiano, G.F., Pighizzini, G., Sannella, D.T. (eds.) MFCS 2015. LNCS, vol. 9235, pp. 348–360. Springer, Heidelberg (2015)
13. Geelen, J., Kabell, K.: The Erdős-Pósa property for matroid circuits. J. Comb. Theor. Ser. B **99**(2), 407–419 (2009)
14. Halin, R.: Tree-partitions of infinite graphs. Discrete Math. **97**(1–3), 203–217 (1991)
15. Kakimura, N., Kawarabayashi, K.: Fixed-parameter tractability for subset feedback set problems with parity constraints. Theor. Comput. Sci. **576**, 61–76 (2015)
16. Kawarabayashi, K.-I., Nakamoto, A.: The Erdös-pósa property for vertex- and edge-disjoint odd cycles in graphs on orientable surfaces. Discrete Math. **307**(6), 764–768 (2007)
17. Král', D., Voss, H.-J.: Edge-disjoint odd cycles in planar graphs. J. Comb. Theor. Ser. B **90**(1), 107–120 (2004)
18. Liu, C.-H.: Packing and covering immersions in 4-edge-connected graphs. CoRR, abs/1505.00867 (2015)
19. Rautenbach, D., Reed, B.A.: The Erdos-Pósa property for odd cycles in highly connected graphs. Combinatorica **21**(2), 267–278 (2001)
20. Raymond, J.-F., Sau, I., Thilikos, D.M.: An edge variant of the Erdős-Pósa property. Discrete Math. **339**(8), 2027–2035 (2016)
21. Reed, B.A., Robertson, N., Seymour, P.D., Thomas, R.: Packing directed circuits. Combinatorica **16**(4), 535–554 (1996)
22. Robertson, N., Seymour, P.D.: Graph minors. V. excluding a planar graph. J. Comb. Theor. Ser. B **41**(2), 92–114 (1986)
23. Seese, D.: Tree-partite graphs and the complexity of algorithms. In: Budach, L. (ed.) Proceedings of Fundamentals of Computation Theory. LNCS, vol. 199, pp. 412–421. Springer, Heidelberg (1985)
24. Wollan, P.: The structure of graphs not admitting a fixed immersion. J. Comb. Theor. Ser. B **110**, 47–66 (2015)

The Maximum Weight Stable Set Problem in (P_6, bull)-Free Graphs

Frédéric Maffray[1] and Lucas Pastor[2(⊠)]

[1] CNRS, Laboratoire G-SCOP, Univ. of Grenoble-Alpes, Grenoble, France
[2] Laboratoire G-SCOP, Univ. of Grenoble-Alpes, Grenoble, France
lucas.pastor@g-scop.grenoble-inp.fr

Abstract. We present a polynomial-time algorithm that finds a maximum weight stable set in a graph that does not contain as an induced subgraph an induced path on six vertices or a bull (the graph with vertices a, b, c, d, e and edges ab, bc, cd, be, ce).

Keywords: Stability · P_6-free · Bull-free · Polynomial time · Algorithm

1 Introduction

In a graph G, a *stable set* (also called *independent* set) is any subset of pairwise non-adjacent vertices. The MAXIMUM STABLE SET PROBLEM (henceforth MSS) is the problem of finding a stable set of maximum size. In the weighted version of this problem, each vertex x of G has a weight $w(x)$, and the weight of any subset of vertices is defined as the total weight of its elements. The MAXIMUM WEIGHT STABLE SET PROBLEM (MWSS) is then the problem of finding a stable set of maximum weight. It is well-known that MSS (and consequently MWSS) is NP-hard in general, even under various restrictions [15].

Given a fixed graph F, a graph G *contains* F when F is isomorphic to an induced subgraph of G. A graph G is said to be F-*free* if it does not contain F. Let us say that F is *special* if every component of F is a tree with no vertex of degree at least four and with at most one vertex of degree three. Alekseev [1] proved that MSS remains NP-complete in the class of F-free graphs whenever F is not special. On the other hand, when F is a special graph, it is still an open problem to know if MSS can be solved in polynomial time in the class of F-free graphs for most instances of F. It is known that MWSS is polynomial-time solvable in the class of F-free graphs when F is any special graph on at most five vertices [2, 20, 22]. Hence the new frontier to explore now is the case where F has six or more vertices.

We denote by P_n the path on n vertices. The complexity (polynomial or not) of MSS in the class of P_6-free graph is still unknown, but it has recently been

---

The authors are partially supported by ANR project STINT (reference ANR-13-BS02-0007).

P. Heggernes (Ed.): WG 2016, LNCS 9941, pp. 85–96, 2016.
DOI: 10.1007/978-3-662-53536-3_8

proved that it is quasi-polynomial [19]. There are several results on the existence of polynomial-time algorithms for MSS in subclasses of P_6-free graphs; see for example [17, 18, 24–26].

The *bull* is the graph with five vertices a, b, c, d, e and edges ab, bc, cd, be, ce (see Fig. 1). Our main result is the following.

Fig. 1. The bull.

Theorem 1. *MWSS can be solved in time $O(n^7)$ for every graph on n vertices in the class of $(P_6, bull)$-free graphs.*

Before presenting the proofs, we recall some closely related results.

Brandstädt and Mosca [7] showed that MWSS can be solved in polynomial time in the class of (odd-hole, bull)-free graphs. This class does not contain the class of $(P_6, bull)$-free graphs, notably because of the graph C_5.

Thomassé, Trotignon and Vuškovič [29] use the decomposition theorem for bull-free trigraphs, due to Chudnovsky [8,9], to prove that MWSS is FPT in the class of bull-free graphs. The bottleneck against polynomiality is a subclass called T_1. It might be that one can prove that MWSS is polynomial in the class of P_6-free graphs in T_1. However the algorithm from [29] runs in $O(n^9)$ time, while our algorithm runs in $O(n^7)$ time and is, we believe, conceptually simpler.

The proof of Theorem 1 works along the following lines. First, we reduce the problem to prime graphs, using modular decomposition (the technical terms will be defined precisely below). Next, we will show that if a prime $(P_6, bull)$-free graph G contains a certain graph G_7, then G has a structure from which we can solve MWSS in polynomial time on G. Finally, we will show that if a prime $(P_6, bull)$-free graph G contains no G_7, then the non-neighborhood of any vertex x is perfect, which implies that a maximum weight stable set containing x can be found in polynomial time, and it suffices to repeat this for every vertex x.

Let us recall some definitions and results we need. Let G be a graph. For each vertex $v \in V(G)$, we denote by $N(v)$ the set of vertices adjacent to v (the *neighbors* of v) in G. For any subset S of $V(G)$ we write $N_S(v)$ instead of $N(v) \cap S$; and for a subgraph H we write $N_H(v)$ instead of $N_{V(H)}(v)$. We denote by $G[S]$ the induced subgraph of G with vertex-set S, and we denote by $N(S)$ the set $\{v \in V(G) \setminus S \mid v$ has a neighbor in $S\}$. The complement of G is denoted by $\overline{G}$. We say that a vertex v is *complete* to S if v is adjacent to every vertex in S, and that v is *anticomplete* to S if v has no neighbor in S. For two sets $S, T \subseteq V(G)$ we say that S is complete to T if every vertex of S is adjacent to every vertex of T, and we say that S is anticomplete to T if no vertex of S is adjacent to any vertex of T.

Let $\omega(G)$ denote the maximum size of a clique in G, and let $\chi(G)$ denote the chromatic number of G (the smallest number of colors needed to color the vertices of G in such a way that no two adjacent vertices receive the same color). A graph G is *perfect* [3–5] if every induced subgraph H of G satisfies $\chi(H) = \omega(H)$. By the Strong Perfect Graph Theorem [10], a graph is perfect if and only if G and $\overline{G}$ contain no induced ℓ-cycle for any odd $\ell \geq 5$.

In a graph G a *homogeneous set* is a set $S \subseteq V(G)$ such that every vertex in $V(G) \setminus S$ is either complete to S or anticomplete to S. A homogeneous set is *proper* if it contains at least two vertices and is different from $V(G)$. A graph is *prime* if it has no proper homogeneous set. Note that prime graphs are connected.

A class of graphs is *hereditary* if, for every graph G in the class, every induced subgraph of G is also in the class. For example, for any family $\mathcal{F}$ of graphs, the class of $\mathcal{F}$-free graphs is hereditary. We will use the following theorem of Lozin and Milanič [21].

Theorem 2 ([21]). *Let $\mathcal{G}$ be a hereditary class of graphs. Suppose that there is a constant $c \geq 1$ such that the MWSS problem can be solved in time $O(|V(G)|^c)$ for every prime graph G in $\mathcal{G}$. Then the MWSS problem can be solved in time $O(|V(G)|^c + |E(G)|)$ for every graph G in $\mathcal{G}$.*

Clearly, the class of (P_6, bull)-free graphs is hereditary. By Theorem 2, in order to prove Theorem 1 it suffices to prove it for prime graphs. This is the object of the following theorem.

Theorem 3. *Let G be a prime (P_6, bull)-free graph, and let x be any vertex in G. Suppose that there is a 5-cycle induced by non-neighbors of x. Then there is a (possibly empty) clique F in G such that the induced subgraph $G \setminus F$ is triangle-free, and such a set F can be found in time $O(n^2)$.*

The proof of Theorem 3 is given in the next section. We close this section by showing how to obtain a proof of Theorem 1 on the basis of Theorem 3.

Our algorithm relies on results concerning graphs of bounded clique-width. We will not develop all the technical aspects concerning the clique-width, but we recall its definition and the results that we use. The concept of clique-width was first introduced in [11]. The *clique-width* of a graph G is defined as the minimum number of labels which are necessary to generate G by using the following operations:

– Create a vertex v labeled by integer i.
– Make the disjoint union of two labeled graphs.
– Join all vertices with label i to all vertices with label j for two labels $i \neq j$.
– Relabel all vertices of label i by label j.

A *c-expression* for a graph G of clique-width c is a sequence of the above four operations that generate G and use at most c different labels. A class of graphs $\mathcal{C}$ has *bounded clique-width* if there exists a constant c such that every graph G in $\mathcal{C}$ has clique-width at most c.

Theorem 4 ([12]). *If a class of graphs C has bounded clique-width c, and there is a function f such that for every graph G in C with n vertices and m edges a c-expression can be found in time $O(f(n,m))$, then the maximum weight stable set problem can be solved in time $O(f(n,m))$ for every graph G in C.*

Theorem 5 ([6]). *The class of $(P_6, triangle)$-free graphs has bounded clique-width c, and a c-expression can be found in time $O(|V(G)|^2)$ for every graph G in this class.*

Hence, as observed in [6], Theorems 4 and 5 imply the following.

Corollary 6 ([6]). *For any $(P_6, triangle)$-free graph G on n vertices one can find a maximum weight stable set of G in time $O(n^2)$.*

A *k-wheel* is a graph that consists of a k-cycle plus a vertex (called the center) adjacent to all vertices of the cycle. The following lemma was proved for $k \geq 7$ in [28]; actually the same proof holds for all $k \geq 6$ as observed in [14].

Lemma 7 ([14,28]). *Let G be a bull-free graph. If G contains a k-wheel for any $k \geq 6$, then G has a proper homogeneous set.*

Note that the bull is a self-complementary graph, so the preceding lemma also says that if G is prime then it does not contain the complementary graph of a k-wheel with $k \geq 6$.

Theorem 8. *Let G be a prime $(P_6, bull)$-free graph on n vertices. Then a maximum weight stable set of G can be found in time $O(n^7)$.*

Proof. Let G be a prime (P_6, bull)-free graph. Let $w : V(G) \to \mathbb{N}$ be a weight function on the vertex set of G. To find the maximum weight stable set in G it is sufficient to compute, for every vertex x of G, a maximum weight stable set containing x. So let x be any vertex in G. We want to compute the weight of a maximum stable set containing x. Clearly it suffices to compute the maximum weight stable set in each component of the induced subgraph $G \backslash (\{x\} \cup N(x))$ and make the sum over all components. Let K be any component of $G \backslash (\{x\} \cup N(x))$. We claim that:

$$\text{Either } K \text{ is perfect or it contains a 5-cycle.} \tag{1}$$

Proof of (1): Suppose that K is not perfect. Note that K contains no odd hole of length at least 7 since G is P_6-free. By the Strong Perfect Graph Theorem K contains an odd antihole C. If C has length at least 7 then $V(C) \cup \{x\}$ induces a wheel in $\overline{G}$, so G has a proper homogeneous set by Lemma 7, a contradiction because G is prime. So C has length 5, i.e., C is a 5-cycle. So (1) holds.

We can test in time $O(n^5)$ if K contains a 5-cycle. This leads to the following two cases.

Suppose that K contains no 5-cycle. Then (1) imples that K is perfect. In that case we can use the algorithms from either [13] or [27], which compute a maximal weight stable set in a bull-free perfect graph in polynomial time. The algorithm from [27] has time complexity $O(n^6)$.

Now suppose that K contains a 5-cycle. Then by Theorem 3 we can find in time $O(n^2)$ a clique F such that $G \setminus F$ is triangle-free. Consider any stable set S in K. If S contains no vertex from F, then S is in the subgraph $G \setminus F$, which is triangle-free. By Corollary 6 we can find a maximum weight stable set S_F in $G \setminus F$ in time $O(n^2)$. If S contains a vertex f from F, then $S \setminus f$ is in the subgraph $G \setminus (\{f\} \cup N(f))$, which, since F is a clique, is a subgraph of $G \setminus F$ and consequently is also triangle-free. By Corollary 6 we can find a maximum weight stable set S'_f in $G \setminus (\{f\} \cup N(f))$ in time $O(n^2)$. Then we set $S_f = S'_f \cup \{f\}$. We do this for every vertex $f \in F$. Now we need only compare the set S_F and the sets S_f (for all $f \in F$) and select the one with the largest weight. This takes time $O(n^3)$ for each component K that contains a 5-cycle.

Repeating the above for each component takes time $O(n^6)$ as the components are disjoint. Repeating this for every vertex x, the total complexity is $O(n^7)$. $\square$

Now Theorem 1 follows directly from Theorems 2 and 8.

2   Proofs

In a graph G, let H be a subgraph of G. For each $k > 0$, a *k-neighbor* of H is any vertex in $V(G) \setminus V(H)$ that has exactly k neighbors in H.

Lemma 9. *Let G be a bull-free graph. Let C be an induced 5-cycle in G, with vertices $c_1, \ldots, c_5$ and edges $c_i c_{i+1}$ for each i modulo 5. Then:*

(i) Every 2-neighbor of C is adjacent to c_i and c_{i+2} for some i.
(ii) Every 3-neighbor of C is adjacent to c_i, c_{i+1} and c_{i+2} for some i.
(iii) Every 5-neighbor of C is adjacent to every k-neighbor with $k \in \{1, 2\}$.
(iv) If C has a 4-neighbor non-adjacent to c_i for some i, then every 1-neighbor of C is adjacent to c_i.
(v) If a non-neighbor of C is adjacent to a k-neighbor of C, then $k \in \{1, 2, 5\}$.

Proof. If either (i) or (ii) fails, there is a vertex x that is either a 2-neighbor adjacent to c_i and c_{i+1} or a 3-neighbor adjacent to c_i, c_{i+1} and c_{i+3} for some i, and then $\{c_{i-1}, c_i, x, c_{i+1}, c_{i+2}\}$ induces a bull.

(iii) Let u be a 5-neighbor of C and x be a k-neighbor of C with $k \in \{1, 2\}$. So for some i the vertex x is adjacent to c_i and maybe to c_{i+2}. Then u is adjacent to x, for otherwise $\{x, c_i, c_{i+1}, u, c_{i+3}\}$ induces a bull.

(iv) Let f be a 4-neighbor of C non-adjacent to c_i. Suppose that there is a 1-neighbor x not adjacent to c_i. So, up to symmetry, x is adjacent to c_{i+1} or c_{i+2}. Then x is adjacent to f, for otherwise $\{x, c_{i+1}, c_{i+2}, f, c_{i-1}\}$ induces a bull; but then $\{x, f, c_{i-2}, c_{i-1}, c_i\}$ induces a bull.

(v) Let z be a non-neighbor of C that is adjacent to a k-neighbor x with $k \in \{3, 4\}$. So there is an integer i such that x is adjacent to c_i and c_{i+1} and not adjacent to c_{i+2}. Then $\{z, x, c_i, c_{i+1}, c_{i+2}\}$ induces a bull. $\square$

An *umbrella* is a graph that consists of a 5-wheel plus a vertex adjacent to the center of the 5-wheel only.

Lemma 10. *Let G be a bull-free graph. If G contains an umbrella, then G has a homogeneous set (that contains the 5-cycle of the umbrella).*

Proof. Let C be the 5-cycle of the umbrella, with vertices $c_1, \ldots, c_5$ and edges $c_i c_{i+1}$ for all i modulo 5. Let A be the set of vertices that are complete to C, and let Z be the set of vertices that are anticomplete to C. Let:

$$A' = \{a \in A \mid a \text{ has a neighbor in } Z\}.$$
$$A'' = \{a \in A \setminus A' \mid a \text{ has a non-neighbor in } A'\}.$$

By the hypothesis that C is part of an umbrella, we have $A' \neq \emptyset$. Let H be the component of $G \setminus (A' \cup A'')$ that contains $V(C)$. We claim that:

$$A' \cup A'' \text{ is complete to } V(H). \tag{2}$$

Proof: Pick any $b \in A' \cup A''$ and $u \in V(H)$, and let us prove that b is adjacent to u. We use the following notation. If $b \in A'$, then b has a neighbor $z \in Z$. If $b \in A''$, then b has a non-neighbor $a' \in A'$, and a' has a neighbor $z \in Z$, and b is not adjacent to z, for otherwise we would have $b \in A'$.

By the definition of H, there is a shortest path $u_0 \cdots u_p$ in H with $u_0 \in V(C)$ and $u_p = u$, and $p \geq 0$. We know that b is adjacent to u_0 by the definition of A. First, we show that b is adjacent to u_1 and finally by induction on $j = 2, \ldots p$, we show that b is adjacent to u_j.

Now suppose that $p \geq 1$. The vertex u_1 is a k-neighbor of C for some $k \geq 1$. If $k \in \{1, 2\}$, then b is adjacent to u_1 by Lemma 9 (iii). Suppose that $k \in \{3, 4\}$. Then there is an integer i such that u_1 is adjacent to c_i and not to c_{i+1}. By Lemma 9 (v), z is not adjacent to u_1. If $b \in A'$, then b is adjacent to u_1, for otherwise $\{z, b, c_{i+1}, c_i, u_1\}$ induces a bull. If $b \in A''$, then, by the preceding sentence we know that a' is adjacent to u_1; and then b is adjacent to u_1, for otherwise $\{z, a', u_1, u_0, b\}$ induces a bull. Suppose that $k = 5$. So $u_1 \in A$. Then u_1 is not adjacent to z, for otherwise we would have $u_1 \in A'$. If $b \in A'$, then b is adjacent to u_1 for otherwise we would have $u_1 \in A''$. If $b \in A''$, then, by the preceding sentence we know that a' is adjacent to u_1; and then b is adjacent to u_1, for otherwise $\{z, a', u_1, u_0, b\}$ induces a bull.

Finally suppose that $p \geq 2$. So $u_2, \ldots, u_p$ are non-neighbors of C. Since $u_2 \in Z$, we have $k \neq 5$, for otherwise we would have $u_1 \in A'$. So there is an integer h such that u_1 is adjacent to c_h and not to c_{h+2}. We may assume up to relabeling that $u_0 = c_h$. It follows that c_{h+2} has no neighbor in $\{u_0, \ldots, u_p\}$. Then, by induction on $j = 2, \ldots, p$, the vertex b is adjacent to u_j, for otherwise $\{c_{h+2}, b, u_{j-2}, u_{j-1}, u_j\}$ induces a bull. So b is adjacent to u. Thus (2) holds.

Let $R = V(G) \setminus (A' \cup A'' \cup V(H))$. By the definition of H, there is no edge between $V(H)$ and R. By (2), $V(H)$ is complete to $A' \cup A''$. Hence $V(H)$ is a homogeneous set that contains $V(C)$, and it is proper since $A' \neq \emptyset$. □

Lemma 11. *Let G be a prime $(P_6, bull)$-free graph. Let C be an induced 5-cycle in G. If a non-neighbor of C is adjacent to a k-neighbor of C, then $k = 2$.*

Proof. Let C have vertices $c_1, \ldots, c_5$ and edges $c_i c_{i+1}$ for each i modulo 5. Suppose that a non-neighbor z of C is adjacent to a k-neighbor x of C. By Lemma 9 (v), we have $k \in \{1, 2, 5\}$. If $k = 1$, say x is adjacent to c_i, then z-x-c_i-c_{i+1}-c_{i+2}-c_{i+3} is an induced P_6 in G. If $k = 5$, then $V(H) \cup \{x, y\}$ induces an umbrella, so, by Lemma 10, G has a proper homogeneous set, a contradiction. So $k = 2$. □

Let G_7 be the graph with vertex-set $\{c_1, \ldots, c_5, d, x\}$ and edge-set $\{c_i c_{i+1} \mid$ for all $i \bmod 5\} \cup \{dc_1, dc_4, dx\}$. See Fig. 2.

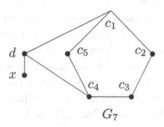

Fig. 2. The graph G_7.

Lemma 12. *Let G be a prime (P_6, bull)-free graph. Assume that G contains a 5-cycle C, with vertices $c_1, \ldots, c_5$ and edges $c_i c_{i+1}$ for all $i \bmod 5$. Moreover assume that C has a non-neighbor x in G. Then:*

(i) *There is a neighbor d of x that is a 2-neighbor of C. And consequently, $V(C) \cup \{d, x\}$ induces a G_7.*

(ii) *C has no 3-neighbor and no 5-neighbor.*

(iii) *If the vertex d from (i) is (up to symmetry) adjacent to c_1 and c_4, then every 4-neighbor of C is non-adjacent to c_5.*

Proof. Since G is prime it is connected, so there is a shortest path from C to x in G. Let x_0-$\cdots$-x_p be such a path, where $x_0 \in V(C)$ and $x_p = x$, and $p \geq 2$. By Lemma 11, x_1 is a 2-neighbor of C, so up to relabeling we may assume that x_1 is adjacent to c_1 and c_4. Then $p = 2$ for otherwise x_3-x_2-x_1-c_1-c_2-c_3 is an induced P_6. So (i) holds with $d = x_1$. Clearly, $\{c_1, \ldots, c_5, x_1, x\}$ induces a G_7.

Therefore we may assume, up to symmetry, that the vertex d from (i) is adjacent to c_1 and c_4.

Suppose that there is a vertex u that is either a 5-neighbor of C or a 4-neighbor adjacent to c_5. In either case we may assume, up to symmetry, that u is adjacent to c_1, c_3 and c_5. Then u is adjacent to d, for otherwise $\{d, c_1, c_5, u, c_3\}$ induces a bull, and u is adjacent to x, for otherwise $\{x, d, c_1, u, c_3\}$ induces a bull. But then u and x contradict Lemma 11. This proves item (iii) and that C has no 5-neighbor.

Finally suppose that C has a 3-neighbor u, adjacent to c_{i-1}, c_i, c_{i+1}; we may assume up to symmetry that $i \in \{5, 1, 2\}$. Let X be the set of vertices that are complete to $\{c_{i-1}, c_{i+1}\}$ and anticomplete to $\{c_{i-2}, c_{i+2}\}$, and let Y be the vertex-set of the component of $G[X]$ that contains c_i and u. Since G is prime, Y is not a homogeneous set, so there is a vertex t in $V(G) \setminus Y$ and vertices y, z in Y such that t is adjacent to y and not to z, and since Y is connected we may choose y and z adjacent. We claim that:

$$t \text{ is adjacent to } c_{i-2} \text{ and } c_{i+2} \text{ and to at least one of } c_{i-1} \text{ and } c_{i+1}. \tag{3}$$

Proof: If t has no neighbor in $\{c_{i-1}, c_{i+1}\}$, then t is adjacent to c_{i-2}, for otherwise $\{t, y, z, c_{i-1}, c_{i-2}\}$ induces a bull, and similarly t is adjacent to c_{i+2}; but then $\{c_{i-1}, c_{i-2}, t, c_{i+2}, c_{i+1}\}$ induces a bull. Hence t has a neighbor in $\{c_{i-1}, c_{i+1}\}$. Suppose that t is adjacent to both c_{i-1} and c_{i+1}. Since t is not in Y it must have a neighbor in $\{c_{i-2}, c_{i+2}\}$, and actually t is complete to $\{c_{i-2}, c_{i+2}\}$, for otherwise t is a 3-neighbor of the 5-cycle induced by $\{z, c_{i-1}, c_{i-2}, c_{i+2}, c_{i+1}\}$ that violates Lemma 9 (ii). Now suppose that t is adjacent to exactly one of c_{i-1}, c_{i+1}, say up to symmetry to c_{i-1}. Then t is adjacent to c_{i-2}, for otherwise $\{c_{i-2}, c_{i-1}, t, y, c_{i+1}\}$ induces a bull, and t is adjacent to c_{i+2}, for otherwise $\{c_{i+2}, c_{i-2}, t, c_{i-1}, z\}$ induces a bull. Thus (3) holds.

Now we claim that:

$$x \text{ has no neighbor in } Y \cup \{t\}. \tag{4}$$

Proof: Suppose that x has a neighbor in Y. Since x also has a non-neighbor c_i in Y, and Y is connected, there are adjacent vertices v, v' in Y such that x is adjacent to v and not to v', and then $\{x, v, v', c_{i-1}, c_{i-2}\}$ induces a bull, a contradiction. So x has no neighbor in Y. In particular x is not adjacent to y, so x has no neighbor in the 5-cycle C_y induced by $\{y, c_{i-1}, c_{i-2}, c_{i+2}, c_{i+1}\}$. By (3), t is a 3- or 4-neighbor of C_y. By Lemma 11, x is not adjacent to t. Thus (4) holds.

Suppose that $i = 5$. By (3), t is adjacent to c_2 and c_3 and, up to symmetry, to c_1. Then d is not adjacent to y, for otherwise $\{x, d, y, c_1, c_2\}$ induces a bull, and d is not adjacent to t, for otherwise $\{x, d, c_1, t, c_3\}$ induces a bull; but then $\{d, c_1, y, t, c_3\}$ induces a bull, a contradiction.

Suppose that $i = 1$. By (3), t is adjacent to c_3 and c_4. Then d is adjacent to y, for otherwise x-d-c_4-c_3-c_2-y is an induced P_6, and similarly d is adjacent to z. Then t is adjacent to d, for otherwise $\{x, d, z, y, t\}$ induces a bull, and t is adjacent to c_2, for otherwise $\{x, d, t, y, c_2\}$ induces a bull; but then $\{x, d, c_4, t, c_2\}$ induces a bull.

Finally suppose that $i = 2$. By (3), t is adjacent to c_4 and c_5. Then d is not adjacent to y, for otherwise $\{x, d, c_1, y, c_3\}$ induces a bull, and d is adjacent to t, for otherwise $\{d, c_4, c_5, t, y\}$ induces a bull; but then $\{x, d, c_4, t, y\}$ induces a bull, a contradiction. $\square$

Theorem 13. *Let G be a prime $(P_6, bull)$-free graph. Suppose that G contains a G_7, with vertex-set $\{c_1, \ldots, c_5, d, x\}$ and edge-set $\{c_i c_{i+1} \mid for~all~i \bmod 5\} \cup \{dc_1, dc_4, dx\}$. Let:*

- C be the 5-cycle induced by $\{c_1, \ldots, c_5\}$;
- F be the set of 4-neighbors of C;
- T be the set of 2-neighbors of C;
- W be the set of 1-neighbors and non-neighbors of C.

Then the following properties hold:

(i) $V(G) = \{c_1, \ldots, c_5\} \cup F \cup T \cup W$.
(ii) F is complete to $\{c_1, \ldots, c_4\}$ and anticomplete to $\{c_5, x, d\}$.
(iii) F is a clique.
(iv) $G \setminus F$ is triangle-free.

Proof. Note that $d \in T$ and $x \in W$. Clearly the sets $\{c_1, \ldots, c_5\}$, F, T, and W are pairwise disjoint subsets of $V(G)$. We observe that item (i) follows directly from the definition of the sets F, T, W and Lemma 12 (ii).

Now we prove item (ii). Consider any $f \in F$. By Lemma 12 (iii), f is non-adjacent to c_5, and consequently f is complete to $\{c_1, \ldots, c_4\}$. Then f is not adjacent to x, for otherwise $\{x, f, c_3, c_4, c_5\}$ induces a bull; and f is not adjacent to d, for otherwise $\{x, d, c_1, f, c_3\}$ induces a bull. Thus (ii) holds.

Now we prove item (iii). Suppose on the contrary that F is not a clique. So $G[F]$ has an anticomponent whose vertex-set F' satisfies $|F'| \geq 2$. Since G is prime, F' is not a homogeneous set, so there are vertices $y, z \in F'$ and a vertex $t \in V(G) \setminus F'$ that is adjacent to y and not to z, and since F' is anticonnected we may choose y and z non-adjacent. By the definition of F', we have $t \notin F$. By (ii), we have $t \notin V(C)$. Therefore, By (i), we have $t \in T \cup W$.

Suppose that $t \in T$, so t is adjacent to c_{i-1} and c_{i+1} for some i in (up to symmetry) $\{1, 2, 5\}$. If $i = 1$, then $\{z, c_2, y, t, c_5\}$ induces a bull. If $i = 2$, then $\{t, c_3, z, c_4, c_5\}$ induces a bull. So $i = 5$. Then t is not adjacent to x, for otherwise $\{x, t, c_1, y, c_3\}$ induces a bull. Then x is a non-neighbor of the 5-cycle induced by $\{c_1, c_2, c_3, c_4, t\}$, and y is a 5-neighbor of that cycle, which contradicts Lemma 12.

Hence $t \in W$. By Lemma 9 (iv), t is anticomplete to $\{c_1, c_2, c_3, c_4\}$. Then t is adjacent to each $u \in \{c_5, d\}$, for otherwise $\{t, y, c_3, c_4, u\}$ induces a bull. So t is a 1-neighbor of C, and by Lemma 11, t is not adjacent to x. But then x-d-t-y-c_3-z is an induced P_6. Thus (iii) holds.

There remains to prove item (iv). Suppose on the contrary that $G \setminus F$ contains a triangle, with vertex-set $R = \{u, v, w\}$. Clearly C and R have at most two common vertices. Moreover:

$$C \text{ and } R \text{ have at most one common vertex.} \tag{5}$$

Proof: Suppose on the contrary that $u, v \in V(C)$, and consequently $w \notin V(C)$. By Lemma 9 (i), w is a k-neighbor of C for some $k \geq 3$. Since $w \notin F$, we have $k \neq 4$, so $k \in \{3, 5\}$; but this contradicts Lemma 12 (ii). So (5) holds.

Suppose that $G \setminus F$ is not connected. Consider the component K of $G \setminus F$ that contains C; then K also contains T. Pick any vertex z in another component. By Lemma 12 (i), the vertex z must have a neighbor in T, a contradiction. Hence $G \setminus F$ is connected. It follows that there is a path from C to R in $G \setminus F$.

Let $P = p_0\text{-}\cdots\text{-}p_\ell$ be a shortest such path, with $p_0 \in V(C)$, $p_\ell = u$, and $\ell \geq 0$. Note that if $\ell \geq 1$, the vertices $p_1, \ldots, p_\ell$ are not in C. We choose R so as to minimize ℓ. Let H be the component of $G[N(u)]$ that contains v and w. Since G is prime, $V(H)$ is not a homogeneous set, so there are two vertices $y, z \in V(H)$ and a vertex $a \in V(G) \setminus V(H)$ such that a is adjacent to y and not to z, and since H is connected we may choose y and z adjacent. By the definition of H, the vertex a is not adjacent to u.

Suppose that $\ell = 0$. So $u = p_0 = c_i$ for some $i \in \{1, \ldots, 5\}$. By (5) the vertices y, z are not in C and are anticomplete to $\{c_{i-1}, c_{i+1}\}$. So, by Lemma 9 (ii), each of y and z is a 1- or 2-neighbor of C. The vertex a is adjacent to c_{i-1}, for otherwise $\{a, y, z, c_i, c_{i-1}\}$ induces a bull; and similarly a is adjacent to c_{i+1}. Note that this implies $a \notin V(C)$. Suppose that a has no neighbor in $\{c_{i-2}, c_{i+2}\}$. Then one of y, z has a neighbor in $\{c_{i-2}, c_{i+2}\}$, for otherwise $z\text{-}y\text{-}a\text{-}c_{i+1}\text{-}c_{i+2}\text{-}c_{i-2}$ is an induced P_6. So assume up to symmetry that one of y, z is adjacent to c_{i+2}. Then both y, z are adjacent to c_{i+2}, for otherwise $\{c_{i+2}, y, z, c_i, c_{i-1}\}$ induces a bull. So y and z are 2-neighbors of C, and they are not adjacent to c_{i-2}. But then $\{a, y, z, c_{i+2}, c_{i-2}\}$ induces a bull, a contradiction. Hence a has a neighbor in $\{c_{i-2}, c_{i+2}\}$. By Lemma 9 (ii) and Lemma 12 (ii), a must be adjacent to both c_{i-2}, c_{i+2}, so a is a 4-neighbor of C. Hence $a \in F$, and $i = 5$, and by (iii) a has no neighbor in $\{d, x\}$. The vertex z is not adjacent to c_2, for otherwise $\{z, c_2, c_1, a, c_4\}$ induces a bull; and similarly z is not adjacent to c_3. Then y is not adjacent to c_2, for otherwise $\{c_4, c_5, z, y, c_2\}$ induces a bull; and similarly y is not adjacent to c_3. So y and z are 1-neighbors of C, and by Lemma 11 they are not adjacent to x. Then d is adjacent to y, for otherwise $\{d, c_1, c_2, a, y\}$ induces a bull, and d is not adjacent to z, for otherwise $\{x, d, z, y, a\}$ induces a bull; but then $z\text{-}y\text{-}d\text{-}c_1\text{-}c_2\text{-}c_3$ is an induced P_6, a contradiction. Therefore $\ell \geq 1$.

We deduce that:

$$\text{Every vertex } c_i \text{ in } C \text{ has at most one neighbor in } \{u, y, z\}. \qquad (6)$$

For otherwise, c_i and two of its neighbors in $\{u, y, z\}$ form a triangle that contradicts the choice of R (the minimality of ℓ). Thus (6) holds.

Suppose that $\ell \geq 2$. By Lemma 11 (applied to p_1 and p_2), p_1 is a 2-neighbor of C, adjacent to c_{i-1} and c_{i+1} for some i. The vertex y has no neighbor c_j in C, for otherwise the path $c_j\text{-}y$ contradicts the choice of P. The vertex p_2 has no neighbor c_j in C, for otherwise the path $c_j\text{-}p_2\text{-}\cdots\text{-}p_\ell$ contradicts the choice of P. Put $p' = p_3$ if $\ell \geq 3$ and $p' = y$ if $\ell = 2$. Then $p'\text{-}p_2\text{-}p_1\text{-}c_{i+1}\text{-}c_{i+2}\text{-}c_{i-2}$ is an induced P_6, a contradiction.

Therefore $\ell = 1$, so $u = p_1$. By (i), and since $u \notin F$, u is either a 1-neighbor or a 2-neighbor of C.

Suppose that u is a 1-neighbor of C, adjacent to c_i for some i. By (6), y and z are not adjacent to c_i. Then a is adjacent to c_i, for otherwise $\{a, y, z, u, c_i\}$ induces a bull. If a has a neighbor in $\{c_{i-1}, c_{i+1}\}$, then, by Lemma 9 (ii) and Lemma 12 (ii), a is a 4-neighbor of C; but then a and u violate Lemma 9 (iv). So a has no neighbor in $\{c_{i-1}, c_{i+1}\}$. Then z is not adjacent to c_{i+1}, for otherwise, by (6), $\{a, y, u, z, c_{i+1}\}$ induces a bull; and z has no neighbor c in $\{c_{i-2}, c_{i+2}\}$, for otherwise, by (6), $\{c_i, u, y, z, c\}$ induces a bull. But then z-u-c_i-c_{i+1}-c_{i+2}-c_{i-2} is an induced P_6, a contradiction.

Therefore u is a 2-neighbor of C, adjacent to c_{i-1} and c_{i+1} for some i. By (6), y and z are anticomplete to $\{c_{i-1}, c_{i+1}\}$. The vertex c_{i+2} has no neighbor in $\{y, z\}$, for otherwise, by (6), $\{c_{i+2}, y, z, u, c_{i-1}\}$ induces a bull. Likewise, c_{i-2} has no neighbor in $\{y, z\}$. The vertex a is adjacent to c_{i-1}, for otherwise $\{a, y, z, u, c_{i-1}\}$ induces a bull, and similarly a is adjacent to c_{i+1}. Then a has a neighbor in $\{c_{i-2}, c_{i+2}\}$, for otherwise z-y-a-c_{i+1}-c_{i+2}-c_{i-2} is an induced P_6. By Lemma 9 (ii) and Lemma 12 (ii), a is a 4-neighbor of C, so $i = 5$, and a has no neighbor in $\{c_5, d, x\}$. Then y is adjacent to c_5, for otherwise $\{y, a, c_3, c_4, c_5\}$ induces a bull; and by (6), z is not adjacent to c_5. But then z-y-c_5-c_4-c_3-c_2 is an induced P_6, a contradiction. This completes the proof of the theorem.   □

Finally, Theorem 3 follows as a direct consequence of Lemma 12 and Theorem 13.

3   Conclusion

In a parallel paper [23], but using different techniques, we proved that the problem of 4-coloring (P_6, bull)-free graphs can be solved in polynomial time. It is not known if there exists a polynomial-time algorithm that determines 4-colorability in the whole class of P_6-free graphs. We note that the class of (P_6, bull)-free graph does not have bounded clique-width, since it contains the class of complements of bipartite graphs, which has unbounded clique-width [16]. Hence the main result of this paper and of [23] cannot be obtained solely with a bounded clique-width argument.

References

1. Alekseev, V.E.: On the local restrictions effect on the complexity of finding the graph independence number. Comb. Algebraic Methods Appl. Math. **132**, 3–13 (1983). Gorkiy Universty Press (in Russian)
2. Alekseev, V.E.: A polynomial algorithm for finding maximum independent sets in fork-free graphs. Discrete Anal. Oper. Res. Ser. **1**(6), 3–19 (1999). (in Russian)
3. Berge, C.: Les problèmes de coloration en théorie des graphes. Publ. Inst. Stat. Univ. Paris **9**, 123–160 (1960)
4. Berge, C.: Färbung von Graphen, deren sämtliche bzw. derenungerade Kreise starr sind (Zusammenfassung). Wiss. Z. MartinLuther Univ. Math. Natur. Reihe (Halle-Wittenberg) **10**, 114–115 (1961)
5. Berge, C.: Graphs. North-Holland, Amsterdam (1985)
6. Brandstädt, A., Klembt, T., Mahfud, S.: P_6- and triangle-free graphs revisited: Structure and bounded clique-width. Discrete Math. Theor. Comput. Sci. **8**, 173–188 (2006)
7. Brandstädt, A., Mosca, R.: Maximum weight independent sets in odd-hole-free graphs without dart or without bull. Graphs Comb. **31**, 1249–1262 (2015)
8. Chudnovsky, M.: The structure of bull-free graphs I: three-edge paths with centers and anticenters. J. Comb. Theor. B **102**, 233–251 (2012)

9. Chudnovsky, M.: The structure of bull-free graphs II and III: a summary. J. Comb. Theor. B **102**, 252–282 (2012)

10. Chudnovsky, M., Robertson, N., Seymour, P., Thomas, R.: The strong perfect graph theorem. Ann. Math. **164**, 51–229 (2006)

11. Courcelle, B., Engelfriet, J., Rozenberg, G.: Handle-rewriting hypergraph grammars. J. Comput. Syst. Sci. **46**, 218–270 (1993)

12. Courcelle, B., Makowsky, J.A., Rotics, U.: Linear time solvable optimization problems on graphs of bounded clique width. Theor. Comput. Syst. **33**, 125–150 (2000)

13. de Figueiredo, C.M.H., Maffray, F.: Optimizing bull-free perfect graphs. SIAM J. Discrete Math. **18**, 226–240 (2004)

14. de Figueiredo, C.M.H., Maffray, F., Porto, O.: On the structure of bull-free perfect graphs. Graphs Comb. **13**, 31–55 (1997)

15. Garey, M.R., Johnson, D.S.: Computers and Intractability: A Guide to the Theory of NP-Completeness. W.H. Freeman, New York (1979)

16. Golumbic, M.C., Rotics, U.: On the clique—width of perfect graph classes. In: Widmayer, P., Neyer, G., Eidenbenz, S. (eds.) WG 1999. LNCS, vol. 1665, pp. 135–147. Springer, Heidelberg (1999). doi:10.1007/3-540-46784-X_14

17. Karthick, T.: Weighted independent sets in a subclass of P_6-free graphs. Discrete Math. **339**, 1412–1418 (2016)

18. Karthick, T., Maffray, F.: Weighted independent sets in classes of P_6-free graphs. Discrete Appl. Math. **209**, 217–226 (2016). doi:10.1016/j.dam.2015.10.015

19. Lokshtanov, D., Pilipczuk, M., van Leeuwen, E.J.: Independence and efficient domination on P_6-free graphs. In: Proceedings of the 27th Annual ACM-SIAM Symposium on Discrete Algorithms, pp. 1784–1803. ACM, New York (2016)

20. Lokshtanov, D., Vatshelle, M., Villanger, Y.: Independent set in P_5-free graphs in polynomial time. In: Proceedings of the 25th Annual ACM-SIAM Symposium on Discrete Algorithms, pp. 570–581. ACM, New York (2014)

21. Lozin, V.V., Milanič, M.: A polynomial algorithm to find an independent set of maximum weight in a fork-free graph. J. Discrete Algorithms **6**, 595–604 (2008)

22. Lozin, V.V., Mosca, R.: Independent sets in extensions of $2K_2$-free graphs. Discrete Appl. Math. **146**, 74–80 (2005)

23. Maffray, F., Pastor, L.: 4-coloring $(P_6,$ bull$)$-free graphs. arXiv:1511.08911. To appear in *Discrete Applied Mathematics*

24. Mosca, R.: Stable sets in certain P_6-free graphs. Discrete Appl. Math. **92**, 177–191 (1999)

25. Mosca, R.: Independent sets in $(P_6,$ diamond$)$-free graphs. Discrete Math. Theor. Comput. Sci. **11**, 125–140 (2009)

26. Mosca, R.: Maximum weight independent sets in $(P_6,$ co-banner$)$-free graphs. Inf. Process. Lett. **113**, 89–93 (2013)

27. Penev, I.: Coloring bull-free perfect graphs. SIAM J. Discrete Math. **26**, 1281–1309 (2012)

28. Reed, B., Sbihi, N.: Recognizing bull-free perfect graphs. Graphs Comb. **11**, 171–178 (1995)

29. Thomassé, S., Trotignon, N., Vušković, K.: A polynomial Turing-kernel for weighted independent set in bull-free graphs. Algorithmica (2015, in press)

Parameterized Power Vertex Cover

Eric Angel[1], Evripidis Bampis[2], Bruno Escoffier[2], and Michael Lampis[3(✉)]

[1] IBISC, Université Evry Val d'Essone, Evry, France
angel@ibisc.fr
[2] Sorbonne Universités, UPMC Univ Paris 06, CNRS, LIP6 UMR 7606, Paris, France
{euripidis.bampis,bruno.escoffier}@upmc.fr
[3] CNRS UMR 7243 and Université Paris Dauphine, Paris, France
michail.lampis@dauphine.fr

Abstract. We study a recently introduced generalization of the VER-
TEX COVER (VC) problem, called POWER VERTEX COVER (PVC). In
this problem, each edge of the input graph is supplied with a positive
integer *demand*. A solution is an assignment of (power) values to the
vertices, so that for each edge one of its endpoints has value as high as
the demand, and the total sum of power values assigned is minimized.

We investigate how this generalization affects the complexity of VER-
TEX COVER from the point of view of *parameterized algorithms*. On the
positive side, when parameterized by the value of the optimal P, we give
an $O^*(1.274^P)$ branching algorithm (O^* is used to hide factors polynomial in the input size), and also an $O^*(1.325^P)$ algorithm for the more
general asymmetric case of the problem, where the demand of each edge
may differ for its two endpoints. When the parameter is the number of
vertices k that receive positive value, we give $O^*(1.619^k)$ and $O^*(k^k)$
algorithms for the symmetric and asymmetric cases respectively, as well
as a simple quadratic kernel for the asymmetric case.

We also show that PVC becomes significantly harder than classical
VC when parameterized by the graph's treewidth t. More specifically,
we prove that unless the ETH is false, there is no $n^{o(t)}$ algorithm for
PVC. We give a method to overcome this hardness by designing an FPT
approximation scheme which obtains a $(1+\epsilon)$-approximation to the optimal solution in time FPT in parameters t and $1/\epsilon$.

1 Introduction

In the classical VERTEX COVER (VC) problem, we are given a graph $G = (V, E)$
and we aim to find a minimum cardinality cover of the edges, i.e. a subset of
the vertices $C \subseteq V$ such that for every edge $e \in E$, at least one of its endpoints
belongs to C. VERTEX COVER is one of the most extensively studied NP-hard
problems in both approximation and parameterized algorithms [13,15].

In this paper, we study a natural generalization of the VC problem, which
we call POWER VERTEX COVER (PVC). In this generalization, we are given an
edge-weighted graph $G = (V, E)$ and we are asked to assign (power) values to
its vertices. We say that an edge e is covered if at least one of its endpoints is

© Springer-Verlag GmbH Germany 2016
P. Heggernes (Ed.): WG 2016, LNCS 9941, pp. 97–108, 2016.
DOI: 10.1007/978-3-662-53536-3_9

assigned a value greater than or equal to the weight of e. The goal is to determine a valuation such that all edges are covered and the sum of all values assigned is minimized. Clearly, if all edge weights are equal to 1, then this problem coincides with VC.

POWER VERTEX COVER was recently introduced in [1], motivated by practical applications in sensor networks (hence the term "power"). The main question posed in [1] was whether this more general problem is harder to approximate than VC. It was then shown that PVC retains enough of the desirable structure of VC to admit a similar 2-approximation algorithm, even for the more general case where the power needed to cover the edge (u, v) is not the same for u and v (a case referred to as DIRECTED POWER VERTEX COVER (DPVC)).

The goal of this paper is to pose a similar question in the context of parameterized complexity: is it possible to leverage known FPT results for VC to obtain FPT algorithms for this more general version? We offer a number of both positive and negative results. Specifically:

- When the parameter is the value of the optimal solution P (and all weights are positive integers), we show an $O^*(1.274^P)$ branching algorithm for PVC, and an $O^*(1.325^P)$ algorithm for DPVC. Thus, in this case, the two problems behave similarly to classical VC.

- When the parameter is the *cardinality* k of the optimal solution, that is, the number of vertices to be assigned non-zero values, we show $O^*(1.619^k)$ and $O^*(k^k)$ algorithms for PVC and DPVC respectively, as well as a simple quadratic (vertex) kernel for DPVC, similar to the classical Buss kernel for VC. This raises the question of whether a kernel of order *linear* in k can be obtained. We give some negative evidence in this direction, by showing that an LP-based approach is very unlikely to succeed. More strongly, we show that, given an optimal *fractional* solution to PVC which assigns value 0 to a vertex, it is NP-hard to decide if an optimal solution exists that does the same.

- When the parameter is the treewidth t of the input graph, we show through an FPT reduction from CLIQUE that there is no $n^{o(t)}$ algorithm for PVC unless the ETH is false. This is essentially tight, since we also supply an $O^*((\Delta + 1)^t)$ algorithm, where Δ is the maximum degree of the graph, and is in stark contrast to VC, which admits an $O^*(2^t)$ algorithm. We complement this hardness result with an FPT approximation scheme, that is, an algorithm which, for any $\epsilon > 0$ returns a $(1 + \epsilon)$-approximate solution while running in time FPT in t and $\frac{1}{\epsilon}$. Specifically, our algorithm runs in time $\left(O(\frac{\log n}{\epsilon})\right)^t n^{O(1)}$.

Our results thus indicate that PVC occupies a very interesting spot in terms of its parameterized complexity. On the one hand, PVC carries over many of the desirable algorithmic properties of VC: branching algorithms and simple kernelization algorithms can be directly applied. On the other, this problem seems to be considerably harder in several (sometimes surprising) respects. In particular, neither the standard treewidth-based DP techniques, nor the Nemhauser-Trotter theorem can be applied to obtain results comparable to those for VC. In fact, in the latter case, the existence of edge weights turns a trivial problem (all vertices with fractional optimal value 0 are placed in the independent set) to an NP-hard one. Yet, despite its added hardness, PVC in fact admits an FPT approximation scheme, a property that is at the moment known for only a handful of other W-hard problems. Because of all these, we view the results of this paper as a first step towards a deeper understanding of a natural generalization of VC that merits further investigation.

Due to space constraints, some proofs are missing or sketched.

Previous Work. As mentioned, PVC and DPVC were introduced in [1], where 2-approximation algorithms were presented for general graphs and it was proved that, like VC, the problem can be solved in polynomial time for bipartite graphs.

VERTEX COVER is one of the most studied problems in FPT algorithms, and the complexity of the fastest algorithm as a function of k has led to a long "race" of improving results, see [3,14] and references therein. The current best result is a $O^*(1.274^k)$-time polynomial-space algorithm. Another direction of intense interest has been kernelization algorithms for VC, with the current best being a kernel with (slightly less than) $2k$ vertices [4,7,9]. Because of the importance of this problem, numerous variations and generalizations have also been thoroughly investigated. These include (among others): WEIGHTED VC (where each vertex has a cost) [14], CONNECTED VC (where the solution is required to be connected) [5,12], PARTIAL VC (where the solution size is fixed and we seek to maximize the number of covered edges) [8,11] and CAPACITATED VC (where each vertex has a capacity of edges it can dominate) [6,8]. Of these, all except PARTIAL VC are FPT when parameterized by k, while all except CAPACITATED VC are FPT when parameterized by the input graph's treewidth t. PARTIAL VC is known to admit an FPT approximation scheme parameterized by k [11], while CAPACITATED VC admits a *bi-criteria* FPT approximation scheme parameterized by t [10], that is, an algorithm that returns a solution that has optimal size, but may violate some capacity constraints by a factor $(1 + \epsilon)$.

In view of the above, and the results of this paper, we observe that PVC displays a different behavior than most VC variants, with CAPACITATED VC being the most similar. Note though, that for PVC we are able to obtain a (much simpler) $(1 + \epsilon)$-approximation for the problem, as opposed to the bi-criteria approximation known for CAPACITATED VC. This is a consequence of a "smoothness" property displayed by one problem and not the other, namely, that any solution that slightly violates the feasibility constraints of PVC can be transformed into a feasible solution with almost the same value. This property separates the two problems, motivating the further study of PVC.

2   Preliminaries

We use standard graph theory terminology. We use n to denote the order of a graph, and Δ to denote its maximum degree. We also use standard parameterized complexity terminology, and refer the reader to related textbook [13] for the definitions of notions such as FPT, kernel, treewidth.

In the DPVC problem we are given a graph $G(V, E)$ and for each edge $(u, v) \in E$ two positive integer values $w_{u,v}$ and $w_{v,u}$. A feasible solution is a function that assigns to each $v \in V$ a value p_v such that for all edges we have either $p_u \geq w_{u,v}$ or $p_v \geq w_{v,u}$. If for all edges we have $w_{u,v} = w_{v,u}$ we say that we have an instance of PVC.

Both of these problems generalize VERTEX COVER, which is the case where $w_{u,v} = 1$ for all edges $(u, v) \in E$. In fact, there are simple cases where the problems are considerable harder.

Theorem 1. PVC *is NP-hard in complete graphs, even if the weights are restricted to* $\{1, 2\}$. *It is even APX-hard in this class of graphs, as hard to approximate as* VC.

As a consequence of the above, PVC is hard on any class of graphs that contains cliques, such as interval graphs. In the remainder we focus on classes that do not contain all cliques, such as graphs of bounded treewidth.

3   Parameterizing by Treewidth

3.1   Hardness for Treewidth

Theorem 2. *If there exists an algorithm which, given an instance* $G(V, E)$ *of* PVC *with treewidth* t, *computes an optimal solution in time* $|V|^{o(t)}$, *then the* ETH *is false. This result holds even if all weights are polynomially bounded in* $|V|$.

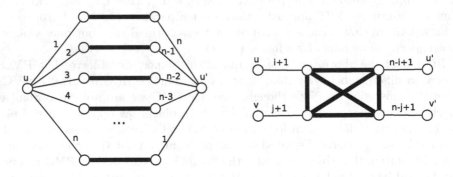

Fig. 1. Main gadgets of Theorem 2. Thick lines represent weight n edges

Proof. We describe a reduction from k-Multicolored Independent Set. In this problem we are given a graph whose vertex set has been partitioned into k cliques $V_1, \ldots, V_k$ and we are asked if this graph contains an independent set of size k. We assume without loss of generality that $|V_1| = |V_2| = \ldots = |V_k| = n$ and that the vertices of each part are numbered $\{1, \ldots, n\}$. It is known that if an algorithm can solve this problem in $n^{o(k)}$ time then the ETH is false.

Our reduction relies on two main gadgets, depicted in Fig. 1. We first describe the choice gadget, depicted on the left side of the figure. This gadget contains two vertices u, u' that will be connected to the rest of the graph. In addition, it contains n independent edges, each of which is given weight n. Each edge has one of its endpoints connected to u and the other to u'. The weights assigned are such that no two edges incident on u have the same weight, and for each internal edge the weight of the edges connecting it to u, u' add up to $n + 1$.

The first step of our construction is to take k independent copies of the choice gadget, and label the high-degree vertices $u_1, \ldots, u_k$ and $u'_1, \ldots, u'_k$. As we will see, the idea of the reduction is that the power assigned to u_i will encode the choice of vertex for the independent set in V_i in the original graph.

We now consider the second gadget of the figure (the checker), which consists of a K_4, all of whose edges have weight n. We complete the construction as follows: for every edge of the original graph, if its endpoints are the i-th vertex of V_c and the j-th vertex of V_d, we add a copy of the checker gadget, where each of the vertices u_c, u'_c, u_d, u'_d is connected to a distinct vertex of the K_4. The weights are $i + 1, n - i + 1, j + 1, n - j + 1$ for the edges incident on u_c, u'_c, u_d, u'_d respectively.

This completes the description of the graph. We now ask if there exists a power vertex cover with total cost at most $k(n^2 + n) + 3mn$, where m is the number of edges of the original graph. Observe that the treewidth of the constructed graph is $2k + O(1)$, because deleting the vertices $u_i, u'_i, i \in \{1, \ldots, k\}$ turns the graph into a disconnected collection of K_2s and K_4s.

First, suppose that the original graph has an independent set of size k. If the independent set contains vertex i from the set V_c, we assign the value i to u_c and $n - i$ to u'_c. Inside each choice gadget, we consider each edge incident on u_c not yet covered, and we assign value n to its other endpoint. Similarly, we consider each edge incident on u'_c not yet covered and assign value n to its other endpoint. Since all weights are distinct and from $\{1, \ldots, n\}$, we will thus select $n - i$ vertices from the uncovered edges incident on u_i and i vertices from the uncovered edges incident on u'_i, thus the total value spent on each choice gadget is $n^2 + n$. To see that this assignment covers also the weight n edges inside the matching, observe that since the edges connecting each to u, u' have total weight $n + 1$, at least one is not covered by u_c or u'_c, thus one of the internal endpoints is taken.

Let us now consider the checker gadgets. Recall that we have one such gadget for every edge. Consider an edge between the i-th vertex of V_c and the j-th vertex of V_d, so that the weights are those depicted in Fig. 1. Because we started from an independent set of G we know that for the values we have assigned at least one of the following is true: $p_{u_c} \neq i$ or $p_{u_d} \neq j$, since these values correspond to the indices of the vertices of the independent set. Suppose without loss of generality that $p_{u_c} \neq i$, therefore $p_{u_c} > i$ or $p_{u_c} < i$. In the first case, the edge connecting u_c to the K_4 is already covered, so we simply assign value n to each of the three vertices of the K_4 not connected to u_c. In the second case, we recall that we have assigned $p_{u'_c} = n - p_{u_c}$ therefore the edge incident on u'_c is covered. Thus, in both cases we can cover all edges of the gadget for a total cost of $3n$. Thus, if we started from an independent set of the original graph we can construct a power vertex cover of total cost $k(n^2 + n) + 3mn$.

For the other direction, suppose that a vertex cover of cost at most $k(n^2 + n) + 3mn$ exists. First, observe that since the checker gadget contains a K_4 of weight n edges, any solution must spend at least $3n$ to cover it. There are m such gadgets, thus the solution spends at most $k(n^2 + n)$ on the remaining vertices.

Consider now the solution restricted to a choice gadget. A first observation is that there exists an optimal solution that assigns all degree 2 vertices values either 0 or n. To see this, suppose that one such vertex has value i, and suppose without loss of generality that it is a neighbor of u. We set its value to 0 and the value of u to $\max\{i, p_u\}$. This is still a feasible solution of the same or lower cost.

Suppose that the optimal solution assigns total value at most $n^2 + n$ to the vertices of a choice gadget. It cannot be using fewer than n degree-two vertices, because then one of the internal weight n edges will be uncovered, thus it spends at least n^2 on such vertices. Furthermore, it cannot be using $n + 1$ such vertices, because then it would have to assign 0 value to u_i, u_i' and some edges would be uncovered. Therefore, the optimal solution uses exactly n degree-two vertices, and assigns total value at most n to u_i, u_i'. We now claim that the total value assigned to u_i, u_i' must be exactly n. To see this, suppose that $p_{u_i} + p_{u_i'} < n$. The total number of edges covered by u_i, u_i' is then strictly less than n. There exist therefore $n + 1$ edges incident on u_i, u_i' which must be covered by other vertices. By pigeonhole principle, two of them must be connected to the same edge. But since we only selected one of the two endpoints of this edge, one of the edges must be uncovered.

Because of the above we can now argue that if the optimal solution has total cost at most $k(n^2 + n) + 3mn$ it must assign value exactly $3n$ to each checker gadget and $n^2 + n$ to each choice gadget. Furthermore, this can only be achieved if $p_{u_c} + p_{u_c'} = n$ for all $c \in \{1, \ldots, k\}$. We can now see that selecting the vertex with index p_{u_c} in V_c in the original graph gives an independent set. To see this, suppose that $p_{u_c} = i$ and $p_{u_d} = j$ and suppose that an edge existed between the corresponding vertices in the original graph. It is not hard to see that in the checker gadget for this edge none of the vertices u_c, u_c', u_d, u_d' covers its incident edge. Therefore, it is impossible to cover everything by spending exactly $3n$ on this gadget. $\square$

3.2   Exact and Approximation Algorithms for Treewidth

In the previous section we showed that PVC is much harder than Vertex Cover, when parameterized by treewidth. This raises the natural question of how one may be able to work around this added complexity. Our first observation is that, using standard techniques, it is possible to obtain FPT algorithms for this problem by adding extra parameters. In particular, if M is the maximum weight of any edge and Δ is the maximum degree of the input graph, we have the following:

Theorem 3. *There exists an algorithm which, given an instance of DPVC and a tree decomposition of width t, computes an optimal solution in time*

$(M + 1)^t n^{O(1)}$. Similarly, there exists an algorithm that performs the same in time $(\Delta + 1)^t n^{O(1)}$.

Theorem 3 indicates that the problem's hardness for treewidth is not purely combinatorial; rather, it stems mostly from the existence of large numbers, which force the natural DP table to grow out of proportion. Using this intuition we are able to state the main algorithmic result of this section which shows that, in a sense, the problem's W-hardness with respect to treewidth is "soft": even if we do not add extra parameters, it is always possible to obtain in FPT time a solution that comes arbitrarily close to the optimal.

Theorem 4. *There exists an algorithm which, given an instance of* DPVC, *$G(V, E)$ and a graph decomposition of G of width t, for any $\epsilon > 0$, produces a $(1+\epsilon)$-approximation of the optimal in time $\left(O(\frac{\log n}{\epsilon})\right)^t n^{O(1)}$. Therefore,* DPVC *admits an FPT approximation scheme parameterized by treewidth.*

Proof (Sketch). The proof relies on two rounding steps. In the first, we deal with the case where the maximum weight is not polynomially bounded in n. In that case, we divide all weights by an appropriate value, so that the maximum weight becomes polynomial in n, while losing a $(1 + \epsilon)$ factor in optimality (this is similar to standard techniques, e.g. for KNAPSACK). We now have an (almost) equivalent instance where $M = n^c$. We now replace all edge weights by setting $w_{u,v} := \lfloor \log_{(1+\epsilon)}(w_{u,v}) \rfloor$. The idea is that this does not significantly affect the feasibility constraints (by more than $(1+\epsilon)$); indeed, a vertex that was receiving value p_v in the old instance may now take value $\lfloor \log_{(1+\epsilon)} p_v \rfloor$ in the new one. We can now modify the algorithm of Theorem 3 to calculate the optimal solution in the new instance. Because the new maximum value is now $\log_{(1+\epsilon)} M$ we get the promised running time. □

4 Parameterizing by Total Power

We focus in this section on the standard parameterization: given an edge-weighted graph G and an integer P (the parameter), we want to determine if there exists a solution of total power at most P. We first focus on PVC and show that it is solvable within time $O^*(1.274^P)$, thus reaching the same bound as VC (when parameterized by the solution value). We then tackle DPVC where a more involved analysis is needed, and we reach time $O^*(1.325^P)$.

4.1 PVC

The algorithm for PVC is based on the following simple property.

Property 1. Consider an edge $e = (u, v)$ of maximum weight. Then, in any optimal solution either $p_u = w_e$ or $p_v = w_e$.

This property can be turned into a branching rule: considering an edge $e = (u,v)$ of maximum weight, then either set $p_u = w_e$ (remove u and incident edges), or set $p_v = w_e$ (remove v and incident edges). This already shows that the problem is FPT, leading to an algorithm in $O^*(2^P)$. To improve this and get the claimed bound, we also use the following reduction rule.

(RR1)   Suppose that there is (u,v) with $w_{u,v} = M$, and the maximum weight of other edges incident to u and v is $B \le M - 1$. Then set $w_{u,v} = B$, and do $P \leftarrow P - (M - B)$.

Property 2. (RR1) is correct.

Now, consider the following branching algorithm.

Algorithm 1
STEP 1: While (RR1) is applicable, apply it;
STEP 2: If $P < 0$ return NO;
If the graph has no edge return YES;
STEP 3: If the maximum weight of edges is 1:
Apply an algorithm in $O^*(1.274^k)$ for VC;
STEP 4: Take two adjacent edges $e = (u,v)$ and $f = (u,w)$ of
maximum weight. Branch as follows:
- either set $p_u = w_e$ (remove u, set $P \leftarrow P - w_e$);
- or set $p_v = p_w = w_e$ (remove v and w, set $P \leftarrow P - 2w_e$)

Theorem 5. *Algorithm 1 solves* PVC *in time* $O^*(1.274^P)$.

4.2   DPVC

For DPVC, the previous simple approach does not work. Indeed, there might be no pair of vertices (u,v) such that both $w_{u,v}$ is the maximum weight of arcs starting at u, and $w_{v,u}$ is the maximum weight of arcs starting at v. If we branch on a pair (u,v), the only thing that we know is that either $p_u \ge w_{u,v}$, or $p_v \ge w_{v,u}$. Setting a constraint $p_u \ge w$ corresponds to the following operation $Adjust(u,w)$.

Definition 1. $Adjust(u,w)$ consists of decreasing weights of each arc starting at u by w, and decreasing P by w.

Of course, if the weight of an arc becomes 0 or negative, then it is removed (as well as the reverse arc). In our algorithm, a typical branching will be to take an edge (u,v), and either to apply $Adjust(u, w_{u,v})$ (and solve recursively), or to apply $Adjust(v, w_{v,u})$ (and solve recursively). Another possibility is to set the power of a vertex u to a certain power w. In this case we must use v to cover (v,u) if $w_{u,v} > w$. Formally:

Definition 2. $Set(u,w)$ consists of (1) setting $p_u = w$, removing u (and incident edges), (2) decreasing P by w, (3) applying $Adjust(v, w_{v,u})$ for all (v,u) such that $w_{u,v} > w$.

Using this it is already easy to show that the problem is solvable in FPT-time $O^*(1.619^P)$. To reach the claimed bound of $O^*(1.325^P)$ we need some more ingredients. Let $M(u)$ be the maximum weight of outgoing arcs from u, and $P(u)$ be the sum of weights of arcs (z, u) (incoming in u). We first define two reduction rules and one branching rule.

(RR2) If there exists u such that $P(u) \leq M(u)$, do $Adjust(v, w_{v,u})$ where v is such that $w_{u,v} = M(u)$.

(RR3) If there is (u, v) with $w_{u,v} = w_{v,u} = 2$, and all other arcs outgoing from u and v have weight 1, then set $w_{u,v} = w_{v,u} = 1$, and do $P \leftarrow P - 1$.

(BR1) If there exists u with $P(u) \geq 5$, branch as follows: either $Set(u, 0)$, or $Adjust(u, 1)$.

Property 3. (RR2) and (RR3) are correct. (BR1) has a branching factor (at worst) $(-1, -5)$.

Now, before giving the whole algorithm, we detail the case where the maximum weight is 2, where a careful case analysis is needed.

Lemma 1. *Let us consider an instance where (1) (RR2) and (RR3) have been extensively applied, and (2) the maximum edge-weight is $w_{u,v} = 2$. Then there is a branching algorithm with branching factor (at worst) $(-1, -5)$ or $(-2, -3)$.*

We are now ready to describe the main algorithm. $N(u)$ denotes the set of neighbors of u, and $N^2(u)$ the set of vertices at distance 2 from u.

Algorithm 2

STEP 1: While (RR2) or (RR3) is applicable, apply them;

STEP 2: If $P < 0$ return NO;
   If the graph has no edge return YES;

STEP 3: If there is a vertex u with $P(u) \geq 5$: apply (BR1);

STEP 4: If there exists (u, v) with either $w_{u,v} + w_{v,u} \geq 6$, or $w_{u,v} = 2$ and $w_{v,u} = 3$: branch by either $Adjust(u, w_{u,v})$, or $Adjust(v, w_{v,u})$;

STEP 5: If there exists (u, v) of weight $w_{u,v} = 3$:
   - if $N^2(u) = \{t\}$ and all arcs $(t, z), z \in N(u)$ have weight 1:
     - either $Adjust(t, 1)$;
     - or $Set(t, 0)$;
   - otherwise:
     - either $Adjust(v, 1)$;
     - or $Set(u, 3)$, and $Set(z, 0)$ for all $z \in N(u)$;

STEP 6: If the maximum weight is at most 2, then branch as in Lemma 1.

Theorem 6. *Algorithm 2 solves DPVC in time $O^*(1.325^P)$.*

Proof. Note that 1.325 corresponds to branching factors $(-1, -5)$ and $(-2, -3)$. We have already seen that (RR2) and (RR3) are sound, and that (BR1) gives a branching factor $(-1, -5)$. For Step 4, if $w_{u,v} + w_{v,u} \geq 6$ this gives in the worst case a branching factor $(-1, -5)$. If $w_{u,v} = 2$ and $w_{v,u} = 3$ the branching factor is $(-2, -3)$. At this point (after Step 4) there cannot remain an edge (u, v) with weight $w_{u,v} \geq 5$, since Step 4 would have been applied. If there is an edge (u, v) with $w_{u,v} = 4$, then either $P(u) \geq 5$ (which is impossible since (BR1) would have been applied), or $P(u) \leq 4$ (which is impossible since (RR2) would have been applied). So, the maximum edge weight after step 4 is at most 3. Thanks to Lemma 1, what remains to do is to focus on Step 5, with $w_{u,v} = 3 = M(u)$. First, note that then $P(u) = 4$ (otherwise (RR2) or (BR1) would have been applied), and $w_{v,u} = 1$ (otherwise Step 4 would have been applied). We consider two cases.

- If $N^2(u) = \{t\}$ and all arcs $(t, z), z \in N(u)$ have weight $w_{t,z} = 1$. As explained in the algorithm, we branch on t: either $p_t \geq 1$, or $p_t = 0$. If $p_t \geq 1$, all arcs $(t, z), z \in N(u)$ are covered, so we need to cover edges incident to u (which are now disconnected from the rest of the graph), and we need at least 2 for this. If $p_t = 0$, we need to optimally cover the edges incident to a vertex in $N(u)$, and we need at least 2 for this. P reduces by at least 3 in one branch, by at least 2 in the other, leading to a branching factor $(-3, -2)$.
- If $|N^2(u)| \geq 2$, or if $N^2(u) = \{t\}$ with at least one arc $(t, z), z \in N(u)$ of weight 2: setting $p_u = 3$ is only interesting if all neighbors of u receive weight 0 - otherwise distributing the power 3 of u on neighbors of u to cover all arcs (v, u) is always at least as good. Then in this case we have to cover arcs between $N(u)$ and $N^2(u)$, so a power at least 2. Then either we have $p_u < 3$ and in this case $p_v \geq 1$, or we set $p_u = 3, p_z = 0$ for neighbors z of u, and we fix at least 2 in $N^2(u)$. In the first branch P reduces by at least 1, in the other by at least 5, leading to a branching factor $(-1, -5)$. $\qquad \square$

5   Parameterizing by the Number k of Vertices

We now consider as parameter the number k of vertices that will receive a positive value in the optimal solution. Note that by definition $k \leq P$; therefore, we expect any FPT algorithm with respect to k to have the same or worse performance than the best algorithm for parameter P.

Theorem 7. PVC *is solvable in time* $O^*(1.619^k)$. DPVC *is solvable in time* $O^*(k^k)$.

Following Theorem 7, a natural question is whether DPVC is solvable in single exponential time with respect to k or not. This does not seem obvious. In particular, it is not clear whether DPVC is solvable in single exponential time *with respect to the number of vertices* n, since the simple brute-force algorithm which guesses the value of each vertex needs $n^{O(n)}$ time.

Interestingly, though we are not able to resolve these questions, we can show that they are actually equivalent.

Theorem 8. *If there exists an* $O^*(\gamma^n)$ *algorithm for* DPVC, *then there exists an* $O^*((4\gamma)^k)$ *algorithm for* DPVC.

6   Kernelization and Linear Programming

Moving to the subject of kernels, we first notice that the same technique as for VC gives a quadratic kernel for DPVC when the parameter is k (and therefore also when the parameter is P):

Theorem 9. *There exists a kernelization algorithm for DPVC that produces a kernel with $O(k^2)$ vertices.*

We observe that the above theorem gives a *bi-kernel* also for PVC. We leave it as an open question whether a pure quadratic kernel exists for PVC.

Let us now consider the question whether the kernel of Theorem 9 could be improved to linear. A way to reach a linear kernel for VC is by means of linear programming. We consider this approach now and show that it seems to fail for the generalization we consider here. Let us consider the following ILP formulation for DPVC, where we have one variable per vertex (x_i is the power of u_i), and one variable $x_{i,j}(i < j)$ per edge (u_i, u_j). $x_{i,j} = 1$ (resp. 0) means that u_i (resp u_j) covers the edge.

$$\begin{cases} Min \ \sum_{i=1}^{n} x_i \\ x_i \geq w_{i,j}x_{i,j}, \forall (u_i, u_j) \in E, i < j \\ x_j \geq w_{j,i}(1 - x_{i,j}), \forall (u_i, u_j) \in E, i < j \\ x_{i,j} \in \{0,1\}, \forall (u_i, u_j) \in E, i < j \\ x_i \geq 0, i = 1, \cdots, n \end{cases}$$

Can we use the relaxation of this ILP to get a linear kernel? Let us focus on PVC, where the relaxation can be written in an equivalent simpler form[1]:

$$\begin{cases} Min \ \sum_{i=1}^{n} x_i \\ x_i + x_j \geq w_{i,j}, \forall (u_i, u_j) \in E, i < j \\ x_i \geq 0, i = 1, \cdots, n \end{cases}$$

Let us call $RPVC$ this LP. We can show that, similarly as for VC, $RPVC$ has the semi-integrality property: in an optimal (extremal) solution x^*, $2x_i^* \in \mathbb{N}$ for all i. However, we *cannot* remove vertices receiving value 0, as in the case of VC. Indeed, there does exist vertices that receive weight 0 in the above relaxation which are in *any* optimal (integer) solution. To see this, consider two edges (u_1, v_1) and (u_2, v_2) both with weight 2, and a vertex v adjacent to all 4 previous vertices with edges of weight 1. Then, there is only one optimal fractional solution, with $p_{u_1} = p_{v_1} = p_{u_2} = p_{v_2} = 1$, and $p_v = 0$. But any (integer) solution has value 5 and gives power 2 to u_1 or v_1, to u_2 or v_2, and weight 1 to v. The difficulty is actually deeper, since we have the following.

Theorem 10. *The following problem is NP-hard: given an instance of PVC, an optimal (extremal) solution x^* of $RPVC$ and i such that $x_i^* = 0$, does there exists an optimal (integer) solution of PVC not containing v_i?*

---

[1] A solution of the relaxation of the former is clearly a solution of the latter. Conversely, if $x_i + x_j \geq w_{i,j}$, set $x_{i,j} = x_i/w_{i,j}$ to get a solution of the former.

References

1. Angel, E., Bampis, E., Chau, V., Kononov, A.: Min-power covering problems. In: Elbassioni, K., et al. (eds.) ISAAC 2015. LNCS, vol. 9472, pp. 367–377. Springer, Heidelberg (2015). doi:10.1007/978-3-662-48971-0_32
2. Bodlaender, H.L., Koster, A.M.C.A.: Combinatorial optimization on graphs of bounded treewidth. Comput. J. **51**(3), 255–269 (2008)
3. Chen, J., Kanj, I.A., Xia, G.: Improved upper bounds for vertex cover. Theor. Comput. Sci. **411**(4042), 3736–3756 (2010)
4. Chlebík, M., Chlebíková, J.: Crown reductions for the minimum weighted vertex cover problem. Discrete Appl. Math. **156**(3), 292–312 (2008)
5. Cygan, M.: Deterministic parameterized connected vertex cover. In: Fomin, F.V., Kaski, P. (eds.) SWAT 2012. LNCS, vol. 7357, pp. 95–106. Springer, Heidelberg (2012)
6. Dom, M., Lokshtanov, D., Saurabh, S., Villanger, Y.: Capacitated domination and covering: a parameterized perspective. In: Grohe, M., Niedermeier, R. (eds.) IWPEC 2008. LNCS, vol. 5018, pp. 78–90. Springer, Heidelberg (2008)
7. Fellows, M.R.: Blow-ups, win/win's, and crown rules: some new directions in *fpt*. In: Bodlaender, H.L. (ed.) WG 2003. LNCS, vol. 2880, pp. 1–12. Springer, Heidelberg (2003)
8. Guo, J., Niedermeier, R., Wernicke, S.: Parameterized complexity of vertex cover variants. Theor. Comput. Syst. **41**(3), 501–520 (2007)
9. Lampis, M.: A kernel of order 2 k-c log k for vertex cover. Inf. Process. Lett. **111**(23–24), 1089–1091 (2011)
10. Lampis, M.: Parameterized approximation schemes using graph widths. In: Esparza, J., Fraigniaud, P., Husfeldt, T., Koutsoupias, E. (eds.) ICALP 2014. LNCS, vol. 8572, pp. 775–786. Springer, Heidelberg (2014)
11. Marx, D.: Parameterized complexity and approximation algorithms. Comput. J. **51**(1), 60–78 (2008)
12. Mölle, D., Richter, S., Rossmanith, P.: Enumerate and expand: improved algorithms for connected vertex cover and tree cover. Theory Comput. Syst. **43**(2), 234–253 (2008)
13. Niedermeier, R.: Invitation to Fixed-Parameter Algorithms. Oxford University Press, Oxford (2006)
14. Niedermeier, R., Rossmanith, P.: On efficient fixed-parameter algorithms for weighted vertex cover. J. Algorithms **47**(2), 63–77 (2003)
15. Williamson, D.P., Shmoys, D.B.: The Design of Approximation Algorithms, 1st edn. Cambridge University Press, New York (2011)

Exhaustive Generation of k-Critical $\mathcal{H}$-Free Graphs

Jan Goedgebeur[1]([⊠]) and Oliver Schaudt[2]

[1] Department of Applied Mathematics, Computer Science and Statistics,
Ghent University, Ghent, Belgium
jan.goedgebeur@ugent.be
[2] Institut Für Informatik, Universität Zu Köln, Cologne, Germany
schaudto@zpr.uni-koeln.de

Abstract. We describe an algorithm for generating all k-critical $\mathcal{H}$-free graphs, based on a method of Hoàng et al. Using this algorithm, we prove that there are only finitely many 4-critical (P_7, C_k)-free graphs, for both $k = 4$ and $k = 5$. We also show that there are only finitely many 4-critical (P_8, C_4)-free graphs. For each case of these cases we also give the complete lists of critical graphs and vertex-critical graphs. These results generalize previous work by Hell and Huang, and yield certifying algorithms for the 3-colorability problem in the respective classes.

Moreover, we prove that for every t, the class of 4-critical planar P_t-free graphs is finite. We also determine all 52 4-critical planar P_7-free graphs. We also prove that every P_{11}-free graph of girth at least five is 3-colorable, and show that this is best possible by determining the smallest 4-chromatic P_{12}-free graph of girth at least five. Moreover, we show that every P_{14}-free graph of girth at least six and every P_{17}-free graph of girth at least seven is 3-colorable. This strengthens results of Golovach et al.

Keywords: Graph coloring · Critical graph · H-free graph · Graph generation

1 Introduction

Given a graph G, a k-coloring is a mapping $c : V(G) \to \{1, \ldots, k\}$ with $c(u) \neq c(v)$ for all edges uv of G. If a k-coloring exists for G, we call G k-colorable. Moreover G is called k-chromatic if it is k-colorable, but not $(k-1)$-colorable.

The graph G is called k-critical if it is k-chromatic, but every proper subgraph of G is $(k-1)$-colorable. For example, the class of 3-critical graphs equals the family of odd cycles. The characterization of k-critical graphs is a notorious problem in graph theory.

To get a grip on this problem, it is common to consider graphs with restricted structure, as follows. Let a graph H and a number k be given. An H-free graph is a graph that does not contain H as an induced subgraph. We say that a graph G is k-critical H-free if G is H-free, k-chromatic, and every H-free proper subgraph of G is $(k-1)$-colorable. If $\mathcal{H}$ is a set of graphs, then we say that a

© Springer-Verlag GmbH Germany 2016
P. Heggernes (Ed.): WG 2016, LNCS 9941, pp. 109–120, 2016.
DOI: 10.1007/978-3-662-53536-3_10

graph G is $\mathcal{H}$-free if G is H-free for each $H \in \mathcal{H}$. The definition of a k-critical $\mathcal{H}$-free graph is analogous.

A notion similar to critical graphs is that of k-vertex-critical graphs: k-chromatic graphs for which every proper induced subgraph is $(k-1)$-colorable. We define k-vertex-critical H-free and k-vertex-critical $\mathcal{H}$-free graphs accordingly. Note that, unlike for critical graphs, the set of k-vertex-critical $\mathcal{H}$-free graphs equals the set of $\mathcal{H}$-free k-vertex-critical graphs.

We remark that every k-critical graph is k-vertex-critical. Moreover, as noted by Hoàng et al. [11], there are finitely many k-critical $\mathcal{H}$-free graphs if and only if there are finitely many k-vertex-critical $\mathcal{H}$-free graphs, for any family of graphs $\mathcal{H}$.

The study of k-critical graphs in a particular graph class received a significant amount of interest in the past decade, which is partly due to the interest in the design of certifying algorithms. Given a decision problem, a solution algorithm is called *certifying* if it provides, together with the yes/no decision, a polynomial time verifiable certificate for this decision. In case of k-colorability for $\mathcal{H}$-free graphs, a canonical certificate would be either a k-coloring or an induced subgraph of the input graph which is (a) not k-colorable and (b) of constant size. However, assertion (b) can only be realized if there is a finite list of $(k+1)$-critical $\mathcal{H}$-free graphs.

Let us now mention some results in this line of research. From [7,12–15] we know that the k-colorability problem remains NP-complete for H-free graphs, unless H is the disjoint union of paths. This motivates the study of graph classes in which some path is forbidden as induced subgraph.

Bruce et al. [5] proved that there are exactly six 4-critical P_5-free graphs, where P_t denotes the path on t vertices. Later, Maffray and Morel [16], by characterizing the 4-vertex-critical P_5-free graphs, designed a linear time algorithm to decide 3-colorability of P_5-free graphs. Randerath et al. [19] have shown that the only 4-critical (P_6, C_3)-free graph is the Grötzsch graph. More recently, Hell and Huang [10] proved that there are four 4-critical (P_6, C_4)-free graphs. They also proved that in general, there are only finitely many k-critical (P_6, C_4)-free graphs.

In a companion paper, we proved the following dichotomy theorem together with Chudnovsky and Zhong for the case of a single forbidden induced subgraph, answering questions of Golovach et al. and Seymour.

Theorem 1 (Chudnovsky et al. [6]). *Let H be a connected graph. There are finitely many 4-critical H-free graphs if and only if H is a subgraph of P_6.*

The main difficulty in the proof of the above theorem is to show that there are only finitely many 4-critical P_6-free graphs, namely 24. A substantial step in this proof, in turn, is to show that there only finitely many 4-critical $(P_6, \text{diamond})$-free graphs. After unsuccessfully trying to prove this by hand, we developed an algorithm to automatize the huge amount of case distinctions, based on a method recently proposed by Hoàng et al. [11].

In the present paper, we thoroughly extend this algorithm in order to derive more characterizations of 4-critical $\mathcal{H}$-free graphs. As a demonstration of the power of this algorithm, we prove the following results.

- There are exactly 6 4-critical (P_7, C_4)-free graphs.
- There are exactly 17 4-critical (P_7, C_5)-free graphs.
- There are exactly 94 4-critical (P_8, C_4)-free graphs.
- There are exactly 52 4-critical planar P_7-free graphs. In addition, using a result of Böhme et al. [1] we show that for every t there are only finitely many 4-critical planar P_t-free graphs.
- Every P_{11}-free graph of girth at least five is 3-colorable and there is a 4-chromatic P_{12}-free graph of girth 5.
- Every P_{14}-free graph of girth at least six is 3-colorable.
- Every P_{17}-free graph of girth at least seven is 3-colorable.

Our results extend and/or strengthen previous results of Hell and Huang [10] and Golovach et al. [9]. Besides these results we see the algorithm as the main contribution of our paper. Its modular design allows to easily implement new expansion rules, which makes it adaptable to closely related problems in this line of research. To this end, we also mention a case that was out of reach with the current algorithm, but where we have a good feeling that there is only a finite set of obstructions.

In the next section we propose a number of lemmas that give necessary conditions for k-critical graphs. Our generation algorithm, which we also present in the next section, is built on these lemmas.

In Sect. 3 we give more details and background on the above mentioned results. The proofs, however, we defer to the full-length version of the paper due to space constraints.

2   A Generic Algorithm to Find All k-Critical $\mathcal{H}$-Free Graphs

We build upon a method recently proposed by Hoàng et al. [11]. With this method they have shown that there is a finite number of 5-critical (P_5, C_5)-free graphs.

The idea is to use necessary conditions for a graph to be critical to generate all critical graphs. The algorithm then performs all remaining case distinctions automatically. In order to deal with more advanced cases, we need to thoroughly alter the approach of Hoàng et al. [11]. We remark that the algorithm presented below is, moreover, a substantial strenghtening of that used in the proof of Theorem 1, and this strenghtening is necessary in order to derive the results of the present paper.

2.1 Preparation

In this section we prove a number of lemmas which will later be used as expansion rules for our generation algorithm. Although none of them is particularly deep, they turn out to be very useful for our purposes.

We use the following notation. The set $N_G(v)$ denotes the neighborhood of a vertex v in G. If it is clear from the context which graph is meant, we abbreviate this to $N(v)$. The graph $G|U$ denotes the subgraph of G induced by the vertex subset $U \subseteq V(G)$. Moreover, for a vertex subset $U \subseteq V(G)$ we denote by $G - U$ the induced subgraph $G|(V(G)\backslash U)$ of G.

Let G be a k-colorable graph. The k-hull of G, which we denote G_k, is the graph obtained from G by making two vertices u and v adjacent if and only if there is no k-coloring of G under which u and v receive the same color. Clearly G_k is a supergraph of G without loops, and G_k is k-colorable.

It is a folklore fact that a k-critical graph cannot contain two distinct vertices u and v with $N(u) \subseteq N(v)$. The following observation is a proper generalization of this fact, and we proved it together with Chudnovsky and Zhong in [6].

Lemma 1 (Chudnovsky et al. [6]). Let $G = (V, E)$ be a k-vertex-critical graph and let U, W be two non-empty disjoint vertex subsets of G. Let $H := (G - U)_{k-1}$. If there exists a homomorphism $\phi : G|U \mapsto H|W$, then $N_G(u)\backslash U \not\subseteq N_H(\phi(u))$ for some $u \in U$.

We make use of Lemma 1 in the following way. Assume that G' is a k-vertex-critical graph and G is a $(k-1)$-colorable induced subgraph of G'. Then pick two disjoint vertex subsets U and W of G, both non-empty, and let $H := (G - U)_{k-1}$. Assume that there is a homomorphism $\phi : G|U \mapsto H|W$ with $N_G(u)\backslash U \subseteq N_H(\phi(u))$ for each $u \in U$. Then we know that there is a vertex $x \in V(G')\backslash V(G)$ adjacent to some $u \in U$ but non-adjacent to $\phi(u)$, in the graph G'. Also x is non-adjacent to $\phi(u)$ in $(G' - U)_{k-1}$.

Recall that any k-critical graph G must have minimum degree at least $k-1$. As otherwise, any $(k-1)$-coloring of $G - u$ can be extended to a $(k-1)$-coloring of G, where u is some vertex in G of degree at most $k-2$. The following is an immediate strengthening.

Lemma 2. Let G be a k-vertex-critical graph and $u \in V(G)$. Then in any $(k-1)$-coloring of $G - u$, the set $N_G(u)$ receives $k-1$ distinct colors.

We make use of Lemma 2 in the following way. Assume that G is a $(k-1)$-colorable graph that is an induced subgraph of some k-vertex-critical graph G'. Suppose that there is a vertex u such that there is no $(k-1)$-coloring of $G - u$ in which the set $N_G(u)$ receives $k-1$ distinct colors. Let us say that ℓ is the maximum number of distinct colors that the set $N_G(u)$ can receive in a $(k-1)$-coloring of $G - u$. Then there must be some vertex $v \in N_{G'}(u)\backslash V(G)$ such that there is a $(k-1)$-coloring of $G'|(V(G - u) \cup \{v\})$ in which the set $N_G(u) \cup \{v\}$ receives $\ell + 1$ distinct colors.

We also need the following fact which is folklore: every cutset of a critical graph contains at least two non-adjacent vertices, where a *cutset* is a vertex subset whose removal increases the number of connected components of the graph.

Lemma 3. *Let G be a k-critical graph, and let X be a clique of G. Then $G - X$ is a connected graph.*

In fact, we only need that a k-critical graph does not have a cutvertex, that is, a cutset of size 1. We use Lemma 3 as follows. Given a $(k-1)$-colorable graph G that is an induced subgraph of some k-vertex-critical graph G'. Suppose that there is a cutvertex u in G. Then there must be some vertex $v \in V(G') \backslash V(G)$ with at least one neighbor in $V(G) \backslash \{u\}$.

The next lemma is custom-made for the case of 4-critical graphs.

Lemma 4. *Let G be a 4-vertex-critical graph. Suppose that there is an induced cycle C where all vertices are of degree three. Then C has odd length, and there is a 3-coloring of $G - V(C)$ for which every member of the set $(\bigcup_{c \in V(C)} N(c)) \backslash V(C)$ receives the same color.*

We use Lemma 4 as follows. Given a 3-colorable graph G that is an induced subgraph of some 4-vertex-critical graph G'. Suppose that there is an induced cycle C in G where all vertices are of degree three. Moreover, suppose that C has even length, or C has odd length and in every 3-coloring of $G - V(C)$ the set $(\bigcup_{c \in V(C)} N(c)) \backslash V(C)$ is not monochromatic. Then there must be some vertex $v \in V(G') \backslash V(G)$ with at least one neighbor in $V(C)$.

The next lemma tells us from which graphs we have to start our enumeration algorithm.

Lemma 5. *Every k-critical P_t-free graph different from K_k contains one of the following graphs as induced subgraph:*

- *an odd hole C_{2s+1}, for $2 \le s \le \lfloor (t-1)/2 \rfloor$, or*
- *an odd antihole $\overline{C_{2s+1}}$, for $3 \le s \le k - 1$.*

2.2 The Enumeration Algorithm

We use Algorithm 1 below to enumerate all $\mathcal{H}$-free k-critical graphs. In order to keep things short, we use the following conventions for a $(k-1)$-colorable graph G.

We call a pair (u, v) of distinct vertices for which $N_G(u) \subseteq N_{(G-u)_{k-1}}(v)$ *similar vertices*. Similarly, we call a 4-tuple (u, v, u', v') of distinct vertices with $uv, u'v' \in E(G)$ such that $N_G(u) \backslash \{v\} \subseteq N_{(G-\{u,v\})_{k-1}}(u')$ and $N_G(v) \backslash \{u\} \subseteq N_{(G-\{u,v\})_{k-1}}(v')$ *similar edges*. Finally, we define *similar triangles* in an analogous fashion.

Recall that a diamond is the graph obtained from K_4 by removing one edge. We define *similar diamonds* in complete analogy to similar triangles.

Let u be a vertex of G for which, in every $(k-1)$-coloring of $G - u$, the set $N_G(u)$ receives at most $k - 2$ distinct colors. Then we call u a *poor vertex*.

Let C be an induced cycle in G such that every vertex of C has degree three. We say that C is a *weak cycle* if C is of even length or, if C has odd length, there is no 3-coloring of $G - V(C)$ for which every member of the set $(\bigcup_{c \in V(C)} N(c)) \backslash V(C)$ receives the same color.

Algorithm 1. Generate $\mathcal{H}$-free k-critical graphs

1: let $\mathcal{F}$ be an empty list
2: Construct(K_k) // i.e. perform Algorithm 2
3: **for all** graphs G mentioned in Lemma 5 **do**
4:     Construct(G) // i.e. perform Algorithm 2
5: **end for**
6: Output $\mathcal{F}$

We now state that Algorithm 1 is correct.

Theorem 2. *Assume that Algorithm 1 terminates, and outputs the list of graphs $\mathcal{F}$. Then $\mathcal{F}$ is the list of all k-critical $\mathcal{H}$-free graphs.*

We implemented this algorithm in C with some further optimizations. We used the program **nauty** [18] to make sure that no isomorphic graphs are accepted. More specifically, we use **nauty** to compute a canonical form of the graphs. We maintain a list of the canonical forms of all non-isomorphic graphs which were generated so far and only accept a graph if it was not generated before (in which case its canonical form is added to the list). More sophisticated isomorphism rejection techniques are known (such as the canonical construction path method [17]), but these methods are not compatible with the destruction of similar *elements* in Algorithm 2. Furthermore isomorphism rejection is not a bottleneck in our program.

Our program does indeed terminate in several cases. More details about this can be found in the next section, Sect. 3. The source code of the program can be downloaded from [8] and in the full version of the paper we describe how we extensively tested the correctness of our implementation.

Due to space constraints, we skip the description of the implementation details and additional optimizations and refer to the full version of the paper.

3   Results

This section describes the main results obtained with our implementation of Algorithm 1. The adjacency lists of all new critical graphs from this section can be found in the appendix of the full paper, and these graphs can also be downloaded from the *House of Graphs* [3] at http://hog.grinvin.org/Critical

---

Algorithm 2. Construct(Graph G)

1: **if** G is $\mathcal{H}$-free AND not generated before **then**
2:    **if** G is not $(k-1)$-colorable **then**
3:       **if** G is k-critical $\mathcal{H}$-free **then**
4:          add G to the list $\mathcal{F}$
5:       **end if**
6:    **else**
7:       **if** G contains similar vertices (u, v) **then**
8:          **for** every graph H obtained from G by attaching a new vertex x and incident edges in all possible ways, such that $ux \in E(H)$, but $vx \notin E((H-u)_{k-1})$ **do**
9:             Construct(H)
10:          **end for**
11:       **else if** G contains a poor vertex u **then**
12:          **for** every graph H obtained from G by attaching a new vertex x and incident edges in all possible ways, such that $ux \in E(H)$ and the maximum number of distinct colors the set $N_H(u)$ recieves in some $(k-1)$-coloring of $H-u$ properly increased **do**
13:             Construct(H)
14:          **end for**
15:       **else if** G contains similar edges (u, v, u', v') **then**
16:          **for** every graph H obtained from G by attaching a new vertex x and incident edges in all possible ways, such that $rx \in E(H)$ and $r'x \notin E((H-\{u,v\})_{k-1})$ for some $r \in \{u, v\}$ **do**
17:             Construct(H)
18:          **end for**
19:       **else if** G contains similar triangles (u, v, w, u', v', w') **then**
20:          **for** every graph H obtained from G by attaching a new vertex x and incident edges in all possible ways, such that $rx \in E(H)$ and $r'x \notin E((H-\{u,v,w\})_{k-1})$ for some $r \in \{u, v, w\}$ **do**
21:             Construct(H)
22:          **end for**
23:       **else if** G contains similar diamonds $(u, v, w, x, u', v', w', x')$ **then**
24:          **for** every graph H obtained from G by attaching a new vertex y and incident edges in all possible ways, such that $ry \in E(H)$ and $r'y \notin E((H-\{u,v,w,x\})_{k-1})$ for some $r \in \{u, v, w, x\}$ **do**
25:             Construct(H)
26:          **end for**
27:       **else if** $k = 4$ and G contains a weak cycle C **then**
28:          **for** every graph H obtained from G by attaching a new vertex x with a neighbor in C and incident edges in all possible ways **do**
29:             Construct(H)
30:          **end for**
31:       **else if** G contains a cutvertex u **then**
32:          **for** every graph H obtained from G by attaching a new vertex x adjacent to some member of $V(G) \setminus \{u\}$ and incident edges in all possible ways **do**
33:             Construct(H)
34:          **end for**
35:       **else**
36:          **for** every graph H obtained from G by attaching a new vertex x and incident edges in all possible ways **do**
37:             Construct(H)
38:          **end for**
39:       **end if**
40:    **end if**
41: **end if**

3.1   4-Critical (P_r, C_s)-Free Graphs

It is known [6] that there is an infinite family of 4-critical P_7-free graphs. However, a careful observation shows that all members of this family contain a C_k for all $k = 3, 4, 5$, and are C_ℓ-free for $\ell = 6, 7$. This motivates the study of 4-critical (P_7, C_k)-free graphs when $k = 3, 4, 5$, since there might be only finitely many of these. We can solve the cases of $k = 4$ and $k = 5$ using Algorithm 1. Moreover, we can also solve the (P_8, C_4)-free case.

Theorem 3. *The following assertions hold.*

(a) *There are exactly 17 4-critical (P_7, C_4)-free graphs.*
(b) *There are exactly 94 4-critical (P_8, C_4)-free graphs.*
(c) *There are exactly 6 4-critical (P_7, C_5)-free graphs.*

The 4-critical (P_7, C_5)-free graphs are shown in Fig. 1. The adjacency lists of all of these graphs can be found in appendix of the full version of the paper.

We also determined that there are exactly 35 4-vertex-critical (P_7, C_4)-free graphs, 164 4-vertex-critical (P_8, C_4)-free graphs, and 27 4-vertex-critical (P_7, C_5)-free graphs.

The Tables 1, 2, and 3 give an overview of the counts of the 4-critical and 4-vertex-critical graphs mentioned in Theorem 3.

We did not succeed in solving the (P_7, C_3)-free case. However, we were able to determine all 4-critical (P_7, C_3)-free graphs with at most 35 vertices. These are shown in Fig. 2, the counts are shown in Table 4 and their adjacency lists can be found in the appendix of the full paper. Since the largest 4-critical (P_7, C_3)-free graph up to 35 vertices has only 16 vertices, we conjecture the following.

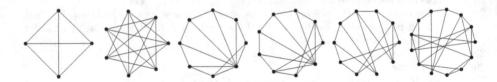

Fig. 1. All 6 4-critical (P_7, C_5)-free graphs

Table 1. Counts of all 4-critical and 4-vertex-critical (P_7, C_4)-free graphs

Vertices	4	6	7	8	9	10	13	Total
Critical graphs	1	1	1	2	3	8	1	17
Vertex-critical graphs	1	1	1	2	4	24	2	35

Table 2. Counts of all 4-critical and 4-vertex-critical (P_8, C_4)-free graphs

Vertices	4	6	7	8	9	10	11	12	13	14	Total
Critical graphs	1	1	1	2	3	15	28	34	8	1	94
Vertex-critical graphs	1	1	1	2	4	33	54	53	14	1	164

Table 3. Counts of all 4-critical and 4-vertex-critical (P_7, C_5)-free graphs

Vertices	4	7	8	9	10	13	Total
Critical graphs	1	1	1	1	1	1	6
Vertex-critical graphs	1	1	1	6	17	1	27

Conjecture 1. The seven graphs from Fig. 2 are the only 4-critical (P_7, C_3)-free graphs.

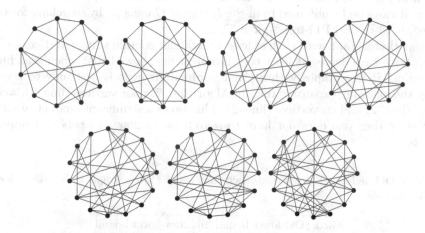

Fig. 2. All seven 4-critical (P_7, C_3)-free graphs with at most 35 vertices

Table 4. Counts of all 7 4-critical (P_7, C_3)-free graphs with at most 35 vertices. The number of 4-vertex-critical graphs is the same

Vertices	11	12	13	15	16	Total
Critical graphs	1	1	2	2	1	7

3.2 4-Critical Graphs with a Given Minimal Girth

In the 1980s, Sumner [20] proved that every (P_5, C_3)-free graph is 3-colorable. This result has been extended in various directions, e.g., Golovach et al. [9] proved that every P_7-free graph of girth at least five is 3-colorable. They mention the Brinkmann graph, constructed by Brinkmann and Meringer in [4], as an example of a P_{10}-free graph of girth five which is not 3-colorable. However, this claim is not entirely correct since the Brinkmann graph contains a P_{12} as an induced subgraph. Using Algorithm 1 we can in fact show that every P_{11}-free graph with girth at least five is 3-colorable. We also improve upon results of Golovach et al. [9] in the cases of girth at least six, and at least seven. These results are also summarized in Table 5.

Theorem 4. *The following assertions hold.*

(a) *Every P_{11}-free graph of girth at least five is 3-colorable.*
(b) *There is a 4-chromatic P_{12}-free graph of girth five, and the smallest such graph has 21 vertices.*
(c) *Every P_{14}-free graph of girth at least six is 3-colorable.*
(d) *Every P_{17}-free graph of girth at least seven is 3-colorable.*

We remark that the 4-critical P_{12}-free graph with girth five mentioned in Theorem 4.(b) is the only such graph up to at least 30 vertices. This graph is shown in Fig. 3 and its adjacency list can be found in the appendix of the full paper. It can also be obtained from the *House of Graphs* [3] by searching for the keywords "4-critical P12-free * girth 5".

Note that a graph with girth at least five cannot contain similar vertices (u, v) which both have degree at least two. Experiments showed that when searching for critical P_t-free graphs with a given minimal girth, It is best only to try to apply the following expansion rules in Algorithm 2: poor vertices (line 2), weak cycles (line 2) and cutvertices (line 2). This saves a significant amount of CPU time since then one does not have to search for similar *elements* or compute *k-hulls*.

Table 5. Old and new lower bounds such that every P_k-free graph with girth at least g is 3-colorable

Girth	Old lower bound [9]	New lower bound
4	P_5-free (exact)	P_5-free (exact)
5	P_7-free	P_{11}-free (exact)
6	P_{10}-free	P_{14}-free
7	P_{12}-free	P_{17}-free

3.3   4-Critical Planar Graphs

In this section we turn to planar critical graphs. Due to the four-color theorem, we may restrict our attention to 4-critical graphs. Let us first note that if an induced path is forbidden, there are only finitely many vertex-critical planar graphs. This implies that there are only finitely many 4-critical planar P_t-free graphs, for all $t \in \mathbb{N}$.

Theorem 5. *For any integer t there are only finitely many vertex-critical graphs that are both planar and P_t-free.*

The proof of Theorem 5 uses the following result of Böhme et al. [1], which we state in a slightly different fashion than in the original paper.

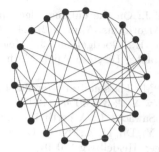

Fig. 3. The smallest 4-critical P_{12}-free graph with girth 5. It has 21 vertices and it is the only 4-critical P_{12}-free graph with girth at least five up to at least 30 vertices.

Theorem 6 (Böhme et al. [1]). For every $k, s, t \in \mathbb{N}$ there is a number $n = n(k, s, t)$ such that every k-connected graph on at least n vertices contains an induced path on t vertices or a subdivision of $K_{2,s}$.

Unfortunately, Theorem 5 does not provide a list of 4-critical graphs, nor yield a useful bound on their order. Using a slight modification of our algorithm, involving the planarity test of Boyer and Myrvold [2], we obtained the following result.

Theorem 7. *There are exactly 52 4-critical planar P_7-free graphs.*

Note that, as mentioned earlier, if the planarity condition is dropped there are infinitely many 4-critical P_7-free graphs. The counts of the 4-critical and 4-vertex-critical planar P_7-free graphs can be found in Table 6. The graphs themselves can be downloaded from http://hog.grinvin.org/Critical.

Table 6. Counts of all planar 4-critical P_7-free graphs

Vertices	4	6	7	8	9	10	11	12	13	Total
Critical graphs	1	1	2	2	14	19	4	6	3	52
Vertex-critical graphs	1	1	6	2	65	347	6	19	15	462

Acknowledgments. Several of the computations for this work were carried out using the Stevin Supercomputer Infrastructure at Ghent University. Jan Goedgebeur is supported by a Postdoctoral Fellowship of the Research Foundation Flanders (FWO).

References

1. Böhme, T., Mohar, B., Škrekovski, R., Stiebitz, M.: Subdivisions of large complete bipartite graphs and long induced paths in k-connected graphs. J. Graph Theor. **45**, 270–274 (2004)

2. Boyer, J.M., Myrvold, W.J.: On the cutting edge: simplified $O(n)$ planarity by edge addition. J. Graph Algorithms. Appl. **8**(2), 241–273 (2004)
3. Brinkmann, G., Coolsaet, K., Goedgebeur, J., Mélot, H.: House of graphs: a database of interesting graphs. Discrete Appl. Math. **161**(1–2), 311–314 (2013). http://hog.grinvin.org/
4. Brinkmann, G., Meringer, M.: The smallest 4-regular 4-chromatic graphs with girth 5. Graph Theor. Notes New York **32**, 40–41 (1997)
5. Bruce, D., Hoàng, C.T., Sawada, J.: A certifying algorithm for 3-colorability of P_5-free graphs. In: Dong, Y., Du, D.-Z., Ibarra, O. (eds.) ISAAC 2009. LNCS, vol. 5878, pp. 594–604. Springer, Heidelberg (2009)
6. Chudnovsky, M., Goedgebeur, J., Schaudt, O., Zhong, M.: Obstructions for three-coloring graphs with one forbidden induced subgraph. In: Proceedings of the Twenty-Seventh Annual ACM-SIAM Symposium on Discrete Algorithms, SODA 2016, 10–12 January 2016, Arlington, VA, USA, pp. 1774–1783 (2016)
7. Emden-Weinert, T., Hougardy, S., Kreuter, B.: Uniquely colourable graphs and the hardness of colouring graphs of large girth. Comb. Probab. Comput. **7**(04), 375–386 (1998)
8. Goedgebeur, J.: Homepage of generator for 4-critical P_t-free graphs. http://caagt.ugent.be/criticalpfree/
9. Golovach, P.A., Paulusma, D., Song, J.: Coloring graphs without short cycles and long induced paths. Discrete Appl. Math. **167**, 107–120 (2014)
10. Hell, P., Huang, S.: Complexity of coloring graphs without paths and cycles. In: Pardo, A., Viola, A. (eds.) LATIN 2014. LNCS, vol. 8392, pp. 538–549. Springer, Heidelberg (2014)
11. Hoàng, C.T., Moore, B., Recoskie, D., Sawada, J., Vatshelle, M.: Constructions of k-critical P_5-free graphs. Discrete Appl. Math. **182**, 91–98 (2015)
12. Holyer, I.: The NP-completeness of edge-coloring. SIAM J. Comput. **10**(4), 718–720 (1981)
13. Kamiński, M., Lozin, V.V.: Coloring edges and vertices of graphs without short or long cycles. Contrib. Discrete Math. **2**, 61–66 (2007)
14. Král', D., Kratochvíl, J., Tuza, Z., Woeginger, G.J.: Complexity of Coloring Graphs without Forbidden Induced Subgraphs. In: Brandstädt, A., Le, V.B. (eds.) WG 2001. LNCS, vol. 2204, pp. 254–262. Springer, Heidelberg (2001)
15. Leven, D., Galil, Z.: NP completeness of finding the chromatic index of regular graphs. J. Algorithms **4**(1), 35–44 (1983)
16. Maffray, F., Morel, G.: On 3-colorable P_5-free graphs. SIAM J. Discrete Math. **26**, 1682–1708 (2012)
17. McKay, B.D.: Isomorph-free exhaustive generation. J. Algorithms **26**(2), 306–324 (1998)
18. McKay, B.D., Piperno, A.: Practical graph isomorphism II. J. Symbolic Comput. **60**, 94–112 (2014)
19. Randerath, B., Schiermeyer, I.: 3-colorability ∈ P for P_6-free graphs. Discrete Appl. Math. **136**(2), 299–313 (2004)
20. Sumner, D.P.: Subtrees of a graph and the chromatic number. In: Chartrand, G. (ed.) Theory and Applications of Graphs, pp. 557–576. John Wiley, New York (1981)

Induced Separation Dimension

Emile Ziedan[1], Deepak Rajendraprasad[1(✉)], Rogers Mathew[2],
Martin Charles Golumbic[1], and Jérémie Dusart[1]

[1] The Caesarea Rothschild Institute, Department of Computer Science,
University of Haifa, Haifa, Israel
deepakmail@gmail.com
[2] Department of Computer Science, Indian Institute of Technology,
Kharagpur, India

Abstract. A linear ordering of the vertices of a graph G *separates* two
edges of G if both the endpoints of one precede both the endpoints of
the other in the order. We call two edges $\{a,b\}$ and $\{c,d\}$ of G *strongly
independent* if the set of endpoints $\{a,b,c,d\}$ induces a $2K_2$ in G. The
induced separation dimension of a graph G is the smallest cardinality of
a family $\mathcal{L}$ of linear orders of $V(G)$ such that every pair of strongly independent edges in G are separated in at least one of the linear orders in $\mathcal{L}$.
For each $k \in \mathbb{N}$, the family of graphs with induced separation dimension
at most k is denoted by ISD(k).

In this article, we initiate a study of this new dimensional parameter. The class ISD(1) or, equivalently, the family of graphs which can
be embedded on a line so that every pair of strongly independent edges
are disjoint line segments, is already an interesting case. On the positive side, we give characterizations for chordal graphs in ISD(1) which
immediately lead to a polynomial time algorithm which determines the
induced separation dimension of chordal graphs. On the negative side, we
show that the recognition problem for ISD(1) is NP-complete for general
graphs. We then briefly study ISD(2) and show that it contains many
important graph classes like outerplanar graphs, chordal graphs, circular
arc graphs and polygon-circle graphs. Finally, we describe two techniques
to construct graphs with large induced separation dimension. The first
one is used to show that the maximum induced separation dimension of
a graph on n vertices is $\Theta(\lg n)$ and the second one is used to construct
AT-free graphs with arbitrarily large induced separation dimension.

1 Introduction

Vertex orderings which meet certain local conditions have turned out to be a
very useful tool in the study of graphs. Perfect elimination orderings of a chordal
graph is perhaps the most striking example. Graph families like comparability
graphs, interval graphs, unit interval graphs, strongly chordal graphs and threshold graphs can be characterized based on the existence of a vertex ordering with
a certain simple property [4,8]. Such orderings are useful not just in providing
structural insights into the family, but also in designing efficient algorithms on

© Springer-Verlag GmbH Germany 2016
P. Heggernes (Ed.): WG 2016, LNCS 9941, pp. 121–132, 2016.
DOI: 10.1007/978-3-662-53536-3_11

those families for problems which are NP-hard on general graphs. In addition, some of these algorithms can be extended to a larger family formed by working with a small family of vertex orderings rather than a single one. Such extensions have resulted in the introduction of many "dimensional" parameters on graphs like boxicity [18], cubicity [18], threshold dimension [7], hypergraph dimension [10], separation dimension [2], etc. and efficient algorithms on families in which one of these dimensions is bounded.

In this article, we use vertex orderings to define a graph parameter, which we call "induced separation dimension", and show that several interesting classes of graphs have a small induced separation dimension.

Let σ be a linear order on the elements of a set U. For two disjoint subsets A and B of U, we say $A \prec_\sigma B$ when every element of A precedes every element of B in σ, i.e., $a \prec_\sigma b$, $\forall (a,b) \in A \times B$. We say that σ separates A and B if either $A \prec_\sigma B$ or $B \prec_\sigma A$.

Definition 1 (Induced separation dimension). Two edges $\{a,b\}$ and $\{c,d\}$ of a graph G are called *strongly independent* if $G[\{a,b,c,d\}]$, the subgraph of G induced on vertices $\{a,b,c,d\}$, is isomorphic to $2K_2$, the disjoint union of two edges. A family $\mathcal{L}$ of linear orders of $V(G)$ is called *weakly separating* if every pair of strongly independent edges in G is separated in at least one order in $\mathcal{L}$. The smallest cardinality of a weakly separating family of linear orders for G is called the *induced separation dimension* of G and is denoted by $\mathrm{isd}(G)$. For each $k \in \mathbb{N}$, the family of graphs with induced separation dimension at most k is denoted by $\mathrm{ISD}(k)$.

For example, one may easily check that a complete graph, a chordless path on at least 5 vertices and a chordless cycle on at least 6 vertices have induced separation dimension, respectively, $0, 1$ and 2. Indeed, a graph G has induced separation dimension 0 if and only if G is $2K_2$-free or, equivalently, if the complement graph $\overline{G}$ is C_4-free. Hence, $\mathrm{ISD}(0) = \{G : G \text{ is } 2K_2\text{-free}\} = \{G : \overline{G} \text{ is } C_4\text{-free}\}$. The family of $2K_2$-free graphs have received considerable attention in literature, resulting in many structural, algorithmic and extremal results [5,6,16]. The left endpoint order of an interval representation of an interval graph separates every pair of strongly independent edges. Hence, interval graphs belong to $\mathrm{ISD}(1)$. Every pair of strongly independent edges in a (rooted) tree is separated either in the DFS pre-order or in the DFS post-order traversal. Thus, trees belong to $\mathrm{ISD}(2)$.

Relation to Separation Dimension. The cardinality of a smallest family $\mathcal{L}$ of linear orders on the vertices of a graph G such that every pair of non-incident edges (two edges with no common endpoints) is separated in at least one of the linear orders in $\mathcal{L}$ is called the *separation dimension* of G [2]. There has been a detailed recent study on the separation dimension of graphs and hypergraphs [1–3]. It follows by definition that the induced separation dimension of a graph is at most its separation dimension. In particular, the induced separation dimension of an n-vertex graph is at most $O(\lg n)$ [3].

But, what we find more interesting is the difference between the two notions. One of the main sources of this difference is that, while separation dimension is a monotone parameter (adding edges cannot decrease the separation dimension of a graph), induced separation dimension is not. Thus, dense graphs, even if highly structured, tend to have large separation dimension. On the other hand, induced separation dimension of some dense but structured graph families is very low. For instance, while separation dimension of cocomparability graphs and chordal graphs is unbounded, their induced separation dimension, as we establish here, is bounded above by 1 and 2 respectively. Their difference is also highlighted by the fact that while the family of graphs with separation dimension 1 has a complete characterization which leads to an easy linear-time recognition algorithm [3], we show here that it is NP-complete to decide whether a graph belongs to ISD(1).

1.1 Results and Organization

We begin by showing that a weakly separating family of linear orders for a graph G corresponds closely with a special family of acyclic orientations of the complement graph $\overline{G}$ (Sect. 2). This characterization is later used to derive both upper and lower bounds on induced separation dimension and also to establish an NP-hardness result.

In Sect. 3, we focus on the graph class ISD(1), i.e., graphs with a single vertex ordering that separates every pair of strongly independent edges. The characterization mentioned above helps us conclude that all cocomparability graphs belong to ISD(1). The same characterization is also used to establish NP-hardness of the recognition problem for ISD(1). We then describe a forbidden configuration for graphs in ISD(1), namely, an asteroidal triple of edges (ATE) and show that a chordal graph belongs to ISD(1) if and only if it is ATE-free. We also note that a tree belongs to ISD(1) if and only if it is a caterpillar with toes.

In Sect. 4, we go one step further and briefly study the graph class ISD(2). The main result here is that ISD(2) contains the class of interval filament graphs. Since the class of interval filament graphs contains many important graph classes like chordal graphs, circular arc graphs and polygon-circle graphs, we conclude that all of them belong to ISD(2). Since chordal graphs belong to ISD(2) and the characterization of chordal graphs in ISD(1) as ATE-free graphs is testable in polynomial time, we get a poly-time algorithm to determine the induced separation dimension of chordal graphs. From the literature on separation dimension, we know that outerplanar graphs belong to ISD(2) and planar graphs belong to ISD(3) [3]. We do not yet know whether planar graphs belong to ISD(2).

Finally, in Sect. 5, we describe two techniques to construct graphs with large induced separation dimension. Using the first one, we construct n-vertex graphs with induced separation dimension at least $\lg n$, showing that the upper bound of $O(\lg n)$ which follows from the relation to separation dimension is tight up to a constant factor. The second construction is used to show that the family of AT-free graphs have unbounded induced separation dimension, in stark contrast to its subfamily of cocomparability graphs.

1.2   Notations and Definitions

All graphs we study here are finite and simple. The vertex set and edge set of a graph G are denoted by $V(G)$ and $E(G)$ respectively. For a graph G and $S \subset V(G)$, the subgraph of G induced on S is denoted by $G[S]$. The complement graph of G is denoted by $\overline{G}$. A graph is called H-free if it has no induced subgraph isomorphic to H. For a vertex v of G, $N(v)$ denotes the set of neighbors of v and $N[v] = N(v) \cup \{v\}$.

The complete graph and the chordless cycle on n vertices are denoted, respectively, by K_n and C_n. The vertex disjoint union of k different copies of a graph is denoted by kG. In particular $2K_2$ denotes two strongly independent edges.

A *cocomparability* graph is an undirected graph that connects pairs of elements that are *not* comparable to each other in a partial order, i.e., the complement of a comparability (transitively orientable) graph. A graph is called *chordal* if it has no induced cycles of size strictly greater than 3. An *interval graph* is an intersection graph of intervals on the real line, and a *unit interval graph* is an intersection graph of unit length intervals on the real line. An independent triple of vertices x, y, z in a graph G is an *asteroidal triple* (*AT*) if, between every pair of vertices in the triple, there is a path that does not contain any neighbor of the third. A graph without asteroidal triples is called an *asteroidal triple-free* (*AT-free*) *graph*. A graph is *outerplanar* if it has a crossing-free embedding in the plane such that all vertices are on the same face. A *caterpillar* is a tree with a dominating path, and a *caterpillar with toes* is a tree with a 2-step dominating path. A 2-*step dominating path* in a graph G is a path P such that every vertex of G is at distance at most 2 from P.

2   Linear Orders and Orientations of the Complement

We start by giving a graph invariant that is equal to the induced separation dimension of the complement graph. This equivalent view will be useful in some of the proofs to come later.

Definition 2 (C_4-transitive orientations). An acyclic orientation of an undirected simple graph G is an assignment of directions to each edge of G so that no directed cycles are formed. A family $\mathcal{O}$ of acyclic orientations of G is called C_4-*transitive* on G if every induced C_4 in G is oriented transitively in at least one orientation in $\mathcal{O}$. The minimum cardinality of a C_4-transitive family of acyclic orientations of G is denoted by $\eta(G)$.

Theorem 3. *For every undirected simple graph G,*

$$\operatorname{isd}(G) = \eta(\overline{G}).$$

Proof. Let $\mathcal{L}$ be a family of linear orders that is weakly separating for G. For every linear order $\sigma \in \mathcal{L}$ we define an orientation O_σ of $\overline{G}$ as follows. An edge $\{u, v\}$ of $\overline{G}$ where $u \prec_\sigma v$ is oriented from u to v (denoted $\overrightarrow{uv}$). This orientation of $\overline{G}$ is obviously acyclic. We claim that the family of acyclic orientations $\{O_\sigma :$ $\sigma \in \mathcal{L}\}$ is C_4-transitive on $\overline{G}$. Let (a, b, c, d) be an induced C_4 in $\overline{G}$. Then the pair of edges ac and bd forms an induced $2K_2$ in G. Let $\sigma \in \mathcal{L}$ be the total order which separates the edges ac and bd of G. That is, we have either $\{a, c\} \prec_\sigma \{b, d\}$ or $\{b, d\} \prec_\sigma \{a, c\}$. In both cases, it is easy to check that O_σ is transitive on the cycle (a, b, c, d).

In the other direction, given a family $\mathcal{O}$ of acyclic orientations that is C_4-transitive on $\overline{G}$, we construct a family of total orders $\{\prec_O : O \in \mathcal{O}\}$ on $V(G)$, where for each $O \in \mathcal{O}$, the total order $\prec_O$ is a linear extension of the transitive closure of O. We claim that $\{\prec_O : O \in \mathcal{O}\}$ is weakly separating for G. Let the pair of edges ab and cd be an induced $2K_2$ in G. Then (a, c, b, d) is an induced C_4 in $\overline{G}$. Let $O \in \mathcal{O}$ be the orientation of $\overline{G}$ which is transitive on (a, c, b, d). There are only two possible transitive orientations for this cycle, namely $\{\overrightarrow{ac}, \overrightarrow{ad}, \overrightarrow{bc}, \overrightarrow{bd}\}$ and the orientation obtained by reversing all the directions in the first one. It is easy to check that $\{a, b\} \prec_O \{c, d\}$ in the first case and $\{c, d\} \prec_O \{a, b\}$ in the second case. □

3   The Graph Class ISD(1)

The following corollary is a restatement of Theorem 3 for ISD(1) and the next one is then immediate.

Corollary 4. *A graph G belongs to ISD(1) if and only if there exists an acyclic orientation of $\overline{G}$ which is transitive on every induced 4-cycle of $\overline{G}$.*

Corollary 5. *The family of cocomparability graphs is contained in ISD(1).*

Remark. The path on 5-vertices P_5 is an interval graph and has a pair of strongly independent edges. Hence, interval graphs and thereby cocomparability graphs are not contained in ISD(0).

Next we use Corollary 4 to show that the recognition problem for ISD(1) is NP-hard. We do this by reducing the 2-coloring problem on 3-uniform hypergraphs to the problem of deciding whether $\eta(G) \leq 1$ for a graph G.

A 3-*uniform hypergraph* H over a set of vertices V is a collection of 3-element subsets of V, called hyperedges. A *proper coloring* of H is a coloring of V so that every hyperedge in H contains vertices of at least two different colors. A hypergraph is called 2-*colorable* if it can be properly colored using 2 colors. It is a result of Lovász from 1973 that testing 2-colorability of 3-uniform hypergraphs is NP-hard [15].

Theorem 6. Problem 1 *below is polynomial-time reducible to* Problem 2.

Problem 1. *Given a 3-uniform hypergraph H, decide whether H is 2-colorable.*
Problem 2. *Given a graph G, decide whether $\eta(G) \leq 1$.*

Proof. Let H contain n vertices $v_1, \ldots, v_n$ and m hyperedges $e_1, \ldots, e_m$. Let L be a bipartite graph on $6m$ vertices with color classes $A = \{a_1, \ldots, a_{3m}\}$ and $B = \{b_1, \ldots, b_{3m}\}$. Vertices a_i and b_j are adjacent in L if and only if $|i - j| \leq 1$. (L is a $3m$-ladder graph). For each $i \in [3m - 1]$, $(a_i, b_i, a_{i+1}, b_{i+1})$ is an induced C_4 in L and these are all the induced C_4's in L. There are only two orientations of L which are transitive on every induced C_4; one which orients every edge from A-side to B-side and the other which orients every edge from B-side to A-side.

To construct G, we first associate a different copy $L(v)$ of the ladder L for each vertex v of H. For each hyperedge $e_l = \{v_i, v_j, v_k\}, i < j < k$, we glue together the three ladders $L(v_i), L(v_j)$ and $L(v_k)$ at their $3l$-th level as follows: the vertex b_{3l} of $L(v_i)$ is identified with the vertex a_{3l} of $L(v_j)$; b_{3l} of $L(v_j)$ with a_{3l} of $L(v_k)$; and b_{3l} of $L(v_k)$ with a_{3k} of $L(v_k)$; forming a 3-cycle. These identifications do not create any new induced 4-cycles since we have chosen to skip 3 levels of the ladder after the modification for each hyperedge. This completes the construction of the graph G given the hypergraph H and it is clearly polynomial time. We complete the proof by showing that $\eta(G) \leq 1$ if and only if H is 2-colorable.

Suppose that H is 2-colorable and let $\phi : V(H) \rightarrow \{0, 1\}$ be a proper coloring of H. Orient the edges of G as follows. If $\phi(v) = 0$, orient every edge of $L(v)$ in G from A-side to B-side and if $\phi(v) = 1$, orient every edge of $L(v)$ from B-side to A-side. Since all the induced 4-cycles in G are subgraphs of the constituent ladders, they are all oriented transitively. All the 3-cycles formed by the hyperedges are oriented acyclically since each of them contains two vertices of different colors. For every longer cycle C (length 4 or more), at least two consecutive edges of C belong to the same ladder and hence C is oriented acyclically. Thus, the above orientation of G is transitive on every induced C_4 and acyclic. Thus $\eta(G) \leq 1$.

In the other direction, suppose $\eta(G) \leq 1$ and let O be an acyclic orientation of G that is transitive on every induced C_4 in G. As noted above, there are only two possible orientations for each ladder that is transitive on every induced C_4. Define a coloring $\phi : V(H) \rightarrow \{0, 1\}$ based on O as follows: $\phi(v) = 0$ if the edges of $L(v)$ in G are oriented from A-side to B-side and $\phi(v) = 1$ otherwise, i.e., if every edge of $L(v)$ is oriented from B-side to A-side. Since O is an acyclic orientation, the 3-cycle corresponding to each hyperedge of H is oriented acyclically in O. That is, every hyperedge contains vertices of both colors under ϕ. Thus, ϕ is a proper 2-coloring of H.   □

Since Problem 1 defined in Theorem 6 is NP-hard [15], so is Problem 2. Moreover, Problem 2 is in NP since the number of induced 4-cycles in a graph is polynomial in the number of vertices. Hence, by Corollary 4, we conclude the following.

Corollary 7. *The recognition problem for* ISD(1) *is NP-complete.*

Next, we give a configuration that is forbidden for graphs in ISD(1). This will turn out to be useful in characterizing trees and chordal graphs in ISD(1). The closed neighborhood of an edge $\{u, v\}$ in a graph G is the set $N[u] \cup N[v]$.

Definition 8 (ATE-free graph). An *asteroidal triple of edges (ATE)* in a graph G is a collection of three edges in G such that, between every pair of them, there exists a path in G which does not contain any vertex in the closed neighborhood of the third edge. A graph without an ATE is called *ATE-free*.

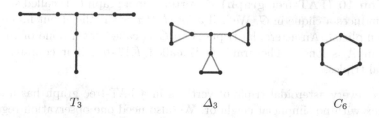

T_3 $\qquad\qquad\qquad\qquad$ Δ_3 $\qquad\qquad\qquad\qquad$ C_6

Fig. 1. Examples of graphs with an asteroidal triple of edges. The three edges which form an asteroidal triple are drawn with thicker lines

Some examples of graphs with an ATE are depicted in Fig. 1. Any ATE-free graph is thus T_3-free, Δ_3-free, C_6-free and so on.

Remark. Note that the three edges of an ATE themselves need not be pairwise strongly independent, as illustrated by the cycle C_6. Nevertheless, one can verify that all AT-free graphs are ATE-free.

Theorem 9. *All graphs in* ISD(1) *are ATE-free.*

Proof. Let $G \in$ ISD(1) and $\prec$ be a single linear order that separates all the strongly independent edges in G. Suppose, for the sake of contradiction, that G contains an ATE. Let aa', bb', and cc' be the three edges forming an ATE in G. Let P_a be the path between bb' and cc' which does not contain any vertex in the closed neighborhood of aa'. P_b and P_c are defined similarly. It is clear that $\prec$ separates the edge xx' from the set $V(P_x)$, for each $x \in \{a, b, c\}$. This demands that no edge of the ATE is completely sandwiched between the endpoints of another. Next we show that one of the above two conditions is violated by $\prec$. This contradiction shall prove the theorem.

Let $S = \{a, a', b, b', c, c'\}$. We can assume, after relabelling if necessary, that $a \prec a', b \prec b', c \prec c'$ and $a \prec b \prec c$. So a is the first vertex of S in $\prec$. The next vertex of S in $\prec$ is not a', since in that case bb' is not separated from $V(P_b)$. Hence, the second vertex from S in $\prec$ is b. The third vertex is not a' for the same reason. Neither is it b' since, in that case bb' is sandwiched between a and a'. Hence, the third vertex is c. The fourth vertex is a' since otherwise either bb' or cc' edge will be sandwiched between a and a'. The fifth vertex has to be b' since otherwise cc' will be sandwiched between b and b'. The sixth vertex is c' by exhaustion. Thus, $a \prec b \prec c \prec a' \prec b' \prec c'$. But in this case, $V(P_b)$ is not separated from bb'. $\qquad\square$

The converse of Theorem 9 is not true in general. We show later that the family of AT-free graphs and thereby the family of ATE-free graphs is not contained in ISD(k) for any constant k. Nevertheless, we show next that the converse of Theorem 9 is true for chordal graphs, i.e., ATE-free chordal graphs belong to ISD(1). We need to define a new notion to streamline the characterization.

Definition 10 (FAT-free graph). A vertex v in a graph G is called simplicial if $N(v)$ induces a clique in G. We call v *lonely* if v is simplicial but no neighbor of v is simplicial. An asteroidal triple A in G is called *fat* if none of the three vertices in A is lonely. The graph G is called *FAT-free* if it contains no fat asteroidal triples.

Hence, every asteroidal triple of vertices in a FAT-free graph has a simplicial vertex with no simplicial neighbor. We also need one observation regarding chordal graphs with an AT.

Oservation 11. *If G is a chordal graph with an asteroidal triple, then G contains an independent set of three simplicial vertices.*

This observation can be verified by looking at a representation of G as an intersection graph of subtrees of a host tree T with the additional property that each node of T corresponds to a different maximal clique in G [12, Theorem 4.8]. Hence, each leaf of T is a subtree in the intersection model. These subtrees correspond to an independent set of simplicial vertices in G. Since G has an AT, the host tree T is not a path and therefore has at least 3 leaves.

Recalling that a *caterpillar* is a tree with a dominating path, we now state and prove a characterization for chordal graphs in ISD(1).

Theorem 12. *For a chordal graph G, the following are equivalent:*

(i) $G \in$ ISD(1).
(ii) G *is ATE-free.*
(iii) G *is FAT-free.*
(iv) G *is an intersection graph of distinct subtrees of a caterpillar.*

The proof is moved to the full version.

Remark. The requirement that the subtrees are *distinct* is essential in Condition (iv) above. The family of graphs which have a representation as the intersection graph of (not necessarily distinct) subtrees of a caterpillar are called *catval* graphs. The graph Δ_3 depicted in Fig. 1 is a catval graph but it has an ATE and therefore cannot be represented as an intersection graph of *distinct* subtrees of a caterpillar. Catval graphs were introduced by Jan Arne Telle in [19] and further studied by Habib, Paul and Telle in [14]. The tolerance version was studied by Eaton and Faubert in [9]. The proof that (iii) $\implies$ (iv) in the above theorem mimics a similar proof in [9].

We conclude this section by specializing the above characterization for trees in ISD(1). Recall that a caterpillar with toes is a tree with a 2-step dominating path.

Theorem 13. *For a tree T the following are equivalent:*

  (i) $T \in \text{ISD}(1)$.
 (ii) T *is ATE-free.*
(iii) T *is T_3-free.*
 (iv) T *is a caterpillar with toes.*

Proof. Theorem 12 establishes the equivalence of (i) and (ii). (ii) $\implies$ (iii) since T_3 contains an ATE. Any longest path in a T_3-free tree is 2-step dominating [13] and thus, (iii) $\implies$ (iv). One can verify easily that (iv) $\implies$ (ii) by a case analysis.   □

Remark. More characterizations of caterpillars with toes can be found in [13, Theorem 3.7].

4   The Graph Class ISD(2)

Since outerplanar graphs have separation dimension at most 2 [3], they also have induced separation dimension at most 2. This bound is tight since C_6 is outerplanar and $\text{isd}(C_6) > 1$. In this section, we show that interval filament graphs, a class introduced by Gavril [11], belongs to ISD(2). Interval filament graphs contain many well known graph classes like chordal graphs, circular-arc graphs (intersection graphs of arcs on a circle), polygon-circle graphs (intersection graphs of a convex polygons inscribed in a circle), etc. Thus, all of the above families belong to ISD(2). Since $\text{isd}(C_6) = 2$, and C_6 is both a circular-arc graph and a polygon-circle graph, both these classes are not contained in ISD(1).

Definition 14 (Interval filament graph [11]). Let $\mathcal{I}$ be a collection of intervals on a horizontal line L embedded in a plane. In the half-plane above L, construct corresponding to each interval $I \in \mathcal{I}$ a curve f_I connecting the two endpoints of I such that f_I remains within the limits of I. The curve f_I is called an *interval filament* above I. A graph is an *interval filament graph* if it has an intersection model consisting of interval filaments.

Theorem 15. *The family of interval filament graphs are contained in* ISD(2).

Proof. Let G be an interval filament graph and $(\mathcal{I}, \mathcal{F})$ be an interval filament intersection model of G. That is, each vertex v of G has an associated interval $I_v \in \mathcal{I}$ on a horizontal line L, and an interval filament $f_v \in \mathcal{F}$ above I_v such that G is the intersection graph of $\mathcal{F}$. Also define $l(v)$ and $r(v)$ to be, respectively, the left and right endpoints of I_v.

Let $\prec_l$ and $\prec_r$ be two linear orders on $V(G)$ such that $l(u) < l(v) \implies u \prec_l v$ and $r(u) < r(v) \implies u \prec_r v$. We argue that any pair of strongly independent edges ab and cd are separated in one of the two permutations above. If two vertices u and v are non-adjacent in G, then the corresponding intervals I_u and I_v are either disjoint or one is contained in the other. Without loss of generality, let a be the vertex with the leftmost left endpoint among $\{a, b, c, d\}$ and c be the vertex with the leftmost left endpoint among $\{c, d\}$. If ab is not separated from cd in $\prec_l$, then $l(a) < l(c) < l(b)$. In this case, since ab is an edge of G, $I_a \cap I_b \neq \emptyset$, hence $I_c \cap I_a \neq \emptyset$ and hence $I_c \subset I_a$. Since c and d are adjacent, $I_c \cap I_d \neq \emptyset$, hence $I_d \cap I_a \neq \emptyset$ and hence $I_d \subset I_a$. Now if I_b is contained in either I_c or I_d, we see that f_b cannot intersect f_a. Thus, I_b is disjoint from I_c and I_d. Moreover since $l(c) < l(b)$ in the case under consideration, we see that I_b has to be to the right of the interval $I_c \cup I_d$. Hence, $\{c, d\} \prec_r \{a, b\}$ in this case. $\square$

Since chordal graphs are interval filament graphs they belong to ISD(2). Hence, a chordal graph G has induced separation dimension either $0, 1$ or 2. It is clear that checking whether $\mathrm{isd}(G) = 0$ can be done in polynomial time. A naive algorithm which tests every triple of edges in G for being an ATE can determine ATE-freeness in poly-time. Hence, by Theorem 12, we can test in poly-time whether $\mathrm{isd}(G) = 1$. In short, we have the following corollary.

Corollary 16. *The induced separation dimension of chordal graphs can be determined in polynomial time.*

5   Graphs with Large Induced Separation Dimension

The separation dimension of an n-vertex graph is at most $O(\lg n)$ [3]. Since induced separation dimension of a graph is at most its separation dimension, we observe that the induced separation dimension of an n-vertex graph is at most $O(\lg n)$. In this section, we construct graphs which show that this upper bound is tight up to a constant factor.

Definition 17 (Bipartite cover). Given a graph G, the bipartite cover B_G of G is the direct product of G with K_2. That is, if $V(G) = [n]$, then the two color classes in $V(B_G)$ are $A = \{a_1, \ldots, a_n\}$ and $B = \{b_1, \ldots, b_n\}$ with a_i adjacent to b_j in B_G if and only if i is adjacent to j in G.

Theorem 18. *For every graph G,*

$$\mathrm{isd}(B_G) \geq \lg \chi(G),$$

where $\chi(G)$ is the chromatic number of G.

Proof. A linear order $\prec$ of $V(B_G)$ is said to *cover* an edge ij of G if the two strongly independent edges $\{a_i b_j, a_j b_i\}$ are separated in $\prec$. The set of edges of G covered by $\prec$ forms a subgraph of G which we denote by $G_\prec$. We now argue that $G_\prec$ is bipartite for any linear order $\prec$. Color a vertex $i \in V(G)$ white if $a_i \prec b_i$ and black otherwise. If an edge ij belongs to $G_\prec$ then $a_i b_j$ and $a_j b_i$ are separated in $\prec$. This happens only if $a_i \prec b_i$ and $a_j \succ b_j$ or vice versa. In both cases i and j are of different color. Hence, we conclude that $G_\prec$ is a bipartite subgraph of G.

Let $\mathcal{L}$ be a family of total orders which separates every pair of strongly independent edges in B_G. For every edge ij in G, the pair of edges $\{a_i b_j, a_j b_i\}$

are strongly independent in B_G. Hence, every edge of G is covered by at least one linear order in $\mathcal{L}$. It is easy to see that at least $\lg \chi$ bipartite graphs are needed to cover all the edges of a χ-chromatic graph. Hence $|\mathcal{L}| \geq \lg \chi(G)$.   □

The bipartite cover of a complete graph is called a *crown graph*. By Theorem 18, we see that the crown graph on $2n$ vertices has induced separation dimension at least $\lg n$. Thus, in general, bipartite graphs have unbounded induced separation dimension.

Another intriguing family is that of AT-free graphs. Since AT-free graphs have a kind of linear structure (dominating pairs) it is tempting to think that their induced separation dimension is at most 1. But we know it is not. The circular ladder CL_k is the graph obtained by taking the Cartesian product of the cycle C_k on $k \geq 3$ vertices with an edge. Orienting a single edge of CL_k forces the orientation on every other edge if we want the orientation to be transitive on each induced C_4. It is easy to check that $\eta(CL_k) \leq 1$ if and only if k is even. Corollary 4 shows that $\mathrm{isd}(\overline{CL_k}) \leq 1$ only when k is even. Notice that for every odd $k \geq 5$, $\overline{CL_k}$ is AT-free (since CL_k is triangle-free) and has induced separation dimension more than 1. Now we amplify this result to show that the induced separation dimension of the family of AT-free graphs is unbounded.

Definition 19 (Double). Given a graph G, the *double D_G* of G is the Cartesian product of G with K_2. That is, D_G consists of two copies of G and a perfect matching of edges between corresponding vertices in the two copies.

Theorem 20. *For every graph G,*

$$\eta(D_G) \geq \lg \chi(G),$$

where $\chi(G)$ is the chromatic number of G.

Proof. To every edge e of G, we associate the induced 4-cycle D_e in D_G formed by the two copies of e and the two matching edges between their endpoints. An acyclic orientation O of D_G is said to *cover* an edge e of G if the associated 4-cycle D_e is oriented transitively by O. The set of edges of G covered by O forms a subgraph of G which we denote by G_O. If G_O contains an odd cycle Z, then it means that O transitively oriented every induced C_4 in the odd circular ladder $D_Z \subset D_G$ which we have observed is impossible. Thus, G_O is bipartite for any acyclic orientation O of D_G.

Let $\mathcal{O}$ be a family of acyclic orientations of D_G such that every induced C_4 in D_G is transitively oriented in at least one orientation in $\mathcal{O}$. Therefore, every edge of G is covered by at least one orientation in $\mathcal{O}$. Hence $|\mathcal{O}| \geq \lg \chi(G)$.   □

If G is triangle free, so is D_G and therefore the maximum size of an independent set in $\overline{D_G}$ is 2 and, in particular, $\overline{D_G}$ is AT-free. There are many classic constructions of families of triangle-free graphs with unbounded chromatic number, Mycielski graphs [17] for instance. If $\mathcal{G}$ is a family of triangle-free graphs with unbounded chromatic number, $\{\overline{D_G} : G \in \mathcal{G}\}$ is a family of AT-free graphs with unbounded induced separation dimension.

References

1. Alon, N., Basavaraju, M., Chandran, L.S., Mathew, R., Rajendraprasad, D.: Separation dimension of bounded degree graphs. SIAM J. Discrete Math. **29**(1), 59–64 (2015)
2. Basavaraju, M., Chandran, L.S., Golumbic, M.C., Mathew, R., Rajendraprasad, D.: Boxicity and separation dimension. In: Kratsch, D., Todinca, I. (eds.) WG 2014. LNCS, vol. 8747, pp. 81–92. Springer, Heidelberg (2014)
3. Basavaraju, M., Chandran, L.S., Golumbic, M.C., Mathew, R., Rajendraprasad, D.: Separation dimension of graphs and hypergraphs. Algorithmica **75**, 187–204 (2015)
4. Brandstädt, A., Le, V.B., Spinrad, J.P.: Graph Classes: A Survey. SIAM, Philadelphia (1999)
5. Broersma, H., Patel, V., Pyatkin, A.: On toughness and Hamiltonicity of $2K_2$-free graphs. J. Graph Theor. **75**(3), 244–255 (2014)
6. Chung, F.R.K., Gyárfás, A., Tuza, Z., Trotter, W.T.: The maximum number of edges in $2K_2$-free graphs of bounded degree. Discrete Math. **81**(2), 129–135 (1990)
7. Chvátal, V., Hammer, P.L.: Aggregation of inequalities in integer programming. Ann. Discret. Math. **1**, 145–162 (1977)
8. Corneil, D.G., Stacho, J.: Vertex ordering characterizations of graphs of bounded asteroidal number. J. Graph Theor. **78**(1), 61–79 (2015)
9. Eaton, N., Faubert, G.: Caterpillar tolerance representations. Bull. Inst. Comb. Appl. **64**, 109–117 (2012)
10. Fishburn, P.C., Trotter, W.T.: Dimensions of hypergraphs. J. Comb. Theor. Ser. B **56**(2), 278–295 (1992)
11. Gavril, F.: Maximum weight independent sets and cliques in intersection graphs of filaments. Inf. Process. Lett. **73**(5), 181–188 (2000)
12. Golumbic, M.C.: Algorithmic Graph Theory and Perfect Graphs. Elsevier, Amsterdam (2004)
13. Golumbic, M.C., Trenk, A.N.: Tolerance Graphs. Cambridge University Press, Cambridge (2004)
14. Habib, M., Paul, C., Telle, J.A.: A linear-time algorithm for recognition of catval graphs. In: Eurocomb 2003: European Conference on Combinatorics, Graphs Theory and Applications (2003)
15. Lovász, L.: Coverings and colorings of hypergraphs. In Proceedings of the 4th Southeastern Conference on Combinatorics, Graph Theory, and Computing, pp. 3–12. Utilitas Mathematica Publishing, Winnipeg (1973)
16. Lozin, V.V., Mosca, R.: Independent sets in extensions of $2K_2$-free graphs. Discrete Appl. Math. **146**(1), 74–80 (2005)
17. Mycielski, J.: Sur le coloriage des graphes. Colloq. Math. **3**, 161–162 (1955)
18. Roberts, F.S.: On the boxicity and cubicity of a graph. In: Recent Progresses in Combinatorics, pp. 301–310. Academic Press, New York (1969)
19. Telle, J.A.: Tree-decompositions of small pathwidth. Discrete Appl. Math. **145**(2), 210–218 (2005)

Tight Bounds for Gomory-Hu-like Cut Counting

Rajesh Chitnis$^{(\boxtimes)}$, Lior Kamma, and Robert Krauthgamer

Weizmann Institute of Science, Rehovot, Israel
{rajesh.chitnis,lior.kamma,robert.krauthgamer}@weizmann.ac.il

Abstract. By a classical result of Gomory and Hu (1961), in every edge-weighted graph $G = (V, E, w)$, the minimum st-cut values, when ranging over all $s, t \in V$, take at most $|V| - 1$ distinct values. That is, these $\binom{|V|}{2}$ instances exhibit *redundancy factor* $\Omega(|V|)$. They further showed how to construct from G a tree (V, E', w') that stores all minimum st-cut values. Motivated by this result, we obtain *tight* bounds for the redundancy factor of several generalizations of the minimum st-cut problem.

1. GROUP-CUT: Consider the minimum (A, B)-cut, ranging over all subsets $A, B \subseteq V$ of given sizes $|A| = \alpha$ and $|B| = \beta$. The redundancy factor is $\Omega_{\alpha,\beta}(|V|)$.
2. MULTIWAY-CUT: Consider the minimum cut separating every two vertices of $S \subseteq V$, ranging over all subsets of a given size $|S| = k$. The redundancy factor is $\Omega_k(|V|)$.
3. MULTICUT: Consider the minimum cut separating every demand-pair in $D \subseteq V \times V$, ranging over collections of $|D| = k$ demand pairs. The redundancy factor is $\Omega_k(|V|^k)$. This result is a bit surprising, as the redundancy factor is much larger than in the first two problems.

A natural application of these bounds is to construct small data structures that stores all relevant cut values, à la the Gomory-Hu tree. We initiate this direction by giving some upper and lower bounds.

1 Introduction

One of the most fundamental combinatorial optimization problems is *minimum st-cut*, where given an edge-weighted graph $G = (V, E, w)$ and two vertices $s, t \in V$, the goal is to find a set of edges of minimum total weight that separates s, t (meaning that removing these edges from G ensures there is no s-t path). This problem was studied extensively, see e.g. the famous minimum-cut/maximum-flow duality [8], and can be solved in polynomial time. It has numerous theoretical applications, such as bipartite matching and edge-disjoint paths, in addition to being extremely useful in many practical settings, including network connectivity, network reliability, and image segmentation, see e.g. [1] for details. Several generalizations of the problem, such as multiway cut, multicut, and k-cut, have been well-studied in operations research and theoretical computer science.

Supported in part by the Israel Science Foundation (grant #897/13). A full version [6] of this extended abstract is available at arXiv:1511.08647.

P. Heggernes (Ed.): WG 2016, LNCS 9941, pp. 133–144, 2016.
DOI: 10.1007/978-3-662-53536-3_12

In every graph $G = (V, E, w)$, there are in total $\binom{|V|}{2}$ instances of the minimum st-cut problem, given by all pairs $s, t \in V$. Potentially, each of these instances could have a different value for the minimum cut. However, the seminal work of Gomory and Hu [9] discovered that *undirected* graphs admit a significantly stronger bound (see also [1, Lemma 8.15] or [7, Sect. 3.5.2]).

Theorem 1 ([9]). *Let $G = (V, E, w)$ be an edge-weighted undirected graph. Then the number of distinct values over all possible $\binom{|V|}{2}$ instances of the minimum st-cut problem is at most $|V| - 1$.*

The beautiful argument of Gomory and Hu shows the existence of a tree $T = (V, E', w')$, usually called a *flow-equivalent tree*, such that for every $s, t \in V$ the minimum st-cut value in T is exactly the same as in G. (They further show how to construct a so-called *cut-equivalent tree*, which has the stronger property that every vertex-partitioning that attains a minimum st-cut in T, also attains a minimum st-cut in G; see Sect. 1.3 for more details on this and related work.) Every G which is a tree (e.g., a path) with distinct edge weights has exactly $|V| - 1$ distinct values, and hence the Gomory-Hu bound is existentially tight.

Another way to view Theorem 1 is that there is always a huge redundancy between the $\binom{|V|}{2}$ minimum st-cut instances in a graph. More precisely, the "redundancy factor", measured as the ratio between the number of instances and the number of distinct optimal values attained by them, is always $\Omega(|V|)$. We study this question of redundancy factor for the following generalizations of minimum st-cut. Let $G = (V, E, w)$ be an undirected edge-weighted graph.

- GROUP-CUT: Given two disjoint sets $A, B \subseteq V$ find a minimum (A, B)-cut, i.e., a set of edges of minimum weight that separates every vertex in A from every vertex in B.
- MULTIWAY-CUT: Given $S \subseteq V$ find a minimum-weight set of edges, whose removal ensures that for every $s \neq s' \in S$ there is no s-s' path.
- MULTICUT: Given $Q \subseteq V \times V$ find a minimum-weight set of edges, whose removal ensures that for every $(q, q') \in Q$ there is no q-q' path.

In order to present our results about the redundancy in these cut problems in a streamlined way, we introduce next the terminology of vertex partitions and demand graphs.

Cut Problems via Demand Graphs. Denote by $\mathrm{Par}(V)$ the set of all partitions of V, where a *partition* of V is, as usual, a collection of pairwise disjoint subsets of V whose union is V. Given a partition $\Pi \in \mathrm{Par}(V)$ and a vertex $v \in V$, denote by $\Pi(v)$ the unique $S \in \Pi$ satisfying $v \in S$. Given a graph $G = (V, E, w)$, define the function $\mathrm{Cut}_G : \mathrm{Par}(V) \to \mathbb{R}^{\geq 0}$ to be $\mathrm{Cut}_G(\Pi) = \sum_{uv \in E \,:\, \Pi(u) \neq \Pi(v)} w(uv)$. We shall usually omit the subscript G, since the graph will be fixed and clear from the context.

Cut problems as above can be defined by specifying the graph G and a collection D of *demands*, which are the vertex pairs that need to be separated. We can view (V, D) as an (undirected and unweighted) *demand graph*, and by slight abuse of notation, D will denote both this graph and its edges. For example, an instance of GROUP-CUT is defined by G and demands that form a complete bipartite graph $K_{A,B}$ (to formally view it as a graph on V, let us add that vertices outside of $A \cup B$ are isolated). We say that partition $\Pi \in \text{Par}(V)$ *agrees* with D if every $uv \in D$ satisfies $\Pi(u) \neq \Pi(v)$. The optimal cut-value for the instance defined by G and D is given by

$$\text{mincut}_G(D) := \min\{\text{Cut}_G(\Pi) : \Pi \in \text{Par}(V) \text{ agrees with } D\}.$$

Redundancy Among Multiple Instances. We study multiple instances on the same graph $G = (V, E, w)$ by considering a family $\mathcal{D}$ of demand graphs. For example, all minimum st-cut instances in a single G corresponds to the family $\mathcal{D}$ of all demands of the form $D = \{(s, t)\}$ (i.e., demand graph with one edge). The collection of optimal cut-values over the entire family $\mathcal{D}$ of instances in a single graph G, is simply $\{\text{mincut}(D) : D \in \mathcal{D}\}$. We are interested in the ratio between the size of this collection as a multiset and its size as a set, i.e., with and without counting multiplicities. Equivalently, we define the *redundancy factor* of a family $\mathcal{D}$ of demand graphs to be

$$\text{redundancy}(\mathcal{D}) := \frac{|\mathcal{D}|}{|\{\text{mincut}(D) : D \in \mathcal{D}\}|},$$

where throughout, $|A|$ denotes the size of A as a *set*, i.e., ignoring multiplicities.

Motivation and Potential Applications. A natural application of the redundancy factor is to construct small data structures that stores all relevant cut values. For the minimum st-cut problem, Gomory and Hu were able to collect all the cut values into a tree on the same vertex set V. This tree can easily support fast query time, or a distributed implementation (labeling scheme) [12].

In addition, large redundancy implies that there is a small collection of cuts that contains a minimum cut for each demand graph. Indeed, first make sure all cut values in G are distinct (e.g., break ties consistently by perturbing edge weights), and then pick for each cut-value in $\{\text{mincut}(D) : D \in \mathcal{D}\}$ just one cut that realizes it. This yields a data structure that reports, given demands $D \in \mathcal{D}$, a vertex partition that forms a minimum cut (see more in Sect. 1.2).

1.1 Main Results

Throughout, we denote $n = |V|$. We use the notation $O_\gamma(\cdot)$ to suppress factors that depend only on γ, and similarly for Ω and Θ.

The GROUP-CUT *problem.* In this problem, the demand graph is a complete bipartite graph $K_{A,B}$ for some subsets $A, B \subset V$. We give a tight bound on the redundancy factor of the family of all instances where A and B are of given sizes α and β, respectively. The special case $\alpha = \beta = 1$ is just all minimum st-cuts in G, and thus recovers the Gomory-Hu bound (Theorem 1).

Theorem 2. *For every graph $G = (V, E, w)$ and for every $\alpha, \beta \in \mathbb{N}$, we have $|\{mincut(K_{A,B}) : |A| = \alpha, |B| = \beta\}| = O_{\alpha,\beta}(n^{\alpha+\beta-1})$, hence the family of (α, β)-group-cuts has redundancy factor $\Omega_{\alpha,\beta}(n)$. Furthermore, this bound is existentially tight (attained by some graph G) for all α, β and n.*

The MULTIWAY-CUT *problem.* In this problem, the demand graph is a complete graph K_S for some subset $S \subseteq V$. We give a tight bound on the redundancy factor of the family of all instances where S is of a given size $k \geq 2$. Again, the Gomory-Hu bound is recovered by the special case $k = 2$.

Theorem 3. *For every graph* $G = (V, E, w)$ *and for every integer* $k \in \mathbb{N}$, *we have* $|\{mincut(K_S) : |S| = k\}| = O_k(n^{k-1})$, *hence the family of* k-*multiway-cuts has redundancy factor* $\Omega_k(n)$. *Furthermore, this bound is existentially tight for all* n *and* k.

The MULTICUT *problem.* In this problem, the demand graph is a collection D of demand pairs. We give a tight bound on the redundancy factor of the family of all instances where D is of a given size $k \in \mathbb{N}$. Again, the Gomory-Hu bound is recovered by the special case $k = 1$.

Theorem 4. *For every graph* $G = (V, E, w)$ *and* $k \in \mathbb{N}$, *we have* $|\{mincut(D) : D \subseteq V \times V, |D| = k\}| = O_k(n^k)$, *and hence the family of* k-*multicuts has redundancy factor* $\Omega_k(n^k)$. *Furthermore, this bound is existentially tight for all* n *and* k.

Theorem 4 is a bit surprising, since it shows a redundancy factor that is polynomial, rather than linear, in n (for fixed α, β and k), so in general MULTICUT has significantly larger redundancy than GROUP-CUT and MULTIWAY-CUT.

1.2 Extensions and Applications

Our main results above actually apply more generally and have algorithmic consequences, as discussed below briefly.

Terminals Version. In this version, the vertices to be separated are limited to a subset $T \subseteq V$ called *terminals*, i.e., we consider only demands inside $T \times T$. All our results above (Theorems 2, 3, and 4) immediately extend to this version of the problem — we simply need to replace $|V|$ by $|T|$ in all the bounds. As an illustration, the terminals version of Theorem 1 states that the $\binom{|T|}{2}$ minimum st-cuts (taken over all $s, t \in T$) attain at most $|T| - 1$ distinct values. (See also [7, Sect. 3.5.2] for this same version.) Extending our proofs to the terminals version is straightforward; for example, in Sect. 2.1 we need to consider polynomials in $|T|$ variables instead of $|V|$ variables.

Data Structures. Flow-equivalent or cut-equivalent trees, such as those constructed by Gomory and Hu [9], may be viewed more generally as succinct data structures that support certain queries, either for the value of an optimal cut, or for its vertex-partition, respectively. Motivated by this view, we define data structures, which we call as evaluation schemes, that preprocess an input graph G, a set of terminals T, and a collection of demand graphs $\mathcal{D}$, so as to answer a cut query given by a demand graph $D \in \mathcal{D}$. The scheme has two flavors, one reports the minimum cut-value, the second reports a corresponding vertex-partition.

In Sect. 5 we initiate the study of such schemes, and provide constructions and lower bounds for some special cases.

Functions Different from Cuts. Recall that the value of the minimum st-cut is $\min\{\mathrm{Cut}_G(X, V \setminus X) : X \subseteq V, \, s \in X, \, t \notin X\}$. Cheng and Hu [5] extended the Gomory-Hu bound (Theorem 1) to a wider class of problems as follows. Instead of a graph G, fix a ground set V and a function $f : 2^V \to \mathbb{R}$. Now for every $s, t \in V$, consider the optimal value $\min\{f(X) : X \subseteq V, |X \cap \{s, t\}| = 1\}$. They showed that ranging over all $s, t \in V$, the number of distinct optimal values is also at most $|V| - 1$. All our results above (Theorems 2, 3, and 4) actually extend to every function $f : \mathrm{Par}(V) \to \mathbb{R}$. However, to keep the notation simple, we opted to present all our results only for the function Cut.

Directed Graphs. What happens if we ask the same questions for the directed variants of the three problems considered previously? Here, an $s \to t$ cut means a set of edges whose removal ensures that no $s \to t$ path exists. Under this definition, we can construct explicit examples for the directed variants of our three problems above where there is no *non-trivial redundancy*, i.e., the number of distinct cut values is asymptotically equal to the total number of instances.

1.3 Related Work

Gomory and Hu [9] showed how to compute a cut-equivalent tree, and in particular a flow-equivalent tree, using $|V| - 1$ minimum st-cut computations on graphs no larger than G. Gusfield [10] has shown a version where all the cut computations are performed on G itself (avoiding contractions). For unweighted graphs, a faster (randomized) algorithm for computing a Gomory-Hu tree which runs in $\tilde{O}(|E| \cdot |V|)$ time was recently given by Bhalgat ct al. [3].

We have mentioned that Cheng and Hu [5] extended Theorem 1 from cuts to an arbitrary function $f : 2^V \to \mathbb{R}$. They further showed how to construct a flow-equivalent tree for this case (but not a cut-equivalent tree). Benczúr [2] showed a function f for which there is no cut-equivalent tree. In addition, he showed that for directed graphs, even flow-equivalent trees do not exist in general.

Another relevant notion here is that of mimicking networks, introduced by Hagerup et al. [11]. A mimicking network for $G = (V, E, w)$ and a terminals set $T \subseteq V$ is a graph $G' = (V', E', w')$ where $T \subset V'$ and for every $X, Y \in T$, the minimum (X, Y)-cut in G and in G' have the exact same value. They showed that every graph has a mimicking network with at most $2^{2^{|T|}}$ vertices. Some improved bounds are known, e.g., for graphs that are planar or have bounded treewidth, as well as some lower bounds [4, 13, 14]. Mimicking networks deal with the GROUP-CUT problem for all $A, B \subset V$; we consider A, B of bounded size, and thus typically achieve much smaller bounds.

2 GROUP-CUT: The Case of Complete Bipartite Demands

This section is devoted to proving Theorem 2. First we give two proofs, one in Sect. 2.1 via polynomials and the second in Sect. 2.2 via matrices, for the bound

$|\{\text{mincut}(K_{A,B}) : |A| = \alpha, |B| = \beta\}| = O_{\alpha,\beta}(n^{\alpha+\beta-1})$. Then in Sect. 2.3 we construct examples of graphs for which this bound is tight. Since $|\{K_{A,B} : |A| = \alpha, |B| = \beta\}| = \binom{n}{\alpha} \cdot \binom{n-\alpha}{\beta} = \Theta_{\alpha,\beta}(n^{\alpha+\beta})$, it follows that the redundancy factor is $\Omega_{\alpha,\beta}(n)$.

2.1 Proof via Polynomials

Let $r = \binom{n}{\alpha}\binom{n-\alpha}{\beta}$ and the set of demand graphs for (α, β)-GROUP-CUT be $\{K_{A_1,B_1}, K_{A_2,B_2}, \ldots, K_{A_r,B_r}\}$. For every vertex $v \in V$ we assign a boolean variable denoted by ϕ_v. Given an instance A, B we can assume that the optimal partition only contains two parts, one which contains A and other which contains B, since we can merge other parts into either of these parts.

Fix some $j \in [r]$. Recall that $\Pi = \{U, V \setminus U\} \in \text{Par}(V)$ agrees with, i.e., is a feasible solution for, the demand graph K_{A_j,B_j} if and only if the following holds: $\Pi(u) \neq \Pi(v)$ whenever $u \in A_j$ and $v \in B_j$ or vice versa. Fix arbitrary $a_j \in A_j$ and $b_j \in B_j$. We associate with the demand graph K_{A_j,B_j} the formal polynomial P_j over the variables $\{\phi_v : v \in V\}$

$$P_j = \prod_{b \in B_j} \left(\phi_{a_j} - \phi_b\right) \cdot \prod_{a \in A_j \setminus \{a_j\}} \left(\phi_a - \phi_{b_j}\right).$$

Note that P_j is a polynomial of degree $\alpha + \beta - 1$. Given $U \subseteq V$, we may think of $\Pi = \{U, V \setminus U\}$ as a vector in $\{0,1\}^n$. We denote by $P_j(\Pi)$ the value of the polynomial P_j (over $\mathbb{F}_2$) when instantiated on Π.

Lemma 1. *A partition Π is feasible for the demand graph K_{A_j,B_j} if and only if $P_j(\Pi) \neq 0$*

Proof. Suppose Π is feasible for the demand graph K_{A_j,B_j}. So $\Pi(u) \neq \Pi(v)$ if $u \in A_j, v \in B_j$ or vice versa. Since every term of P_j contains one variable from each of A_j and B_j, it follows that $P_j(\Pi) \neq 0$.

Conversely, assume $P_j(\Pi) \neq 0$. Let $u \in A_j$. Since $\Pi(u) \neq \Pi(b_j)$ and $\Pi(b_j) \neq \Pi(a_j)$ it follows that $\Pi(u) = \Pi(a_j)$. Similarly for every $v \in B_j$, $\Pi(v) = \Pi(b_j)$. Therefore, it follows that $\Pi(u) \neq \Pi(v)$ whenever $u \in A_j$ and $v \in B_j$ or vice versa, i.e., Π is feasible for K_{A_j,B_j}. □

Next we show that the polynomials corresponding to demand graphs with distinct values under mincut are linearly independent.

Lemma 2. *Reorder the demand graphs such that $\text{mincut}(K_{A_1,B_1}) < \cdots < \text{mincut}(K_{A_q,B_q})$. Then the polynomials $P_1, \ldots, P_q$ are linearly independent.*

Proof. Let $\Pi_1, \ldots, \Pi_q$ be the optimal partitions for the instances corresponding to the demand graphs $K_{A_1,B_1}, \ldots, K_{A_q,B_q}$ respectively, i.e., for each $i \in [q]$ we have that $\text{mincut}(K_{A_i,B_i}) = \text{Cut}(\Pi_i)$. Since $\text{mincut}(K_{A_i,B_i}) < \text{mincut}(K_{A_j,B_j})$ whenever $i < j$, it follows that Π_i is not feasible for the demand graph K_{A_j,B_j} for all $i < j$.

Suppose that the polynomials $P_1, P_2, \ldots, P_q$ are not linearly independent. Then there exist constants $\lambda_1, \ldots, \lambda_q \in \mathbb{R}$ which are not all zero such that $P = \sum_{j \in [q]} \lambda_j P_j$ is the zero polynomial. We will now show that each of the constants $\lambda_1, \lambda_2, \ldots, \lambda_q$ is zero, leading to a contradiction. Instantiate P on Π_1. Recall that Π_1 is not feasible for any K_{A_i, B_i} with $i \geq 2$. Therefore, by Lemma 1, we have that $P_i(\Pi_1) = 0$ for all $i \geq 2$. Therefore $\lambda_1 P_1(\Pi_1) = 0$. Since Π_1 is an (optimal) feasible partition for instance corresponding to K_{A_1, B_1}, applying Lemma 1 we get that $P_1(\Pi_1) \neq 0$. This implies $\lambda_1 = 0$. Hence, we have $P = \sum_{2 \leq j \leq q} \lambda_j P_j$ is the zero polynomial. Now instantiate P on Π_2 to obtain $\lambda_2 = 0$ via a similar argument as above. In the last step, we will get that $\lambda_{q-1} = 0$ and hence $P = \lambda_q P_q$ is the zero polynomial. Instantiating on Π_q gives $0 = P(\Pi_q) = \lambda_q P_q(\Pi_q)$. Since Π_q is (optimal) feasible partition for the demand graph K_{A_q, B_q} it follows that $P_q(\Pi_q) \neq 0$, and hence $\lambda_q = 0$. $\qquad\square$

Note that each of the polynomials $P_1, P_2, \ldots, P_q$ is contained in the vector space of polynomials with n variables and degree $\leq \alpha + \beta - 1$. This vector space is spanned by $\{\prod_{v \in V} \phi_v^{r_v} : \sum_{v \in V} r_v \leq \alpha + \beta - 1\}$ and therefore is of dimension $\binom{n + (\alpha + \beta - 1)}{\alpha + \beta - 1} = O_{\alpha, \beta}(n^{\alpha + \beta - 1})$. From Lemma 2 and the fact that size of any set of linearly independent elements is at most the size of a basis, it follows that $\left| \{\texttt{mincut}(K_{A,B}) : |A| = \alpha, |B| = \beta\} \right| = O_{\alpha, \beta}(n^{\alpha + \beta - 1})$.

2.2 Proof via Matrices

We shall prove the (slightly stronger) bound that $| \{\texttt{mincut}(K_{A,B}) : |A| \leq \alpha, |B| \leq \beta\}| = O_{\alpha, \beta}(n^{\alpha + \beta - 1})$. Let $\texttt{Par}_2(V) \subseteq \texttt{Par}(V)$ be the set of partitions of V into exactly two parts. Let $\mathcal{Q} := \{(A, B) : |A| \leq \alpha, \ |B| \leq \beta\}$. Consider the matrix $\mathcal{M}$ over $\mathbb{F}_2$ with $|\mathcal{Q}|$ rows (one for each element from $\mathcal{Q}$) and $|\texttt{Par}_2(V)| = 2^n$ columns (one for each partition Π of V into two parts). We now define the entries of $\mathcal{M}$. Given $(A, B) \in \mathcal{Q}$ and $\Pi \in \texttt{Par}_2(V)$, we set $\mathcal{M}_{(A,B), \Pi} = 1$ if and only if the partition $\Pi \in \texttt{Par}_2(V)$ agrees with the demand graph $K_{A,B}$, which is equivalent to saying that $\Pi(u) \neq \Pi(v)$ whenever $u \in A$ and $v \in B$ or vice versa. Fix a vertex $v_0 \in V$, and consider the set $\mathcal{R} := \{(A, B) \in \mathcal{Q} : v_0 \in A \cup B\}$.

Proposition 1. *Over $\mathbb{F}_2$, the row space of $\mathcal{M}$ is spanned by the rows corresponding to elements from $\mathcal{R}$.*

Proof. Consider $(A, B) \in \mathcal{Q}$ and $\Pi \in \texttt{Par}_2(V)$. If $v_0 \in A \cup B$ then $(A, B) \in \mathcal{R}$. Henceforth we assume that $v_0 \notin A \cup B$. Let

$$L(\Pi) := \mathcal{M}_{(A,B), \Pi} + \sum_{A' \subset A} \mathcal{M}_{(v_0 \cup A', B), \Pi} + \sum_{B' \subset B} \mathcal{M}_{(A, B' \cup v_0), \Pi},$$

where addition is over $\mathbb{F}_2$. Note that $(v_0 \cup A', B), (A, B' \cup v_0) \in \mathcal{R}$ for every $A' \subset A$ and $B' \subset B$, and therefore it is enough to show that $L(\Pi) \equiv 0 \pmod 2$.

Assume first that $\mathcal{M}_{(A,B), \Pi} = 1$, i.e. Π agrees with the demand graph $K_{A,B}$. Without loss of generality assume that $\Pi(v_0) = \Pi(a)$ for some $a \in A$. Then we have $\mathcal{M}_{(v_0 \cup A', B), \Pi} = 1$ for all $A' \subset A$, and $\mathcal{M}_{(A, v_0 \cup B'), \Pi} = 0$ for all $B' \subset B$. So, $L(\Pi) = 1 + (2^{|A|} - 1) \equiv 0 \pmod{2}$.

Otherwise, we have $\mathcal{M}_{(A,B),\Pi} = 0$. If for every $v \in A \cup B$ it holds that $\Pi(v) \neq \Pi(v_0)$ then

$$L(\Pi) = 0 + \mathcal{M}_{(v_0, B), \Pi} + \mathcal{M}_{(A, v_0), \Pi} = 1 + 1 \equiv 0 \pmod{2}.$$

Hence suppose that there exists $v \in A \cup B$ such that $\Pi(v) = \Pi(v_0)$. Without loss of generality, assume $v \in A$. Then $\mathcal{M}_{(A, B' + v_0), \Pi} = 0$ for all $B' \subset B$. Note that if $A_1, A_2 \subset A$ satisfy $\mathcal{M}_{(v_0 \cup A_1, B), \Pi} = 1 = \mathcal{M}_{(v_0 \cup A_2, B), \Pi}$, then $\mathcal{M}_{(v_0 \cup A_1 \cup A_2, B), \Pi} = 1$. Hence there is an inclusion-wise maximal set $A^* \subseteq A$ such that $\mathcal{M}_{(v_0 \cup A^*, B), \Pi} = 1$. Since $\mathcal{M}_{(A,B),\Pi} = 0$, we conclude that $A^* \subset A$. If $A^* = \emptyset$, then since $\Pi(v) = \Pi(v_0)$, we conclude that there exists $v' \in B$ such that $\Pi(v') = \Pi(v_0)$. Then $\mathcal{M}_{(A' + v_0, B), \Pi} = 0$ for all $A' \subset A$, and $L(\Pi) = 0$. Otherwise, $|A^*| \geq 1$, and therefore

$$L(\Pi) = \mathcal{M}_{(A,B),\Pi} + \sum_{A' \subset A} \mathcal{M}_{(v_0 \cup A', B), \Pi} = \sum_{A' \subseteq A^*} \mathcal{M}_{(v_0 \cup A', B), \Pi} = 2^{|A^*|} \equiv 0 \bmod(2)$$

□

An argument similar to Lemma 2 shows that rows corresponding to demand graphs with distinct values under mincut are linearly independent. Hence, we have $\left| \{\mathrm{mincut}(K_{A,B}) : |A| \leq \alpha, |B| \leq \beta\} \right| \leq \mathrm{rank}(\mathcal{M}) \leq |\mathcal{R}|$, where the last inequality follows since $\mathcal{R}$ spans the row space of $\mathcal{M}$. We now obtain the final bound

$$|\mathcal{R}| = \sum_{i \leq \alpha - 1, j \leq \beta} \binom{n-1}{i} \cdot \binom{n-i-1}{j} + \sum_{j \leq \beta - 1, i \leq \alpha} \binom{n-1}{j} \cdot \binom{n-j-1}{i}$$

$$= \sum_{i \leq \alpha - 1, j \leq \beta} O_{i,j}(n^{i+j}) + \sum_{j \leq \beta - 1, i \leq \alpha} O_{i,j}(n^{i+j})$$

$$= O_{\alpha, \beta}(n^{\alpha + \beta - 1})$$

2.3 Lower Bound on Number of Distinct Cuts for (α, β)-Group-Cut

We now turn to prove that the bound given in Theorem 2 is existentially tight. To this end, we construct an infinite family $G_n^{\alpha, \beta}$ of graphs satisfying $|\{\mathrm{mincut}(K_{A,B}) : |A| = \alpha, |B| = \beta\}| \geq \Omega_{\alpha, \beta}(n^{\alpha + \beta - 1})$.

Let $n, \alpha, \beta \in \mathbb{N}$ be such that n is odd, and both α and $\beta - 1$ divide $(n - 3)/2$. We define a graph $G_n^{\alpha, \beta}$ on n vertices as follows. $G_n^{\alpha, \beta}$ is composed of two graphs that share a common vertex H_n^α and J_n^β defined below.

- H_n^α has $(n+1)/2$ vertices, and is given by α parallel paths $P_1, \ldots, P_\alpha$ between two designated vertices s, t, each path having $(n-3)/2\alpha$ internal vertices. The edge weights are given by distinct powers of 2, monotonically decreasing from s to t. All edges in H_n^α incident on t have ∞ weight (see Fig. 1).

- J_n^β has $(n+1)/2$ vertices, and is given by $(\beta - 1)$ parallel paths $Q_1, \ldots, Q_{\beta-1}$, between t and a designated vertex u, each having $(n - 3)/2(\beta - 1)$ internal vertices. As in H_n^α, edge weights are given by distinct powers of 2, monotonically decreasing from t to u, and all of which are strictly smaller than the weights of H_n^α. All edges in J_n^β incident on u have ∞ weight.

The following proposition implies the desired lower bound.

Proposition 2. $|\{mincut(K_{A,B}) : |A| = \alpha, |B| = \beta\}| \geq \Omega_{\alpha,\beta}(n^{\alpha+\beta-1})$.

Proof. Pick one internal vertex from each P_i for $i \in [\alpha]$ to form A. Similarly for $\beta - 1$ elements in B, we pick one internal vertex from each Q_j for $j \in [\beta]$. In addition, $s \in B$ (as demonstrated in Fig. 1). We claim that every such choice of A, B gives a distinct value for the minimum (A, B)-cut.

Indeed, for $i \in [\alpha]$ let a_i be the unique element in $A \cap P_i$. In order to separate A from B, we need to separate a_i from s. This implies that at least one edge on the segment of P_i between s and a_i has to be in the cut. By monotonicity of weights and minimality of the cut, this must be the edge incident to a_i. Similarly, for every $b \in B \setminus \{s\}$, the left edge incident to b must be cut. It is easy to see (as demonstrated in Fig. 1) that this set of edges also separates A and B.

By the choice of weights, each such cut has a unique value, and therefore $|\{mincut(K_{A,B}) : |A| = \alpha, |B| = \beta\}| \geq ((n - 3)/2\alpha)^\alpha((n - 3)/2(\beta - 1))^{\beta-1} = \Omega_{\alpha,\beta}(n^{\alpha+\beta-1})$. $\qquad\square$

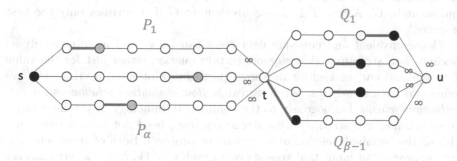

Fig. 1. The graph $G_n^{\alpha,\beta}$ used in the lower bound of Sect. 2.3. The left part of the graph is H_n^α, consisting of α parallel s-t paths. The right part of the graph is J_n^β, consisting of $(\beta - 1)$ parallel t-u paths. The gray vertices are in A, and the black ones are in B. The red edges represent the minimum cut for this choice of A and B.

3   MULTIWAY-CUT: The Case of Clique Demands

Our tight bounds for MULTIWAY-CUT are described in Theorem 3. First we show that for every graph $G = (V, E, w)$ we have $|\{mincut(K_S) : |S| = k\}| = O_k(n^{k-1})$. The proof technique for this upper bound is quite similar to that from

Sect. 2.2. We also show that this bound is tight for paths (with specially chosen edge-weights). Hence, the redundancy factor is $\Omega_k(n)$, since $|\{K_S : |S| = k\}| = \binom{n}{k} = \Theta_k(n^k)$. We refer to the full version [6] for the technical details.

4   MULTICUT: The Case of Demands with Fixed Number of Edges

Our tight bounds for MULTICUT are described in Theorem 4. First we show that $|\{\texttt{mincut}(D) : D \subseteq V \times V, |D| = k\}| = O_k(n^k)$. The proof technique for this upper bound is quite similar to that from Sect. 2.1. We also show that this bound is tight for graphs which are perfect matchings (with specially chosen edge-weights). Since $|\{D : D \subseteq V \times V, |D| = k\}| = \binom{\binom{n}{2}}{k} = \Theta_k(n^{2k})$, it follows that the redundancy factor is $\Omega_k(n^k)$. We refer to the full version [6] for the technical details.

5   Evaluation Schemes: Constructing Succinct Data Structures

Gomory and Hu [9] showed that for every undirected edge-weighted graph $G = (V, E, w)$ there is a tree $\mathcal{T} = (V, E', w')$ that represents the minimum st-cuts exactly both in terms of the *cut-values* and in terms of their *vertex-partitions*. The common terminology, probably due to Benczúr [2], is to say that $\mathcal{T}$ is cut-equivalent to G. A tree $\mathcal{T}$ is flow-equivalent to G if it satisfies only the first property.[1]

Flow-equivalent and cut-equivalent trees can be viewed more generally as succinct data structures that support certain queries, either just for the value of an optimal cut, or also for its vertex-partition. Motivated by this view, we define two data structures, which we call a *flow-evaluation scheme* and a *cut-evaluation scheme* (analogously to the common terminology in the literature). These schemes are arbitrary data structures (e.g., need not form a tree), and address the terminals version of a certain cut problem. Both of these schemes, first preprocess an input that consists of a graph $G = (V, E, w)$, a terminals set $T \subset V$, and a collection of demand graphs $\mathcal{D}$. The preprocessed data can then be used (without further access to G) to answer a cut query given by a demand graph $D \in \mathcal{D}$. The answer of a *flow-evaluation scheme* is the corresponding minimum cut-value $\texttt{mincut}(D)$. A *cut-evaluation scheme* will also give a vertex-partition that attains this cut-value $\texttt{mincut}(D)$.

A natural goal is to provide succinct constructions and lower bounds for the storage and query time of flow-evaluation schemes and cut-equivalent schemes,

---

[1] A tree $\mathcal{T}$ is flow-equivalent to G if for every $s, t \in V$ the minimum st-cut value in $\mathcal{T}$ is exactly the same as in G. We say $\mathcal{T}$ is cut-equivalent to G if, in addition, every vertex partition that attains a minimum st-cut in $\mathcal{T}$, also attains a minimum st-cut in G.

for the three cut problems studied in this paper, viz. GROUP-CUT, MULTIWAY-CUT and MULTICUT. In order to analyze the storage size (in terms of bits) of such data structures, we henceforth assume that all edge weights are integers. Our bounds on the number of distinct cut values naturally lead to the following construction of cut-evaluation schemes for the aforementioned three problems. We state below the result for GROUP-CUT (proof deferred to the full version [6]); similar results also hold for the MULTIWAY-CUT and MULTICUT problems.

Theorem 5. *For every $\alpha, \beta \in \mathbb{N}$ there is a cut-evaluation scheme such that for every graph $G = (V, E, w)$, and a set of terminals $T \subseteq V$, the scheme uses a storage of $O_{\alpha,\beta}(|T|^{\alpha+\beta-1} \cdot (|T| + \log W))$ bits, where $W = \sum_{e \in E} w(e)$, to answer every (α, β)-GROUP-CUT query in time $O_{\alpha,\beta}(|T|^{\alpha+\beta-1})$.*

The result of Theorem 5 is especially meaningful for the case where $|T|$ is much smaller than n. For large $|T|$, say $T = V$, the graph G itself serves as a cut-evaluation scheme of size $O(n^2 \log W)$.

We do not know whether the upper bound in Theorem 5 is tight, and proving lower bounds for the storage size of such schemes is left as an interesting open question. However, for $(2,1)$-GROUP-CUT with $V = T$ and edge weights bounded by $n^{O(1)}$, we can prove that simply storing the graph G using $O(n^2 \log n)$ bits is essentially optimal, even for the weaker notion of flow-evaluation schemes.

5.1 Lower Bound on Flow-Evaluation Schemes for $(2,1)$-GROUP-CUT

Using an information-theoretic argument, we can show the following lower bound (proof deferred to [6]) on the storage required by any flow-evaluation scheme for $(2,1)$-GROUP-CUT. We remark that similar arguments give a lower bound of $\Omega(n^3 \log n)$ by allowing weights which are exponential in n^3.

Theorem 6. *For every $n \geq 3$, a flow-evaluation scheme for $(2,1)$-GROUP-CUT on graphs with n terminals (in which $T = V$) and with edge-weights bounded by a polynomial in n requires storage of $\Omega(n^2 \log n)$ bits.*

6 Future Directions

A natural direction for future work is to construct better data structures for the problems discussed in this paper. Our tight bounds on the number of distinct cut values (redundancy factor) yield straightforward schemes with improved storage requirement, as described in Sect. 5. But one may potentially improve these schemes in several respects. First, our storage requirement is a factor $|T|$ larger than the number of distinct cut values. The latter number could possibly be the "right bound", and it is important to prove it is a lower bound for required storage; we only proved this for $(2,1)$-GROUP-CUT. Second, it is desirable to achieve fast query time, say sublinear in $|T|$ or perhaps even constant. Third, one may ask for a distributed version of the data structure (i.e., a labeling scheme) that can report the same cut values; this would extend the known results [12] for minimum st-cuts. All these improvements require better understanding of the structure of the optimal partitions (those that attain minimum cut values). Such structure is known for minimum st-cuts, where the Gomory-Hu tree essentially shows the existence of a family of minimum st-cuts, one for each $s, t \in V$, which is laminar.

Another very interesting question is to explore approximation to the minimum cut, i.e., versions of the above problems where we only seek for each instance a cut within a small factor of the optimal. For instance, the cut values of (α, β)-GROUP-CUT can be easily approximated within factor $\alpha \cdot \beta$ using Gomory-Hu trees, which requires storage that is linear in $|T|$, much below the aforementioned "right bound" of $|T|^{\alpha+\beta-1}$.

References

1. Ahuja, R.K., Magnanti, T.L., Orlin, J.B.: Network Flows - Theory, Algorithms and Applications. Prentice Hall, Upper Saddle River (1993)
2. Benczúr, A.A.: Counterexamples for directed and node capacitated cut-trees. SIAM J. Comput. **24**(3), 505–510 (1995)
3. Bhalgat, A., Hariharan, R., Kavitha, T., Panigrahi, D.: An $\tilde{O}(mn)$ Gomory-Hu tree construction algorithm for unweighted graphs. In: STOC 2007, pp. 605–614 (2007)
4. Chaudhuri, S., Subrahmanyam, K.V., Wagner, F., Zaroliagis, C.D.: Computing mimicking networks. Algorithmica **26**, 31–49 (2000)
5. Cheng, C., Hu, T.: Ancestor tree for arbitrary multi-terminal cut functions. Ann. Oper. Res. **33**(3), 199–213 (1991)
6. Chitnis, R., Kamma, L., Krauthgamer, R.: Tight bounds for Gomory-Hu-like cut counting. CoRR abs/1511.08647 (2015)
7. Cook, W.J., Cunningham, W.H., Pulleyblank, W.R., Schrijver, A.: Combinatorial Optimization. Wiley, New York (1998)
8. Ford, L.R., Fulkerson, D.R.: Maximal flow through a network. Can. J. Math. **8**(3), 399–404 (1956)
9. Gomory, R.E., Hu, T.C.: Multi-terminal network flows. J. Soc. Ind. Appl. Math. **9**, 551–570 (1961)
10. Gusfield, D.: Very simple methods for all pairs network flow analysis. SIAM J. Comput. **19**(1), 143–155 (1990)
11. Hagerup, T., Katajainen, J., Nishimura, N., Ragde, P.: Characterizing multiterminal flow networks and computing flows in networks of small treewidth. J. Comput. Syst. Sci. **57**(3), 366–375 (1998)
12. Katz, M., Katz, N.A., Korman, A., Peleg, D.: Labeling schemes for flow and connectivity. SIAM J. Comput. **34**(1), 23–40 (2005)
13. Khan, A., Raghavendra, P.: On mimicking networks representing minimum terminal cuts. Inf. Process. Lett. **114**(7), 365–371 (2014)
14. Krauthgamer, R., Rika, I.: Mimicking networks and succinct representations of terminal cuts. In: SODA 2013, pp. 1789–1799. SIAM (2013)

Eccentricity Approximating Trees
Extended Abstract

Feodor F. Dragan[1]($^{(\boxtimes)}$), Ekkehard Köhler[2], and Hend Alrasheed[1]

[1] Algorithmic Research Laboratory, Department of Computer Science,
Kent State University, Kent, OH 44242, USA
`dragan@cs.kent.edu, halrashe@kent.edu`
[2] Mathematisches Institut, Brandenburgische Technische Universität,
03013 Cottbus, Germany
`ekkehard.koehler@b-tu.de`

Abstract. Using the characteristic property of chordal graphs that they are the intersection graphs of subtrees of a tree, Erich Prisner showed that every chordal graph admits an eccentricity 2-approximating spanning tree. That is, every chordal graph G has a spanning tree T such that $ecc_T(v) - ecc_G(v) \leq 2$ for every vertex v, where $ecc_G(v)$ ($ecc_T(v)$) is the eccentricity of a vertex v in G (in T, respectively). Using only metric properties of graphs, we extend that result to a much larger family of graphs containing among others chordal graphs and the underlying graphs of 7-systolic complexes. Furthermore, based on our approach, we propose two heuristics for constructing eccentricity k-approximating trees with small values of k for general unweighted graphs. We validate those heuristics on a set of real-world networks and demonstrate that all those networks have very good eccentricity approximating trees.

1 Introduction

All graphs $G = (V, E)$ occurring in this paper are connected, finite, unweighted, undirected, loopless and without multiple edges. The *length* of a path from a vertex v to a vertex u is the number of edges in the path. The *distance* $d_G(u, v)$ between two vertices u and v is the length of a shortest path connecting u and v in G. If no confusion arises, we will omit subindex G. The *interval* $I(u, v)$ between u and v consists of all vertices on shortest (u, v)-paths, that is, it consists of all vertices (metrically) between u and v: $I(u, v) = \{x \in V : d_G(u, x) + d_G(x, v) = d_G(u, v)\}$. The *eccentricity* $ecc_G(v)$ of a vertex v in G is defined by $\max_{u \in V} d_G(u, v)$, i.e., it is the distance to a most distant vertex. The maximum value of the eccentricity represents the graph's *diameter*: $diam(G) = \max_{u \in V} ecc_G(u) = \max_{u,v \in V} d_G(u, v)$. The minimum value of the eccentricity represents the graph's *radius*: $rad(G) = \min_{u \in V} ecc_G(u)$. The set of vertices with minimum eccentricity forms the *center* $C(G)$ of a graph G, i.e., $C(G) = \{u \in V : ecc_G(u) = rad(G)\}$.

A spanning tree T of a graph G with $d_T(u, v) - d_G(u, v) \leq k$, for all $u, v \in V$, is known as an *additive tree spanner* of G [9] and, if it exists for a small integer k,

© Springer-Verlag GmbH Germany 2016
P. Heggernes (Ed.): WG 2016, LNCS 9941, pp. 145–157, 2016.
DOI: 10.1007/978-3-662-53536-3_13

then it gives a good approximation of all distances in G by the distances in T. Many optimization problems involving distances in graphs are known to be NP-hard in general but have efficient solutions in simpler metric spaces, with well-understood metric structures, including trees. A solution to such an optimization problem obtained for a tree spanner T of G usually serves as a good approximate solution to the problem in G.

In [13], the new notion of eccentricity approximating spanning trees was introduced by Prisner. A spanning tree T of a graph G is called an *eccentricity k-approximating spanning tree* if $ecc_T(v) - ecc_G(v) \leq k$ holds for all $v \in V$. Such a tree tries to approximately preserve only distances from each vertex v to its most distant vertices and can tolerate larger increases to nearby vertices. They are important in applications where vertices measure their degree of centrality by means of their eccentricity and would tolerate a small surplus to the actual eccentricities [13]. Note also that Nandakumar and Parthasarasthy considered in [11] eccentricity-preserving spanning trees (i.e., eccentricity 0-approximating spanning trees) and showed that a graph G has an eccentricity 0-approximating spanning tree if and only if: (a) either $diam(G) = 2rad(G)$ and $|C(G)| = 1$, or $diam(G) = 2rad(G) - 1$, $|C(G)| = 2$, and those two center vertices are adjacent; (b) every vertex $u \in V \setminus C(G)$ has a neighbor v such that $ecc_G(v) < ecc_G(u)$.

Every additive tree k-spanner is clearly eccentricity k-approximating. Therefore, eccentricity k-approximating spanning trees can be found in every interval graph for $k = 2$ [9,10,12], and in every asteroidal-triple–free graph [9], strongly chordal graph [3] and dually chordal graph [3] for $k = 3$. On the other hand, although for every k there is a chordal graph without a tree k-spanner [9,12], yet as Prisner demonstrated in [13], every chordal graph has an eccentricity 2-approximating spanning tree, i.e., with the slightly weaker concept of eccentricity-approximation, one can be successful even for chordal graphs.

Unfortunately, the method used by Prisner in [13] heavily relies on a characteristic property of chordal graphs (*chordal graphs are exactly the intersection graphs of subtrees of a tree*) and is hardly extendable to larger families of graphs.

In this paper we present a new proof of the result of [13] using only metric properties of chordal graphs (see Theorem 9 and Corollary 3). This allows us to extend the result to a much larger family of graphs which includes not only chordal graphs but also other families of graphs known from the literature.

It is known [4,15] that every chordal graph satisfies the following two metric properties:

α_1-**metric:** if $v \in I(u,w)$ and $w \in I(v,x)$ are adjacent, then $d_G(u,x) \geq d_G(u,v) + d_G(v,x) - 1 = d_G(u,v) + d_G(w,x)$.

triangle condition: for any three vertices u, v, w with $1 = d_G(v,w) < d_G(u,v) = d_G(u,w)$ there exists a common neighbor x of v and w such that $d_G(u,x) = d_G(u,v) - 1$.

A graph G satisfying the α_1-metric property is called an α_1-*metric graph*. If an α_1-metric graph G satisfies also the triangle condition then G is called an (α_1, Δ)-*metric graph*. We prove that every (α_1, Δ)-metric graph $G = (V, E)$ has an eccentricity 2-approximating spanning tree and that such a tree can be constructed in $\mathcal{O}(|V||E|)$ total time. As a consequence, we get that the underlying graph of every 7-systolic complex (and, hence, every chordal graph) has an eccentricity 2-approximating spanning tree.

The paper is organized as follows. In Sect. 2, we present additional notions and notations and some auxiliary results. In Sect. 3, some useful properties of the eccentricity function on (α_1, Δ)-metric graphs are described. Our eccentricity approximating spanning tree is constructed and analyzed in Sect. 4. In Sect. 5, the algorithm for the construction of an eccentricity approximating spanning tree developed in Sect. 4 for (α_1, Δ)-metric graphs is generalized and validated on some real-world networks. Our experiments show that all those real-world networks have very good eccentricity approximating trees.

Due to space limitations some proofs are omitted, they can be found in the full journal version of the paper [1].

2   Preliminaries

For a graph $G = (V, E)$, we use $n = |V|$ and $m = |E|$ to denote the cardinality of the vertex set and the edge set of G. We denote an *induced cycle* of length k by C_k (i.e., it has k vertices) and by W_k an *induced wheel* of size k which is a C_k with one extra vertex universal to C_k. For a vertex v of G, $N_G(v) = \{u \in V : uv \in E\}$ is called the *open neighborhood*, and $N_G[v] = N_G(v) \cup \{v\}$ the *closed neighborhood* of v. The distance between a vertex v and a set $S \subseteq V$ is defined as $d_G(v, S) = \min_{u \in S} d_G(u, v)$ and the set of furthest (most distant) vertices from v is denoted by $F(v) = \{u \in V : d_G(u, v) = ecc_G(v)\}$.

An induced subgraph of G (or the corresponding vertex set A) is called *convex* if for each pair of vertices $u, v \in A$ it includes the interval $I(v, u)$ of G between u, v. An induced subgraph H of G is called *isometric* if the distance between any pair of vertices in H is the same as their distance in G. In particular, convex subgraphs are isometric. The *disk* $D(x, r)$ with center x and radius $r \geq 0$ consists of all vertices of G at distance at most r from x. In particular, the unit disk $D(x, 1) = N[x]$ comprises x and the neighborhood $N(x)$. For an edge $e = xy$ of a graph G, let $D(e, r) := D(x, r) \cup D(y, r)$.

By the definition of α_1-metric graphs clearly, such a graph cannot contain any isometric cycles of length $k > 5$ and any induced cycle of length 4. The following results characterize α_1-metric graphs and the class of chordal graphs within the class of α_1-metric graphs. Recall that a graph is *chordal* if all its induced cycles are of length 3.

Theorem 1 ([15]). *G is chordal if and only if it is an α_1-metric graph not containing any induced subgraphs isomorphic to cycle C_5 and wheel $W_k, k \geq 5$.*

Theorem 2 ([15]). *G is an α_1-metric graph if and only if all disks $D(v, k)$ ($v \in V$, $k \geq 1$) of G are convex and G does not contain the graph W_6^{++} (see Fig. 1) as an isometric subgraph.*

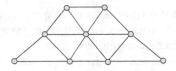

Fig. 1. Forbidden isometric subgraph W_6^{++}.

Theorem 3 ([8,14]). *All disks $D(v,k)$ $(v \in V$, $k \geq 1)$ of a graph G are convex if and only if G does not contain isometric cycles of length $k > 5$, and for any two vertices x, y the neighbors of x in the interval $I(x, y)$ are pairwise adjacent.*

A graph G is called a *bridged graph* if all isometric cycles of G have length three [8]. The class of bridged graphs is a natural generalization of the class of chordal graphs. They can be characterized in the following way.

Theorem 4 ([8,14]). *$G = (V, E)$ is a bridged graph if and only if the disks $D(v,k)$ and $D(e,k)$ are convex for all $v \in V$, $e \in E$, and $k \geq 1$.*

As a consequence of Theorems 2, 3 and 4 we obtain the following equivalences.

Lemma 1. *For a graph $G = (V, E)$ the following statements are equivalent:*

(a) G is an α_1-metric graph not containing an induced C_5;
(b) G is a bridged graph not containing W_6^{++} as an isometric subgraph;
(c) The disks $D(v,k)$ and $D(e,k)$ of G are convex for all $v \in V$, $e \in E$, and $k \geq 1$, and G does not contain W_6^{++} as an isometric subgraph.

As we will show now the class of (α_1, Δ)-metric graphs contains all graphs described in Lemma 1. An induced C_5 is called *suspended* in G if there is a vertex in G which is adjacent to all vertices of the C_5.

Theorem 5. *A graph G is (α_1, Δ)-metric if and only if it is an α_1-metric graph where for each induced C_5 there is a vertex $v \in V$ such that $C_5 \subseteq N(v)$, i.e., every induced C_5 is suspended.*

We will also need the following fact.

Lemma 2. *Let $G = (V, E)$ be an (α_1, Δ)-metric graph, let K be a complete subgraph of G, and let v be a vertex of G. If for every vertex $z \in K$, $d(z, v) = k$ holds, then there is a vertex v' at distance $k - 1$ from v which is adjacent to every vertex of K.*

We note here, without going into the rich theory of systolic complexes, that the underlying graph of any 7-systolic complex is nothing else than a bridged graph not containing a 6-wheel W_6 as an induced (equivalently, isometric) subgraph (see [6] for this fact and a relation of 7-systolic complexes with CAT(0) complexes). Hence, the class of (α_1, Δ)-metric graphs contains the underlying graphs of 7-systolic complexes.

3   Eccentricity Function on (α_1, Δ)-Metric Graphs

In what follows, by $C(G)$ we denote not only the set of all central vertices of G but also the subgraph of G induced by this set. We say that the eccentricity function $ecc_G(v)$ on G is *unimodal* if every vertex $u \in V \setminus C(G)$ has a neighbor v such that $ecc_G(v) < ecc_G(u)$. In other words, every local minimum of the eccentricity function $ecc_G(v)$ is a global minimum on G. It this section we will often omit subindex G since we deal only with a graph G here. A spanning tree T of G will be built only in the next section.

In this section, we will show that the eccentricity function $ecc_G(v)$ on an (α_1, Δ)-metric graph G is almost unimodal and that the radius of the center $C(G)$ of G is at most 2. Recall that for every graph G, $diam(G) \leq 2rad(G)$.

Lemma 3. *Let G be an α_1-metric graph and x be its arbitrary vertex with $ecc(x) \geq rad(G) + 1$. Then, for every vertex $z \in F(x)$ and every neighbor v of x in $I(x, z)$, $ecc(v) \leq ecc(x)$ holds.*

Proof. Assume, by way of contradiction, that $ecc(v) > ecc(x)$ and consider an arbitrary vertex $u \in F(v)$. Since x and v are adjacent, necessarily, $d(v, u) = ecc(v) = ecc(x) + 1 = d(u, x) + 1$, i.e., $x \in I(v, u)$. By the α_1-metric property, $d(u, z) \geq d(u, x) + d(v, z) = ecc(v) - 1 + ecc(x) - 1 = 2ecc(x) - 1 \geq 2rad(G) + 1$. The latter gives a contradiction to $d(u, z) \leq diam(G) \leq 2rad(G)$. $\square$

Theorem 6. *Let G be an (α_1, Δ)-metric graph and x be an arbitrary vertex of G. If (i) $ecc(x) > rad(G) + 1$ or (ii) $ecc(x) = rad(G) + 1$ and $diam(G) < 2rad(G)$, then there must exist a neighbor v of x with $ecc(v) < ecc(x)$.*

Proof. Define for a neighbor v of x a set $S_v := \{z \in F(x) : v \in I(x, z)\}$ of vertices that are most distant from x and have v on a shortest path from x. Choose a neighbor v of x which maximizes $|S_v|$. We claim that $ecc(v) < ecc(x)$. We know, by Lemma 3, that $ecc(v) \leq ecc(x)$. Assume $ecc(v) = ecc(x)$ and consider an arbitrary vertex $u \in F(v)$.

Suppose first that $x \in I(v, u)$. Then, by the α_1-metric property, $d(u, z) \geq d(u, x) + d(v, z) = 2ecc(x) - 2$ holds for every $z \in S_v$. Hence, if $ecc(x) > rad(G) + 1$ then $d(u, z) > 2rad(G)$ and thus a contradiction to $d(u, z) \leq diam(G) \leq 2rad(G)$ arises. If, on the other hand, case (ii) applies, i.e., $ecc(x) = rad(G) + 1$ and $diam(G) < 2rad(G)$, then it follows that $d(u, z) \geq 2rad(G) > diam(G)$ and again a contradiction arises.

Now consider the case that $x \notin I(v, u)$. Then $ecc(v) = ecc(x)$ implies that $d(u, x) = d(u, v)$ and $u \in F(x)$. By the triangle condition, there must exist a common neighbor w of x and v such that $w \in I(x, u) \cap I(v, u)$. Since u belongs to S_w but not to S_v, then, by the maximality of $|S_v|$, there must exist a vertex $z \in F(x)$ which is in S_v but not in S_w. Thus, $d(w, z) > d(v, z)$ and $v \in I(w, z)$ must hold. Now, the α_1-metric property applied to $v \in I(w, z)$ and $w \in I(v, u)$ gives $d(u, z) \geq d(u, w) + d(v, z) = 2ecc(x) - 2$. As before we get $d(u, z) > 2rad(G) \geq diam(G)$, if $ecc(x) > rad(G) + 1$ (case (i)), and $d(u, z) \geq 2rad(G) > diam(G)$, if $ecc(x) = rad(G) + 1$ and $diam(G) < 2rad(G)$ (case (ii)). These contradictions complete the proof. $\square$

For each vertex $v \in V \setminus C(G)$ of a graph G we can define a parameter $loc(v) = \min\{d(v, x) : x \in V, ecc(x) < ecc(v)\}$ and call it the *locality* of v. We define the locality of any vertex from $C(G)$ to be 1. Theorem 6 says that if a vertex v with $loc(v) > 1$ exists in an (α_1, Δ)-metric graph G then $diam(G) = 2rad(G)$ and $ecc(v) = rad(G) + 1$. That is, only in the case that $diam(G) = 2rad(G)$ the eccentricity function can be not unimodal on G.

Observe that the center $C(G)$ of a graph $G = (V, E)$ can be represented as the intersection of all the disks of G of radius $rad(G)$, i.e., $C(G) = \bigcap \{D(v, rad(G)) : v \in V\}$. Consequently, the center $C(G)$ of an α_1-metric graph G is convex (in particular, it is connected), as the intersection of convex sets is always a convex set. In general, any set $\mathcal{C}_{\leq i}(G) := \{z \in V : ecc(z) \leq rad(G) + i\}$ is a convex set of G as $\mathcal{C}_{\leq i}(G) = \bigcap \{D(v, rad(G) + i) : v \in V\}$.

Corollary 1. *In an α_1-metric graph G, all sets $\mathcal{C}_{\leq i}(G)$, $i \in \{0, \ldots, diam(G) - rad(G)\}$, are convex. In particular, $C(G)$ of an α_1-metric graph G is convex.*

The following result gives bounds on the diameter and the radius of the center of an (α_1, Δ)-metric graph. Previously it was known that the diameter (the radius) of the center of a chordal graph is at most 3 (at most 2, respectively) [5].

Theorem 7. *For an (α_1, Δ)-metric graph G, $rad(C(G)) \leq 2$.*

Proof. Assume, by way of contradiction, that there are vertices $s, t \in C(G)$ such that $d(s, t) = 4$. Consider an arbitrary shortest path $P = (s = x_1, x_2, x_3, x_4, x_5 = t)$. Since $C(G)$ is convex any shortest path connecting s and t is in $C(G)$.

First we claim that for any vertex $u \in F(x_3)$ all vertices of P are at distance $r := d(u, x_3) = rad(G)$ from u. As $x_i \in C(G)$, we know that $d(u, x_i) \leq r$ ($1 \leq i \leq 5$). Assume $d(u, x_i) = r - 1$, $d(u, x_{i+1}) = r$, and $i \leq 2$. Then, the α_1-metric property applied to $x_i \in I(u, x_{i+1})$ and $x_{i+1} \in I(x_i, x_{i+3})$ gives $d(x_{i+3}, u) \geq r - 1 + 2 = r + 1$ which is a contradiction to $d(u, x_{i+3}) \leq r$. So, $d(u, x_1) = d(u, x_2) = r$. By symmetry, also $d(u, x_4) = d(u, x_5) = r$.

By the triangle condition, there must exist vertices v and w at distance $r - 1$ from u such that $vx_1, vx_2, wx_4, wx_5 \in E$. We claim that x_3 is adjacent to neither v nor w. Assume, without loss of generality, that $vx_3 \in E$. Then, $d(x_5, x_1) = 4$ implies $d(x_5, v) = 3$ and therefore $x_3 \in I(x_5, v)$. Now, the α_1-metric property applied to $x_3 \in I(x_5, v)$ and $v \in I(u, x_3)$ gives $d(x_5, u) \geq r - 1 + 2 = r + 1$ which is impossible. So, $vx_3, wx_3 \notin E$.

Obviously, $vx_4, wx_2 \notin E$. If $d(x_4, v) = 3$ then $x_2 \in I(x_4, v)$. Thus, by $v \in I(x_2, u)$ and the α_1-metric property, we would get $d(x_4, u) \geq r - 1 + 2 = r + 1$ which, again, is impossible. Thus, $d(x_4, v) = 2$ must hold. Since, by Theorem 5, every induced C_5 is suspended in G and, further, G cannot contain an induced C_4, we can choose a vertex $y \in N(v) \cap N(x_4)$ which is adjacent both to x_2 and x_3 as well. If $d(y, u) = r$ then again $y \in I(v, x_5)$ and $v \in I(u, y)$ will imply $d(x_5, u) \geq r - 1 + 2 = r + 1$, which is impossible. So, $d(y, u) = r - 1$ must hold and, by the convexity of disks, y must be adjacent to w.

All the above holds for every shortest path $P = (s = x_1, x_2, x_3, x_4, x_5 = t)$ connecting vertices s and t. Now, assume that P is chosen in such a way that among all vertices in $I(s,t)$ that are at distance 2 from s (we will call this set of vertices $S_2(s,t)$) the vertex x_3 has the minimum number of furthest vertices, i.e., $|F(x_3)|$ is as small as possible. Observe that, by convexity of the center, $S_2(s,t) \subseteq C(G)$. As y also belongs to $S_2(s,t)$ and has u at distance $r-1$, by the choice of x_3, there must exist a vertex $u' \in F(y)$ which is at distance $r-1$ from x_3. Applying the previous arguments to the path $P' := (s = x_1, x_2, y, x_4, x_5 = t)$, we will have $d(x_i, u') = d(y, u') = r$ for $i = 1, 2, 4, 5$, and get two more vertices v' and w' at distance $r-1$ from u' such that $v'x_1, v'x_2, w'x_4, w'x_5 \in E$ and $v'y, w'y \notin E$. By the convexity of disk $D(u', r-1)$, also $v'x_3, w'x_3 \in E$. Now consider the disk $D(x_2, 2)$. Since w, w' are in the disk and x_5 is not, vertices w and w' must be adjacent. But then vertices y, x_3, w', w form a forbidden induced cycle C_4.

The obtained contradictions show that a shortest path P of length 4 cannot exist in $C(G)$, i.e., $diam(C(G)) \leq 3$. As $C(G)$ is a convex set of G, the subgraph of G induced by $C(G)$ is also an α_1-metric graph. According to [15], $diam(G) \geq 2rad(G) - 2$ holds for every α_1-metric graph G. Hence, for a graph induced by $C(G)$ we have $3 \geq diam(C(G)) \geq 2rad(C(G)) - 2$, i.e., $rad(C(G)) \leq 2$.   □

Corollary 2 ([5]). *For a chordal graph G, $rad(C(G)) \leq 2$.*

For our next arguments we need a generalization of the set $S_2(s,t)$, as used in the proof of Theorem 7. We define a *slice* of the interval $I(u,v)$ from u to v for $0 \leq k \leq d(u,v)$ to be the set $S_k(u,v) = \{w \in I(u,v) : d(w,u) = k\}$.

Theorem 8. *Let G be an (α_1, Δ)-metric graph. Then, in every slice $S_k(u,v)$ there is a vertex x that is universal to that slice, i.e., $S_k(u,v) \subseteq N[x]$. In particular, if $diam(G) = 2rad(G)$, then $diam(C(G)) \leq 2$ and $rad(C(G)) \leq 1$.*

4 Eccentricity Approximating Spanning Tree Construction

It this section, we construct an eccentricity approximating spanning tree and analyze its quality for (α_1, Δ)-metric graphs. Here, we will use sub-indices G and T to indicate whether the distances or the eccentricities are considered in G or in T. $I(u,v)$ will always mean the interval between vertices u and v in G.

4.1 Tree Construction for Unimodal Eccentricity Functions

First consider the case when the eccentricity function on G is unimodal, that is, every non-central vertex of G has a neighbor with smaller eccentricity. We will need the following lemmas.

Lemma 4 ([7]). *Let G be an arbitrary graph. The eccentricity function on G is unimodal if and only if, for every vertex v of G, $ecc_G(v) = d_G(v, C(G)) + rad(G)$.*

Lemma 5 ([2]). *Let G be an arbitrary α_1-metric graph. Let x, y, v, u be vertices of G such that $v \in I(x, y)$, $x \in I(v, u)$, and x and v are adjacent. Then $d(u, y) = d(u, x) + d(v, y)$ holds if and only if there exist a neighbor x' of x in $I(x, u)$ and a neighbor v' of v in $I(v, y)$ with $d_G(x', v') = 2$; in particular, x' and v' lie on a common shortest path of G between u and y.*

We construct a spanning tree T of G as follows. First find the center $C(G)$ of G and pick an arbitrary central vertex c of the graph $C(G)$, i.e., $c \in C(C(G))$. Compute a *breadth-first-search tree* T' of $C(G)$ started at c. Expand this tree T' to a spanning tree T of G by identifying for every vertex $v \in V \setminus C(G)$ its parent vertex in the following way: among all neighbors x of v with $ecc_G(x) = ecc_G(v) - 1$ pick that vertex which is closest to c in G (break ties arbitrarily).

Lemma 6. *Let G be an (α_1, Δ)-metric graph whose eccentricity function is unimodal. Then, for a tree T constructed as described above and every vertex v of G, $d_G(v, c) = d_T(v, c)$ holds, i.e., T is a shortest-path-tree of G started at c.*

Proof. Let v be an arbitrary vertex of G and let v' be a vertex of $C(G)$ closest to v in T. By Lemma 4 and by the construction of T, $d_G(v, v') = d_T(v, v')$ and v' is a vertex of $C(G)$ closest to v in G. By the construction of T', also $d_G(c, v') = d_T(c, v')$ (note that, as $C(G)$ is a convex subgraph of G, clearly, $d_{C(G)}(x, y) = d_G(x, y)$ for every pair x, y of $C(G)$). So, in the tree T, we have $d_T(c, v') + d_T(v', v) = d_T(v, c)$. If $d_G(c, v') + d_G(v', v) = d_G(v, c)$, then $d_G(v, c) = d_T(v, c)$, and we are done. Assume, therefore, that $d_G(c, v') + d_G(v', v) > d_G(v, c)$ and among all vertices that fulfill this inequality, let v be the one that is closest to $C(G)$. Consider the neighbor x of v' on the path in T from v' to v. We have $x \in I(v', v)$ and, by Lemma 4, $ecc_G(x) = rad(G) + 1$. Note that $x = v$ is possible.

If $v' \notin I(x, c)$ then $d_G(x, c) \leq d_G(v', c)$. By the convexity of $C(G)$, x with $ecc_G(x) = rad(G) + 1$ cannot be on a shortest path between two central vertices c and v'. Hence, $d_G(x, c) = d_G(v', c)$ holds. By the triangle condition, there must exist a common neighbor y of v' and x which is at distance $d_G(v', c) - 1$ from c. Since $y \in I(v', c)$, by the convexity of $C(G)$, $ecc_G(y) = rad(G)$. But then, as y is closer to c than v' is, vertex x cannot choose v' as its parent in T, since y is a better choice.

If $v' \in I(x, c)$ then, by the α_1-metric property, $d_G(c, v') + d_G(x, v) \leq d_G(v, c)$. As $d_G(c, v') + d_G(v', v) > d_G(v, c)$, we have $d_G(c, v') + d_G(x, v) = d_G(v, c)$. By Lemma 5, there must exist a neighbor x' of x in $I(x, v)$ and a neighbor v'' of v' in $I(v', c)$ with $d_G(x', v'') = 2$. Denote by w a common neighbor of x' and v''. We have $d_G(x, c) > d_G(w, c)$. Set $k := d_G(v, v') = d_G(v, C(G)) = ecc_G(v) - rad(G)$. Let $P_T := (x = a_1, \ldots, a_k = v)$ be the path in T between x and v. Let $P_G := (w = b_1, x' = b_2, \ldots, b_k = v)$ be a shortest path of G between w and v which shares a longest suffix with P_T, that is, $a_j = b_j$ for all $j > i$, $a_i \neq b_i$, and i is minimal under these conditions. Note that $i = 1$ and $a_2 = b_2 = v$ is possible. By Lemma 4, $ecc_G(a_i) = ecc_G(b_i) = rad(G) + i = ecc_G(a_{i+1}) - 1$.

Since v is a vertex closest to $C(G)$ fulfilling inequality $d_G(c, v') + d_G(v', v) > d_G(v, c)$, for vertex a_i $(i < k)$, the equation $d_G(c, v') + d_G(v', a_i) = d_G(a_i, c)$ holds. Hence, $d_G(c, x) + d_G(x, a_i) = d_G(a_i, c)$. Also, by Lemma 5,

$d_G(c, w) + d_G(w, b_i) = d_G(b_i, c)$. Consequently, $d_G(x, c) > d_G(w, c)$ and $d_G(x, a_i) = d_G(w, b_i)$ imply $d_G(a_i, c) > d_G(b_i, c)$. Therefore, vertex a_{i+1} cannot choose a_i as its parent in T, since b_i is a better choice.

The obtained contradictions prove that $d_G(c, v') + d_G(v', v) = d_G(v, c)$ and hence $d_G(v, c) = d_T(v, c)$. □

Lemma 7. *Let G be an (α_1, Δ)-metric graph whose eccentricity function is unimodal. Then, for a tree T constructed as described above and for every vertex v of G, $ecc_T(v) \leq ecc_G(v) + rad(C(G))$ holds.*

Proof. Let v be an arbitrary vertex of G, v' be a vertex of $C(G)$ closest to v in T, and u be a vertex most distant from v in T, i.e., $ecc_T(v) = d_T(v, u)$. By Lemma 4 and by the construction of T, $d_G(v, v') = d_T(v, v')$ and v' is a vertex of $C(G)$ closest to v in G. We have $ecc_T(v) = d_T(v, u) \leq d_T(v, v') + d_T(v', c) + d_T(c, u)$, where $c \in C(C(G))$ is the root of the tree T (see the construction of T). Since $d_G(v, v') = d_T(v, v')$, $d_T(v', c) = d_G(v', c) \leq rad(C(G))$, and $d_T(c, u) = d_G(c, u) \leq rad(G)$ (by Lemma 6 and the fact that $c \in C(C(G))$), we obtain $ecc_T(v) \leq d_G(v, v') + rad(C(G)) + rad(G) = ecc_G(v) + rad(C(G))$, as $d_G(v, v') + rad(G) = d_G(v, C(G)) + rad(G) = ecc_G(v)$ by Lemma 4. □

4.2   Construction for Eccentricity Functions that Are Not Unimodal

Consider now the case when the eccentricity function on G is not unimodal, that is, there is at least one vertex $v \notin C(G)$ in G which has no neighbor with smaller eccentricity. By Theorem 6, $ecc_G(v) = rad(G) + 1$, $diam(G) = 2rad(G)$ and v has a neighbor with the eccentricity equal to $ecc_G(v)$. We will need the following weaker version of Lemma 4.

Lemma 8. *Let $G = (V, E)$ be an (α_1, Δ)-metric graph. Let v be an arbitrary vertex of G and v' be an arbitrary vertex of $C(G)$ closest to v. Then,*

$$d_G(v, C(G)) + rad(G) - 1 \leq ecc_G(v) \leq d_G(v, C(G)) + rad(G).$$

Furthermore, there is a shortest path $P := (v' = x_0, x_1, \ldots, x_\ell = v)$, connecting v with v', for which the following holds:

(a) if $ecc_G(v) = d_G(v, C(G)) + rad(G)$ then $ecc_G(x_i) = d_G(x_i, C(G)) + rad(G) = i + rad(G)$ for each $i \in \{0, \ldots, \ell\}$;

(b) if $ecc_G(v) = d_G(v, C(G)) + rad(G) - 1$ then $ecc_G(x_i) = d_G(x_i, C(G)) - 1 + rad(G) = i - 1 + rad(G)$ for each $i \in \{3, \ldots, \ell\}$ and $ecc_G(x_1) = ecc_G(x_2) = rad(G) + 1$.

In particular, if $ecc_G(v) = rad(G) + 1$ then $d_G(v, C(G)) \leq 2$.

Now we are ready to construct an eccentricity approximating spanning tree T of G for the case when the eccentricity function is not unimodal. We know that $diam(G) = 2rad(G)$ in this case and, therefore, $C(G) \subseteq S_{rad(G)}(x, y)$ for any diametral pair of vertices x and y, i.e., for x, y with $d_G(x, y) = diam(G)$.

By Theorem 8 and since $C(G)$ is convex, there is a vertex $c \in C(G)$ such that $C(G) \subseteq N[c]$. First we find such a vertex c in $C(G)$ and build a tree T' by making c adjacent with every other vertex of $C(G)$. Then, we expand this tree T' to a spanning tree T of G by identifying for every vertex $v \in V \setminus C(G)$ its parent vertex in the following way: if v has a neighbor with eccentricity less than $ecc_G(v)$, then among all such neighbors pick that vertex which is closest to c in G (break ties arbitrarily); if v has no neighbors with eccentricity less than $ecc_G(v)$ (i.e., $ecc_G(v) = rad(G) + 1$ by Theorem 6), then among all neighbors x of v with $ecc_G(x) = ecc_G(v) = rad(G)+1$ pick again that vertex which is closest to c in G (break ties arbitrarily).

Lemma 9. *Let G be an (α_1, Δ)-metric graph whose eccentricity function is not unimodal. Then, for a tree T constructed as described above and every vertex v of G, $d_T(v, c) = d_G(v, c)$ holds.*

Proof. Assume, by way of contradiction, that $d_G(v, c) < k := d_T(v, c)$ and let v be a vertex with such a condition that has smallest eccentricity $ecc_G(v)$. We may assume that $ecc_G(v) > rad(G) + 1$. Indeed, every v with $ecc_G(v) = rad(G) + 1$ either has a neighbor in $C(G)$ or has a neighbor with a neighbor in $C(G)$ (see Lemma 8). Therefore, if $d_G(v, c) < d_T(v, c)$ then, by the construction of T, necessarily $d_G(v, c) = 2$, $d_T(v, c) = 3$ and the neighbor x of v on the path of T between v and c must have the eccentricity equal to $rad(G) + 1 = ecc_G(v)$. But then, for a common neighbor w of v and c in G, $ecc_G(w) \leq rad(G) + 1$ must hold and hence vertex v cannot choose x as its parent in T, since w is a better choice.

So, let $ecc_G(v) > rad(G) + 1$. By Lemma 8, there must exist a shortest path in G between v and c such that the neighbor w of v on this path has eccentricity $ecc_G(w) = ecc_G(v) - 1$. Hence, by the construction of T, $ecc_G(x) = ecc_G(v) - 1$ must hold for the neighbor x of v on the path of T between v and c. By the minimality of $ecc_G(v)$, we have $d_G(x, c) = d_T(x, c) = k - 1$. Since $d_G(w, c) = d_G(v, c) - 1 < k - 1$, a contradiction arises; again v cannot choose x as its parent in T, since w is a better choice. □

Lemma 10. *Let G be an (α_1, Δ)-metric graph with $diam(G) = 2rad(G)$. Then, for a tree T constructed as described above and every vertex v of G, $ecc_T(v) \leq ecc_G(v) + 2$ holds.*

Proof. Let v be an arbitrary vertex of G and u be a vertex most distant from v in T, i.e., $ecc_T(v) = d_T(v, u)$. We have $ecc_T(v) = d_T(v, u) \leq d_T(v, c) + d_T(c, u) = d_G(v, c) + d_G(c, u) \leq d_G(v, c) + rad(G) \leq d_G(v, C(G)) + 1 + rad(G) \leq ecc_G(v) + 2$ since $d_G(c, u) \leq ecc_G(c) = rad(G)$, $d_G(v, c) \leq d_G(v, C(G)) + 1$ (recall that $C(G) \subseteq N[c]$), and $d_G(v, C(G)) - 1 + rad(G) \leq ecc_G(v)$ (by Lemma 8). □

Our main result is the following theorem. It combines Theorem 7, Lemmas 7 and 10; the complexity follows straightforward.

Theorem 9. *Every (α_1, Δ)-metric graph $G = (V, E)$ has an eccentricity 2-approximating spanning tree. Furthermore, such a tree can be constructed in $\mathcal{O}(|V||E|)$ total time.*

As a consequence we have the following corollary. Note that the result of Corollary 3 (and hence of Theorem 9) is sharp as there are chordal graphs that do not have any eccentricity 1-approximating spanning tree [13].

Corollary 3. *The underlying graph of every 7-systolic complex has an eccentricity 2-approximating spanning tree. In particular, every chordal graph has an eccentricity 2-approximating spanning tree.*

5   Experimental Results for Some Real-World Networks

We say that a tree T is an *eccentricity k-approximating tree* for a graph G if for every vertex v of G, $|ecc_T(v) - ecc_G(v)| \leq k$ holds. If T is a spanning tree, then $ecc_T(v) \geq ecc_G(v)$, for all $v \in V$, and this new definition agrees with the definition of an eccentricity k-approximating *spanning* tree.

Table 1. A spanning tree T constructed by heuristic EAST: for each vertex $u \in V$, $k(u) = ecc_T(u) - ecc_G(u)$; $k_{max} = \max_{u \in V} k(u)$; $k_{avg} = \frac{1}{n} \sum_{u \in V} k(u)$. A tree T' constructed by heuristic EAT: for each vertex $u \in V$, $k(u) = ecc_{T'}(u) - ecc_G(u)$; $k_{max} = \max_{u \in V} k(u)$; $k_{min} = \min_{u \in V} k(u)$; $k_{avg} = \frac{1}{n} \sum_{u \in V} k(u)$

Network	$diam(G)$	k_{max} of T	k_{avg} of T	$[k_{min}, k_{max}]$ of T'	k_{avg} of T'
EMAIL	8	3	1.774	$[-1, 0]$	-0.0009
FACEBOOK	8	2	0.69	$[0, 0]$	0
DUTCH-ELITE	22	6	2.083	$[-1, 0]$	-0.771
JAZZ	6	2	1.742	$[-1, 0]$	-0.015
EVA	18	2	0.575	$[-1, 0]$	-0.36
AS-GRAPH-1	9	2	0.64	$[0, 1]$	0.62
AS-GRAPH-2	11	3	1.272	$[0, 1]$	0.949
AS-GRAPH-3	9	2	0.312	$[0, 1]$	0.248
E-COLI-PI	5	2	0.769	$[0, 1]$	0.595
YEAST-PI	12	4	0.972	$[-1, 0]$	-0.168
MACAQUE-BRAIN-1	4	1	0.222	$[0, 0]$	0
MACAQUE-BRAIN-2	4	2	1.489	$[-1, 0]$	-0.003
E-COLI-METABOLIC	16	4	1.132	$[-1, 0]$	-0.624
C-ELEGANS-METABOLIC	7	1	0.349	$[0, 1]$	0.342
YEAST-TRANSCRIPTION	9	3	1.121	$[0, 1]$	0.019
US-AIRLINES	6	0	0	$[0, 0]$	0
POWER-GRID	46	4	1.409	$[-3, 0]$	-1.309
WORD-ADJACENCY	5	1	0.411	$[0, 1]$	0.152
FOOD	4	2	1.629	$[-1, 0]$	-0.015

Based on what we learned from (α_1, Δ)-metric graphs in Sect. 4, we propose two heuristics for constructing eccentricity approximating trees in general graphs and analyze their performance on a set of real-world networks. Both heuristics try to mimic the construction for (α_1, Δ)-metric graphs that we used in Sect. 4. For more details on the data-set and the experiments see the full journal version of the paper [1].

Our first heuristic, named *EAST*, constructs an *E*ccentricity *A*pproximating *S*panning *T*ree T_{EAST} as a shortest-path-tree starting at a vertex $c \in C(C(G))$. We identify an arbitrary vertex $c \in C(C(G))$ as the root of T_{EAST}, and for each other vertex v of G define its parent in T_{EAST} as follows: among all neighbors of v in $I(v, c)$ choose a vertex with minimum eccentricity (break ties arbitrarily).

Our second heuristic, named *EAT*, constructs for a graph G an *E*ccentricity *A*pproximating *T*ree T_{EAT} (not necessarily a spanning tree; it may have a few edges not present in graph G) as follows. We again identify an arbitrary vertex $c \in C(C(G))$ as the root of T_{EAT} and make it adjacent in T_{EAT} to all other vertices of $C(G)$ (clearly, some of these edges might not be contained in G). Then, for each vertex $v \in V \setminus C(G)$, we find a vertex u with $ecc_G(u) < ecc_G(v)$ which is closest to v, and if there is more than one such vertex, we pick the one which is closest to c. In other words, among all vertices $\{u \in V : d_G(u, v) = loc(v)$ and $ecc_G(u) < ecc_G(v)\}$, we choose a vertex u which is closest to c (break ties arbitrarily). Such a vertex u becomes the parent of v in T_{EAT}. Clearly, if $loc(v) > 1$ then edge uv of T_{EAT} is not present in G.

We tested both heuristics on a set of real-world networks. Experimental results obtained are presented in Table 1. See the full journal version of the paper [1] for more details. It turns out that the eccentricity terrain of each of those networks resembles the eccentricity terrain of a tree.

References

1. http://www.cs.kent.edu/~dragan/EccApprTree--journal.pdf
2. Bandelt, H.-J., Chepoi, V.: 1-hyperbolic graphs. SIAM J. Discret. Math. **16**, 323–334 (2003)
3. Brandstädt, A., Chepoi, V., Dragan, F.F.: Distance approximating trees for chordal and dually chordal graphs. J. Algorithms **30**, 166–184 (1999)
4. Chepoi, V.: Some d-convexity properties in triangulated graphs. Math. Res. **87**, 164–177, Ştiinţa, Chişinău (1986). (Russian)
5. Chepoi, V.: Centers of triangulated graphs. Math. Notes **43**, 143–151 (1988)
6. Chepoi, V.: Graphs of some CAT(0) complexes. Adv. Appl. Math. **24**, 125–179 (2000)
7. Dragan, F.F.: Centers of graphs and the Helly property (in Russian). Ph.D. thesis, Moldova State University (1989)
8. Farber, M., Jamison, R.E.: On local convexity in graphs. Discret. Math. **66**, 231–247 (1987)
9. Kratsch, D., Le, H.-O., Müller, H., Prisner, E., Wagner, D.: Additive tree spanners. SIAM J. Discret. Math. **17**, 332–340 (2003)
10. Madanlal, M.S., Vankatesan, G., Pandu Rangan, C.: Tree 3-spanners on interval, permutation and regularbipartite graphs. Inf. Process. Lett. **59**, 97–102 (1996)

11. Nandakumar, R., Parthasarathy, K.: Eccentricity-preserving spanning trees. J. Math. Phys. Sci. **24**, 33–36 (1990)
12. Prisner, E.: Distance approximating spanning trees. In: Reischuk, R., Morvan, M. (eds.) STACS 1997. LNCS, vol. 1200, pp. 499–510. Springer, Heidelberg (1997)
13. Prisner, E.: Eccentricity-approximating trees in chordal graphs. Discret. Math. **220**, 263–269 (2000)
14. Soltan, V.P., Chepoi, V.D.: Conditions for invariance of set diameters under d-convexification in a graph. Cybernetics **19**, 750–756 (1983). (Russian, English transl.)
15. Yushmanov, S.V., Chepoi, V.: A general method of investigation of metric graph-properties related to the eccentricity. In: Mathematical Problems in Cybernetics, vol. 3, pp. 217–232. Nauka, Moscow (1991). (Russian)

Drawing Planar Graphs with Prescribed Face Areas

Linda Kleist[✉]

Institut für Mathematik, TU Berlin, Berlin, Germany
kleist@math.tu-berlin.de

Abstract. We study drawings of planar graphs such that every inner face has a prescribed area. A plane graph is *area-universal* if for every area assignment on the inner faces, there exists a straight-line drawing realizing the assigned areas. It is known that not all plane graphs are area-universal. The only counterexample in literature is the octahedron graph.

We give a counting argument that allows to prove non-area-universality for a large class of triangulations. Moreover, we relax the straight-line property of the drawings, namely we allow the edges to bend. We show that one bend per edge is enough to realize any face area assignment of every plane graph. For plane bipartite graphs, it suffices that half of the edges have a bend.

1 Introduction

Planar graphs link graph theory and geometry. Since various real-life problems are connected to embedded graphs on the surface of our planet, planar graphs and their representations have many practical applications: in the manufacture of chips and electrical circuits, in the design of network infrastructure such as roads, subway, and utility lines. Other applications are in cartography, geography, and visualization. Consequently, there is a large body of theoretical and applied work on representations of planar graphs with special features [11].

One direction is the representation of planar graphs with given areas. Proportional contact representations, so-called cartograms, are studied when areas are assigned to the vertices [2,5].

We are interested in drawings of plane graphs such that the inner faces have prescribed face areas. Let $G = (V, E)$ be a plane graph and F' the set of its inner faces. A *redrawing* of G is a drawing such that the set of inner faces remains. We denote the set of all redrawings of G by $\mathcal{D}$.

A *face area assignment* is a function $A : F' \to \mathbb{R}^+$. Let $a : F' \times \mathcal{D} \to \mathbb{R}^+$ be a function measuring the area of an inner face f in a specified redrawing D of G. A redrawing D of G is *A-realizing* if $a(f, D) = A(f)$ for each face $f \in F'$. The graph G is *area-universal* if for every face area assignment A there exists an A-realizing straight-line redrawing of G. The graph G is *equiareal* if there exists a straight-line redrawing D of G such that $a(f, D) = 1$ for each face $f \in F'$.

© Springer-Verlag GmbH Germany 2016
P. Heggernes (Ed.): WG 2016, LNCS 9941, pp. 158–170, 2016.
DOI: 10.1007/978-3-662-53536-3_14

1.1   Previous Work and Our Contribution

Ringel [10] can be seen as the initiator of the study of drawings of plane graphs with prescribed face areas. He gave examples of equiareal graphs and proved that not every plane graph is equiareal. Moreover, Ringel conjectured that every plane 3-regular graph is equiareal. This turned out to be true when Thomassen [12] proved something stronger: Every plane graph with maximum degree 3 is area-universal. There are further area-universal graph classes. A straight-forward result is the class of *planar 3-trees*, also known as *stacked triangulations*. Biedl and Ruiz Velázquez [3] proved that planar 3-trees have realizing drawings with rational coordinates if the face areas are rational.

To the best of our knowledge, this is the state of the art. Clearly, many interesting questions remain open. In this paper, we study two directions, namely negative and positive results. In terms of negative results, all non-area-universal graphs in literature contain the octahedron graph as a subgraph. This leads to the following questions:

- Is the octahedron graph the only minimal counterexample (by taking subgraphs)? Is it the only 4-connected counterexample?
- Are highly connected planar graphs area-universal?

The answers to these questions are negative. In Sect. 2 we give a broad class of triangulations which are not area-universal. This class contains many 4-connected graphs. Additionally, we show that the 5-connected icosahedron graph is not area-universal. Hence, high connectivity does not imply area-universality.

In terms of positive results, we investigate relaxations:

- What drawings can realize all face area assignments for every plane graph?

In Sect. 3 we show that every plane graph has a drawing realizing any face area assignment such that each edge has at most one bend. Moreover, every plane bipartite graph has a drawing realizing any face area assignment such that at most half of the edges have a bend.

2   Non-area-Universal Graphs

In this section we discuss non-area-universality. It is known that the octahedron graph is not area-universal [10]. There exist two proofs. In both proofs, the used area assignment is similar to ours; however, the concepts behind the proofs are different. Ringel [10] shows that the system of equations has no rational solution. A different proof relies on a classical geometric result on the area of a triangle inscribed into a triangle. The proof is similar to [1] (for details on the geometric result see [4]).

In contrast, we give a simple counting argument. Interestingly, the ideas can be extended to show that every plane Eulerian triangulation is not area-universal. Indeed, we use the fact that the dual graph of a plane triangulation is bipartite, and hence has an inner face 2-coloring. An *inner face 2-coloring* of a triangulation is a coloring of the inner faces with white and black such that every inner edge is incident to a white and black face. For an inner face 2-coloring of the octahedron graph, see Fig. 1.

Theorem 1. *Every plane Eulerian triangulation (on more than 3 vertices) is not area-universal.*

Proof. Let T be a plane triangulation on n vertices with an inner face 2-coloring. We denote the set of white faces by W and the set of black faces by B. Recall that the number of inner faces of a triangulation with n vertices is $2n - 5$. Without loss of generality $|W| > |B|$, that is $|W| \geq n - 2$.

We show that for sufficiently small $\varepsilon > 0, T$ has no straight-line drawing realizing the following area assignment with a total area of 1:

$$A(f) := \begin{cases} \varepsilon & \text{if } f \in W, \\ \delta_\varepsilon := \frac{1 - \varepsilon|W|}{|B|} & \text{else.} \end{cases}$$

The idea is to show that an A-realizing drawing has contradicting properties. One property is that every white face needs a big angle of almost π. Another property is that every inner vertex can have at most one such big angle, and every outer vertex has none. Since the number of white faces exceeds the number of inner vertices this gives a contradiction. To illustrate the idea of the proof, we start with the degenerate case of $\varepsilon = 0$.

We suppose, for the purpose of contradiction, that there exists an A-realizing drawing D of T. Since every inner edge e is incident to a black face, e has positive length in D; otherwise, the area of the black face cannot be realized. This yields the main property of D: One angle of every white face is of size π, i.e., a *big* angle. The area of a white face vanishes only if a vertex lies on a non-incident edge. Since all edges have positive length, the vertex must lie on an inner point of the non-incident edge. We *assign* (the big angle of) this white face to the vertex with the big angle. Clearly, only inner vertices may have big angles. Recall that T has $n - 2$ white faces and $n - 3$ inner vertices. By the pigeonhole principle, in every assignment of big angles to inner vertices, there exists a vertex v which is assigned to two big angles, see Fig. 1. Vertex v is also incident to at least 2 black faces which are separating the white faces. However, due to the two big angles, no space remains in order to realize the area of the black faces incident to v. Consequently, D is not a realizing drawing, thus a contradiction.

For the case $\varepsilon > 0$, we suppose that there exists a realizing straight-line drawing. Since triangles are affine equivalent, any two realizing straight-line drawings are affine equivalent. Hence, we suppose that there exists a straight-line drawing within an equilateral triangle with area 1 and sidelength L. It has the following properties:

- The longest edge is of length L.
- Each edge is incident to a black face and, hence, at least of length $s := {}^{2\delta_\varepsilon}/L$.
- The height of a white face is bounded by ${}^{2\varepsilon}/s$.

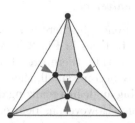

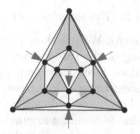

Fig. 1. The octahedron graph with an inner face 2-coloring indicating the area assignment. In every assignment of white faces to inner vertices, one vertex is assigned to two faces. (Color figure online)

Fig. 2. The icosahedron graph with an inner face 2-coloring indicating an area assignment which cannot be realized by any straight-line drawing. (Color figure online)

- There exists a small α'_ε, continuously decreasing with ε, such that a white face angle is either *tiny* (at most α'_ε) or *big* (at least $\alpha_\varepsilon := \pi - 2\alpha'_\varepsilon$).
- Each white face has a big angle.
- There exists β_ε such that a black face angle is at least of size β_ε.

Recall that $2\alpha_0 + 2\beta_0 > 2\pi$. By continuity of α_ε and β_ε, the intermediate value theorem implies the existence of $\epsilon > 0$ with

$$2\alpha_\varepsilon + 2\beta_\varepsilon > 2\pi.$$

Consequently, in every realizing drawing no inner vertex may realize the big angles of two white faces. However, the number of white faces exceeds the number of inner vertices. This is a contradiction and, hence, establishes the proof. □

Remark 1. It suffices to choose ε in the order of n^{-3}.

Remark 2. Indeed, a triangulation has an inner face 2-coloring if and only if it has a face 2-coloring. This stems from the fact that the number of edges of a Eulerian inner triangulation of a quadrangle is divisible by 3.

Remark 3. This construction implies that one cannot hope for drawings realizing the areas up to a constant factor. If ε is small enough, then there is no c such that a drawing of an Eulerian triangulation fulfills $1/c \cdot A(f) \le a(f, D) \le c \cdot A(f)$ for all inner faces f.

Note that the graphs of Theorem 1 are at most 4-connected. With similar ideas (but more work) we can show that the 5-connected icosahedron graph is not area-universal. Here, we only sketch the proof idea.

Theorem 2. *The icosahedron graph is not area-universal.*

Proof (Sketch). We can show that for $\varepsilon > 0$ small enough, there is no straight-line drawing of the icosahedron in which the white faces in Fig. 2 have area ε and the black faces have area $\delta_\varepsilon := \frac{1-10\varepsilon}{9}$. Note that there are three (red) edges adjacent to two white faces. These edges may become small, therefore, the adjacent white triangles do not necessarily need a big angle in a realizing drawing. Hence, a simple counting argument is not sufficient. However, as before, each of the four white triangles with three black edges needs a big angle. In every big angle assignment of these four white triangles, there is a red *special* edge whose vertices are both assigned. Moreover, the angle between two black edges in a white face, is either *tiny* (almost 0) or *big* (almost π). A case distinction on the size of the (green) angles opposite to the red special edge yields the result.   □

3   Drawing Planar Graphs with Bends

We now aim for realizing each face area assignment of every plane graph. As discussed in Sect. 2, this is impossible for straight-line drawings. Therefore, we relax the straight-line property by allowing the edges to have bends. A drawing of a plane graph is a *k-bend drawing* if each edge is a concatenation of at most $k + 1$ segments. These drawings are also called *polyline* drawings. In this section we show that one bend per edge is sufficient.

Theorem 3. *Let G be a plane graph and $A \colon F' \to \mathbb{R}^+$ a face area assignment. Then, there exists an A-realizing 1-bend redrawing of G.*

Proof. Without loss of generality, we assume that G is a plane triangulation: If G is not a triangulation, there exists a triangulation T such that G is an induced subgraph. For each face of G, partition the assigned area between its subfaces in T and obtain the area assignment A' of T. Given an A'-realizing 1-bend redrawing of T, delete the artificial vertices and edges. The result is an A-realizing 1-bend redrawing of G.

We construct the final drawing of G in four steps (see definitions below):

1. Take a $\bot$-contact representation $\mathcal{C}$ which yields a rectangular layout $\mathcal{L}$.
2. Obtain a weak equivalent rectangular layout $\mathcal{L}'$ realizing the areas.
3. Define a degenerate drawing $D_\bot$.
4. Construct a non-degenerate drawing from $D_\bot$.

The steps are visualized for the octahedron graph in Fig. 3.

In the first step, we construct a $\bot$-contact representation $\mathcal{C}$ of G. A $\bot$-shape is the union of a horizontal and vertical segment such that the lower end of the vertical segment lies in the horizontal segment. We call this point of intersection the *heart* of the $\bot$-shape. Each of the other three ends of the segments is an *end* of the $\bot$-shape. A $\bot$-*contact representation* of a graph $G = (V, E)$ is a family of $\bot$-shapes $\{\bot_v : v \in V\}$ where $\bot_u$ and $\bot_v$ intersect if and only if $(u, v) \in E$. Moreover, if $\bot_u$ and $\bot_v$ intersect then the intersection must consist of a single point which is an end of $\bot_u$ or $\bot_v$. The point is the *contact point* of $\bot_u$ and $\bot_v$.

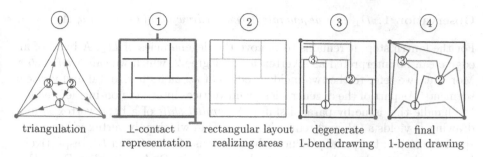

Fig. 3. Construction of a 1-bend drawing realizing the prescribed areas in 4 steps.

Lemma 1 ([8]). *Every plane triangulation has a ⊥-contact representation C such that each inner face is represented by a rectangular region.*

Following the ideas of [8], C can be constructed as described in detail in Sect. 3.1 in [2]. Observe that the ⊥-shapes of two outer vertices may be pruned to segments such that the outer face is the complement of a rectangle. The segments of the ⊥-shapes of inner vertices partition the rectangle into finitely many rectangles; such a partition is called a *rectangular layout*. By Lemma 1, C yields a rectangular layout $\mathcal{L}$ in which every rectangle r corresponds to a face f_r of G.

In the second step, we want to achieve correct areas in a weak-equivalent layout. The maximal segments of a rectangular layout yield a segment contact graph. Two rectangular layouts are *weak-equivalent* if their segment contact graphs are isomorphic. We apply the following lemma.

Lemma 2. *For every rectangular layout with area assignment w on the inner rectangles, there exists a weak-equivalent layout realizing the areas of w.*

This lemma has several variants and proofs; we refer to [5,6,13]. For each rectangle r corresponding to the face f_r, we set $w(r) := A(f_r)$. By Lemma 2, we obtain a weak-equivalent rectangular layout $\mathcal{L}'$ in which the area of each rectangle r is $w(r)$. Due to the weak-equivalence of $\mathcal{L}$ and $\mathcal{L}'$, the layout $\mathcal{L}'$ can be viewed as a ⊥-contact representation C', which now realizes the areas.

In the third step, we obtain a (degenerate) 1-bend drawing $D_\perp$ of G from C':

- Place each vertex v in the heart of $\perp_v$; (for a pruned ⊥-shape, the heart coincides with the bottom or left end of the remaining segment),
- The edges are supported by the segments of C' in the following way: If two vertices u and v share an edge, $\perp_u$ and $\perp_v$ have a point of contact in which a vertical and a horizontal segment meet. We define the edges to run from the heart of one ⊥-shape (along the horizontal segment) to the contact point and then (along the vertical segment) to the heart of the other ⊥-shape.

Up to the fact that two edges may intersect interiorly (but do not cross), the properties of a plane redrawing of G are fulfilled. We call such an embedding *degenerate drawing*. By construction, each edge consists of a horizontal and a vertical segment and, hence, has at most (and in general exactly) one bend.

Observation 1. $D_\perp$ *is a degenerate 1-bend redrawing of G realizing A.*

For the fourth step, it remains to remove the degeneracies of $D_\perp$. A bend of an edge e can be interpreted as a vertex w_e of degree 2, which we call *bend vertex*. Moreover, we refer to the two incident edges of w_e as its horizontal and vertical segments. As part of the degeneracies, bend vertices intersect non-incident edges. We handle this issue by parallel shifts. A *parallel shift* of a bend vertex w in a drawing D yields a (planar) redrawing D' of D in which only vertex w has a new special position: Let p and p' denote the positions of w in D and D', respectively. Let ℓ be the line through the two neighbors of w in D. A redrawing D' of D is obtained by a parallel shift of w if p' lies on the line ℓ' parallel to ℓ through p, see Fig. 4. A bend vertex w is *shiftable* if there exists a parallel shift of w.

Observation 2. *A parallel shift of a bend vertex keeps all face areas invariant.*

The $\perp$-contact representation induces a coloring and an orientation of the inner edges: each edge corresponds to a contact point of two $\perp$-shapes. Orient the edge such that it is an outgoing edge for the vertex belonging to the $\perp$-shape whose end is the contact point. Color the edge red, blue, or green, if the contact point is the top end, left end, or right end of the $\perp$-shape, respectively. (Such a coloring and orientation is a *Schnyder wood* of G.) We analyze the typical situation for a vertex v in $D_\perp$, see Fig. 5. By construction, v has three outgoing edges such that all incoming edges partially run on one of these outgoing edges. Observe that the vertical segments of the incoming blue and green edges are free of segments touching it from the top: Due to the $\perp$-contact representation, every horizontal segment has exactly one vertical segment touching it from above. Moreover, the vertex is placed on this intersection point. Thus, the bend vertex of the lowest green and blue incoming edges is shiftable topwards. Due to the fact that every rectangle has positive area, some space is guaranteed. Therefore, we can parallel shift the bend vertex such that the edge is free of degeneracies. In particular, the bend vertex does not intersect non-incident edges anymore. Hence, the bend vertex of the second lowest incoming edge of v becomes shiftable. We iterate this process for all v such that all blue and green bend vertices do not intersect non-incident edges. Afterwards, only red bend vertices are involved in degeneracies.

For every vertex, we consider the incoming red edges, which have either a left or a right bend with respect to the orientation. Consider the rightmost right bend (and likewise the leftmost left bend) vertex. Its horizontal segment is free to the bottom since it is rightmost (leftmost) and the vertical segment is free to both sides, since by the first step there is no green or blue bend vertex on a red segment. Consequently, the rightmost right bend (leftmost left) vertex is shiftable to the bottom. We shift it parallel downwards such that no new degeneracies are introduced. Hence, the number of degeneracies decreased. Moreover, this process achieved that the second rightmost (leftmost) bend vertex becomes shiftable. By iterating, we remove all degeneracies. Finally, we have a 1-bend drawing realizing the areas prescribed by A. □

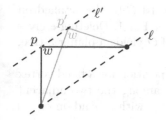
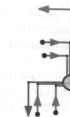
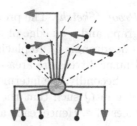

Fig. 4. A parallel shift of w.   **Fig. 5.** The neighborhood of a typical vertex in the degenerate drawing $D_\perp$ and in the final 1-bend drawing. (Color figure online)

Knowing that one bend per edge is always sufficient, we wonder how many bends may be necessary to realize a face area assignment of a plane graph? We construct a family of lower bound examples with the octahedron graph.

Lemma 3. *For every $k \geq 3$, there is a graph G on $n := 4k - 6$ vertices and a face area assignment A, such that in every A-realizing polyline drawing of G at least $1/12$ of the edges have a bend.*

Proof. Recall that, by Theorem 1, the octahedron graph is not area-universal. Hence, there are area assignments such that at least one of its twelve edges needs a bend. We call such an area assignment *bad*.

Now, we construct a graph for each $k \geq 3$. Take a triangulation on k vertices with an inner face 2-coloring. Without loss of generality, assume that $k-2$ inner faces are white. *Stack* an octahedron graph into each white face f, i.e., identify three outer vertices of a plane octahedron graph and with the vertices of f. This yields a graph G on $k+3(k-2) = n$ vertices, consisting of $k-2$ octahedron graphs which are pairwisely edge-disjoint. Let A be a face area assignment such that the inner faces of the stacked octahedron graphs obtain a bad area assignment and the remaining black faces receive some arbitrary value.

Consider an A-realizing polyline drawing D of G. By Theorem 1, each of the octahedron-subgraphs has least one edge with a bend in D. By the edge-disjointness, every edge with a bend can satisfy only one of the octahedron graphs. Hence, in each of the octahedron graphs at least one of the twelve edges has a bend. This implies the claim.   □

Corollary 1. *For a plane graph G with area assignment A, let $B_k(G, A)$ denote the minimum number of bends in a k-bend drawing realizing A. For all $k \geq 1$*

$$1/12|E| \leq \max_G \max_A \{B_k(G, A)\} \leq |E|.$$

Note that the octahedron graph may not yield a better lower bound.

Proposition 1. *For every face area assignment of the octahedron graph, there exists a realizing drawing with at most one bend in total.*

Proof (Sketch). The proof consists of two ideas. Firstly, let G be the octahedron graph and e an edge of the triangle of inner vertices, see Fig. 1. Deleting e gives a quadrangle Q. Replacing e by the other diagonal $\bar{e}$ of Q yields a planar 3-tree. Planar 3-trees are area-universal.

Secondly, the intermediate value theorem asserts a position for a bend vertex of e in Q such that the area of Q is arbitrarily split among the two adjacent faces of e. Hence, each area assignment of G is realizable with a bend on e. □

Observation 3. *The last argument shows that every 2-degenerate quadrangulation is area-universal since it can be constructed by iteratively inserting a degree 2 vertex in a quadrangle.*

3.1 Fewer Bends for Planar Bipartite Graphs

Now, we improve the number of sufficient bends for plane bipartite graphs. We show that at most half of the edges need a bend. Interestingly, no plane bipartite graph is known that needs a bend.

Theorem 4. *Let $G = (X \cup Y, E)$ be a plane bipartite graph and $A \colon F' \to \mathbb{R}^+$ a face area assignment. Then, there exists an A-realizing 1-bend redrawing of G with at most $|E|/2$ bends.*

Proof. First, we assume that G is a quadrangulation. The proof consists of four steps, illustrated in Fig. 6. The main difference to the proof of Theorem 3 lies in Step 1. Steps 3 and 4 are relatively more involved.

1. Take a segment contact representation $\mathcal{C}$ yielding a rectangular layout $\mathcal{L}$.
2. Obtain a weak equivalent rectangular layout $\mathcal{L}'$ realizing the areas.
3. Define a degenerate drawing D.
4. Construct a non-degenerate drawing from D by parallel shifts.

The first step is to take a *segment contact representation* of $G = (X \cup Y, E)$. This is a family $\{s_v | v \in X \cup Y\}$ of segments satisfying the following conditions: s_v is vertical if $v \in X$ and horizontal if $v \in Y$. Two segments s_v and s_u intersect if and only if $(u, v) \in E$. If two segments intersect, the point of intersection is an end of at least one of the two segments.

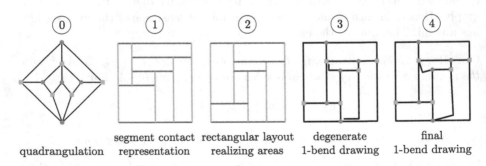

| quadrangulation | segment contact representation | rectangular layout realizing areas | degenerate 1-bend drawing | final 1-bend drawing |

Fig. 6. Construction of a 1-bend drawing realizing the prescribed areas in 4 steps.

Lemma 4 ([7,9]). *Every plane quadrangulation has a segment contact represen-tation such that each inner face is represented by a rectangle.*

Let $\mathcal{C}$ be a segment contact representation of G. The segments of $\mathcal{C}$ partition a rectangle into rectangles, and hence, $\mathcal{C}$ yields a rectangular layout $\mathcal{L}$. In the second step, we obtain a weak equivalent rectangular layout $\mathcal{L}'$ realizing the areas by Lemma 2; let $\mathcal{C}'$ denote the corresponding segment contact representation. In the third step, we define a degenerate drawing D from $\mathcal{C}'$. The challenge is to place the vertices such that, firstly, we save one bend per vertex, and secondly, the degeneracies can be removed by parallel shifts. We distinguish two cases depending on the minimal degree δ of G. Note that in a quadrangulation δ is 2 or 3.

<u>Case 1:</u> $\delta(G) = 3$. Since every segment has only two endpoints but at least three contacts, every segment has an inner contact point. We construct D as follows:

- $v \in X$ is placed on topmost inner contact point of the vertical segment s_v.
- $v \in Y$ is placed on leftmost inner contact point of the horizontal segment s_v.
- $e = (v, w) \in E$ is supported by the segments s_v and s_w in C: e runs from v along s_v to the contact point of s_v and s_w, and then along s_w to w.

Observation 4. *D is a degenerate (orthogonal) 1-bend redrawing of G realizing the areas prescribed by A. The number of bends is at most $|E| - |V|$.*

For the number of bends, observe that by placing the vertex on an inner contact point, the corresponding edge has no bend. Hence, we save one bend per vertex and the number of bends is at most $|E| - |V|$.

In the fourth step, we remove the degeneracies; again, by parallel shifts of bend vertices. Indeed, we iterate twice through all vertices. For a vertex $v \in X$ ($v \in Y$) with a vertical (horizontal) segment s_v, let $\mathcal{B}_1(s_v)$ denote the set of bend vertices on s_v with a horizontal (vertical) segment touching s_v from the right (bottom); we exclude the endpoints of s_v. Likewise $\mathcal{B}_2(s_v)$, denotes the set of bend vertices on s_v not in $\mathcal{B}_1(s_v)$.

Loop 1: For each $v \in X$ ($v \in Y$), do: while $\mathcal{B}_1(s_v)$ is not empty, choose the topmost (leftmost) bend vertex $b \in \mathcal{B}_1(s_v)$, parallel shift b. Delete b from $\mathcal{B}_1(s_v)$.
Loop 2: For each $v \in X$ ($v \in Y$), do: while $\mathcal{B}_2(s_v)$ is not empty, choose the topmost (leftmost) bend vertex $b \in \mathcal{B}_2(s_v)$, parallel shift b. Delete b from $\mathcal{B}_2(s_v)$.

In order to prove that Loop 1 is possible, consider a vertex $v \in X$ with vertical segment s_v, see Fig. 7. We need to argue that the topmost $b \in \mathcal{B}_1(s_v)$ is shiftable. By definition, each $w \in Y$ is placed on the leftmost contact point of s_w. Hence, the horizontal segment of b is the leftmost part of some s_w and therefore free of bend vertices. Since b is the leftmost bend vertex in $\mathcal{B}_1(s_v)$, the vertical segment of s_v is free to the right. Therefore, b is shiftable down-rightwards. Moreover, after shifting b, the second topmost bend vertex becomes shiftable. Consequently, by the order from top to bottom, the horizontal segment of each bend vertex is free to the right if considered. The argument for $v \in Y$ is analogous. After Loop 1, every segment is free to one side in the following way:

Every vertical segment is free to the right and every horizontal segment is free to the bottom.

In Loop 2, a considered bend vertex $b \in \mathcal{B}_2(s_v)$ is shiftable due to the order and the 1-side-freeness guaranteed by Loop 1. Now, no bend vertex sits on a non-incident edge. Consequently, this process yields a non-degenerate 1-bend drawing of G which realizes the prescribed areas and has at most $|E| - |V|$ bends.

<u>Case 2:</u> $\delta(G) = 2$. If it exists, choose an inner vertex of degree 2 and remove the segment s_v in $\mathcal{C}'$. This results in a quadrangulation where two old faces are unified to a new face. Assign the sum of the two old face areas to the new face. Delete inner vertices of degree 2 until all inner vertices are of degree at least 3. This yields a graph G' with area assignment A'. Proceed with G' as in Case 1 with some extra care. If an outer vertex is of degree 2, we injectively place the outer vertices of degree 2 on incident endpoints of their segments. Moreover, we make the parallel shifts small enough, such that the following *special* property is fulfilled in an A'-realizing drawing of G': up to a tiny ε with $2^n \varepsilon \ll A_{\min}$, each face f of G' contains an axis-aligned rectangle with area $A'(f) - \varepsilon$, where $A_{\min} := \min_{f \in F'(G)} A(f)$. We use the special property to reinsert the degree 2 vertices in reverse order of deletion and obtain a sequence of drawings G'_i. We use the invariant that G'_k is a non-degenerate drawing where each face area is realized by an axis-aligned rectangle up to $2^k \varepsilon$. Consider the $(k+1)$th vertex v of degree 2 and the face f in G_k where v must be inserted. Assume f has area $a_1 + a_2$ and must be split into two faces f_1 and f_2 with area a_1 and a_2, where $a_1, a_2 \geq A_{\min}$. By the invariant, f contains an axis-aligned rectangle R of area $a_1 + a_2 - 2^k \varepsilon$, see Fig. 8. Assume that $v \in X$. By the intermediate value theorem, there exists a vertical segment s within R such that s dissects f into two parts of area a_1 and a_2, respectively. Place v on one endpoint of s and a bend vertex b on the other endpoint of s. Note that the areas of f_1 and f_2 are realized by a rectangle up to $2^k \varepsilon$. In order to remove the degeneracies, use parallel shifts of v and b which are small enough to guarantee that f_i contains a rectangle of area $a_i - 2^{k+1} \varepsilon$. This ends Case 2.

If G is not a quadrangulation, then we consider a quadrangulation Q with G as an induced subgraph. For each face in G, dividing its area assignment among its subfaces in Q yields A'. Clearly, an A'-realizing 1-bend drawing of Q induces an A-realizing 1-bend drawing of G. However, the number of bends may

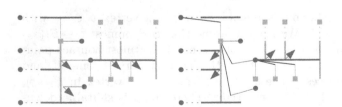

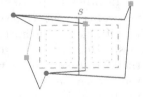

Fig. 7. Schematic neighborhood of vertices in D and after Loop 1.

Fig. 8. Inserting degree 2 vertices in Case 2.

exceed $|E|/2$. Therefore, we ensure to save one bend per vertex by placing the vertices on inner contact points which belong to edges of G. To do so, delete all segments belonging to artificial vertices in C. If necessary, remove vertices of low degree iteratively as in Case 2. Afterwards, place vertices, remove degeneracies and reinsert vertices of G with low degree as in Case 1 and Case 2. Note that degree 1 vertices may also appear, but are no problem to be reinserted.

A planar bipartite graph has at most $(2|V|-4)$ edges. Therefore, the number of edges with bends is at most $|E|-|V| \leq |V|-4$ and without bends at least $|V|$. Consequently, in all cases the number of bends is less than $|E|/2$.   □

If two adjacent contact points around a corner are chosen, one can save an additional bend. For the cube graph this saves 4 bends and shows that it is not only area-universal (partial 3-tree, cubic), but also *convex area-universal*, i.e. for every face area assignment there exists a realizing drawing with convex faces.

Proposition 2. *The cube graph is convex area-universal.*

Proof. Assign the doubled area to the four boundary faces. Theorem 4 and Fig. 9 show the existence of a 1-bend drawing realizing the perturbed face areas with one bend on each outer edge. Replacing the boundary edges by segments halves the area of the boundary faces and gives a realizing straight-line drawing with convex faces.   □

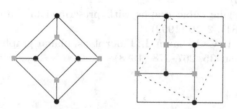

Fig. 9. The cube graph is convex area-universal.

4   Open Questions

Various interesting questions remain. We want to emphasize three of them:

- Are plane bipartite graphs area-universal?
- Are 3-connected plane bipartite graphs convex area-universal?
- How many bends are necessary and sufficient to realize arbitrary prescribed areas for all planar graphs?

Acknowledgments. We thank Stefan Felsner und Udo Hoffmann for discussions about the problem and helpful comments on drafts of this manuscript.

References

1. Alam, M.J., Biedl, T., Felsner, S., Kaufmann, M., Kobourov, S.G.: Proportional contact representations of planar graphs. In: Speckmann, B. (ed.) GD 2011. LNCS, vol. 7034, pp. 26–38. Springer, Heidelberg (2011)
2. Alam, M.J., Biedl, T., Felsner, S., Kaufmann, M., Kobourov, S.G., Ueckert, T.: Computing cartograms with optimal complexity. Discret. Comput. Geom. **50**(3), 784–810 (2013)
3. Biedl, T.C., Velázquez, L.E.R.: Drawing planar 3-trees with given face areas. Comput. Geom. **46**(3), 276–285 (2013)
4. Debrunner, H.: Aufgabe 260. Elemente der Mathematik 12 (1957)
5. Eppstein, D., Mumford, E., Speckmann, B., Verbeek, K.: Area-universal and constrained rectangular layouts. SIAM J. Comput. **41**(3), 537–564 (2012)
6. Felsner, S.: Exploiting air-pressure to map floorplans on point sets. J. Graph Algorithms Appl. **18**(2), 233–252 (2014)
7. de Fraysseix, H., Ossona de Mendez, P.: Representation of planar graphs. J. Intuitive Geom. **63**, 109–117 (1991)
8. de Fraysseix, H., Ossona de Mendez, P., Rosenstiehl, P.: On triangle contact graphs. Comb. Probab. Comput. **3**, 233–246 (1994)
9. Hartman, I.A., Newman, I., Ziv, R.: On grid intersection graphs. Discret. Math. **87**(1), 41–52 (1991)
10. Ringel, G.: Equiareal graphs. In: Contemporary Methods in Graph Theory, in honour of Prof. Dr. K. Wagner, pp. 503–505 (1990)
11. Tamassia, R.: Handbook of Graph Drawing and Visualization. CRC Press, Boca Raton (2013)
12. Thomassen, C.: Plane cubic graphs with prescribed face areas. Comb. Probab. Comput. **1**(371–381), 2–10 (1992)
13. Wimer, S., Koren, I., Cederbaum, I.: Floorplans, planar graphs, and layouts. IEEE Trans. Circuits Syst. **35**, 267–278 (1988)

Vertex Cover Structural Parameterization Revisited

Fedor V. Fomin and Torstein J.F. Strømme[⊠]

Department of Informatics, University of Bergen, Bergen, Norway
{fedor.fomin,torstein.stromme}@ii.uib.no

Abstract. A pseudoforest is a graph whose connected components have at most one cycle. Let X be a pseudoforest modulator of graph G, i.e. a vertex subset of G such that $G - X$ is a pseudoforest. We show that VERTEX COVER admits a polynomial kernel being parameterized by the size of the pseudoforest modulator. In other words, we provide a polynomial time algorithm that for an input graph G and integer k, outputs a graph G' and integer k', such that G' has $\mathcal{O}(|X|^{12})$ vertices and G has a vertex cover of size k if and only if G' has vertex cover of size k'. We complement our findings by proving that there is no polynomial kernel for VERTEX COVER parameterized by the size of a modulator to a *mock forest* (a graph where no cycles share a vertex) unless NP $\subseteq$ coNP/poly. In particular, this also rules out polynomial kernels when parameterized by the size of a modulator to cactus graphs.

1 Introduction

Kernelization is a fundamental algorithmic methodology rooted in parameterized complexity. It also serves as a rigorous mathematical tool for analyzing certain polynomial-time preprocessing or data-reductions algorithms. In this paper we provide new kernelization algorithm for "structural" parameterization of VERTEX COVER.

In the VERTEX COVER problem, we are given as input a graph G and a positive integer k, and are asked if there exists a set S of at most k vertices in G such that every edge in G is adjacent to at least one of the vertices in S; such an S is called a *vertex cover* of G. As a part of a general program on kernelization with structural parameterization, Jansen and Bodlaender [9] initiated the study of kernelization for VERTEX COVER with "refined" parameterization by showing that it admits a polynomial kernel when parameterized by the size of a feedback vertex set, i.e. a forest-modulator. Since a feedback vertex set can be significantly smaller than a vertex cover, in various situations such a kernel can be preferable.

It is a very natural question if the kernelization result of Jansen and Bodlaender can be extended to parameters which are "stronger" than the size of a feedback vertex set. Forests are exactly the graphs of treewidth one and a natural direction of such an extension would be to explore the parameterization by a constant

Supported by Rigorous Theory of Preprocessing, ERC Advanced Investigator Grant 267959.

© Springer-Verlag GmbH Germany 2016
P. Heggernes (Ed.): WG 2016, LNCS 9941, pp. 171–182, 2016.
DOI: 10.1007/978-3-662-53536-3_15

treewidth modulator. However, as it was shown by Cygan et al. [2], for each $t \geq 2$, VERTEX COVER does not admit a polynomial kernel being parameterized by the size of the treewidth t modulator unless NP $\subseteq$ coNP /poly. This result was further strenghtened by Jansen to also hold for a modulators to outerplanar graphs [8]. Since the result of Jansen rules out polynomial kernels for VERTEX COVER even when parameterized by outerplanar modulators, the next natural step in the study of polynomial kernelization for VERTEX COVER is to see if the problem admits a polynomial kernel when parameterized by a modulator to some subclasses of outerplanar graphs. Towards this end, Majumdar et al. [10] obtain a polynomial kernel for VERTEX COVER parameterized by the size of a degree-2 modulator.

In this work we show that VERTEX COVER admits a polynomial kernel when the parameter is the size of a pseudoforest modulator. More precisely, a *pseudoforest* is an undirected graph in which every connected component has at most one cycle. In a graph G, a vertex set X is a *pseudoforest modulator* if the graph $G - X$ obtained from G by deleting X is a pseudoforest. We define the following problem.

VERTEX COVER/PSEUDOFOREST MODULATOR (VC/PFM)
Input: A simple undirected graph G, a pseudoforest-modulator set $X \subseteq V(G)$ such that $G - X$ is a pseudoforest, integer k.
Parameter: Size of a pseudoforest modulator $|X|$.
Question: Does G contain a vertex cover of size at most k?

Our Results. We show that VC/PFM admits a polynomial kernel with $\mathcal{O}(|X|^{12})$ vertices. Since every feedback vertex set is a pseudoforest-modulator and every degree-2-modulator is also a pseudoforest-modulator, our result extends the borders of polynomial kernelization for VERTEX COVER established by Jansen and Bodlaender [9] and by Majumdar et al. [10].

We complement our kernelization algorithm with a lower bound. Let us observe that the works of Cygan et al. and Jansen [2,8] does not rule out the existence of polynomial kernels when the problem is parameterized by the size of a modulator to some proper subclass of outerplanar graphs, such as cactus graphs, i.e. graphs where every 2-connected component is a cycle. We refine the known lower bounds by proving that a polynomial kernel for VERTEX COVER parameterized by the size of mock forest modulator would imply NP $\subseteq$ coNP /poly. (*Mock forest* is a graph with no two cycles sharing a vertex and thus is outerplanar.) Since a mock forest is also a cactus graph, this rules out polynomial kernels parameterized by the size of a modulator to this class as well.

While we state our kernelization result assuming that a pseudoforest modulator is given as a part of the input, this condition can be omitted. There are several approximation algorithms for pseudoforest modulator. For example, computing a modulator to a pseudoforest is a special case of the $\mathcal{F}$-DELETION problem considered in [4], and there is a randomized constant factor approximation algorithm of running time $\mathcal{O}(nm)$. Also since pseudoforests of a graph form independent sets of a bicircular matroid, it follows from the generic framework of Fujito [7] that there is a deterministic polynomial time 2-approximation algorithm for pseudoforest modulator.

The proof of our main result is constructive and consists of several reduction rules. While many steps follow Jansen and Bodlaender [9], the essential part of the proof is different. Our algorithm is based on a novel combinatorial result about maximum independent sets in pseudotrees (Lemma 4), which is also interesting in its own.

The remaining part of the paper is organized as follows. In Sect. 2 we develop the kernelization algorithm for VERTEX COVER/PSEUDOFOREST MODULATOR, which is the main content of this paper. The section containing the proof of Lemma 4 is quite technical and is found in the full version [5]. We obtain lower bounds for VERTEX COVER parameterized by the vertex deletion distances to mock forests, in Sect. 3. Graph theoretic notions and standard definitions are found in [5] together with proofs of lemmata marked with ($\star$).

2   Kernelization

This section contains the kernelization of VERTEX COVER/PSEUDOFOREST MODULATOR and is the main section of the paper. We will first develop a kernel for INDEPENDENT SET/PSEUDOFOREST MODULATOR, and then by the immediate correspondence between the VERTEX COVER and INDEPENDENT SET problems the kernel for VC/PFM will follow. For the remainder of this section, we will thus focus on the INDEPENDENT SET/PSEUDOFOREST MODULATOR problem:

INDEPENDENT SET/PSEUDOFOREST MODULATOR (IS/PFM)
Input: A simple undirected graph G, a pseudoforest-modulator set $X \subseteq V(G)$ such that $G - X$ is a pseudoforest, integer k.
Parameter: Size of a pseudoforest modulator $|X|$.
Question: Does G contain an independent set of size at least k?

Throughout the section, let $F := G - X$ be the induced subgraph remaining after the modulator X has been removed from G. Note that F is a pseudoforest.

We say that (G, X, k) is a *yes-instance* of IS/PFM if there exists an independent set I of G such that $|I| \geq k$. We say it is a *no-instance* if there is no such set.

Definition 1 *(Conflicts). Let (G, X, k) be an instance of IS/PFM where $F' \subseteq F$ is a subgraph of the pseudoforest F and $X' \subseteq X$ is a subset of the modulator X. Then the number of* conflicts *induced by X' on F' is defined as $\mathrm{CONF}_{F'}(X') := \alpha(F') - \alpha(F' - N_G(X'))$.*

Choosing X' to be in the independent set I of G may prevent some vertices in F' from being included in same set I. In particular, no vertex $v \in V(F') \cap N_G(X')$ can be chosen to be in I. In light of this, the term $\mathrm{CONF}_{F'}(X')$ can be understood

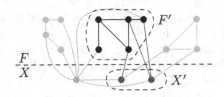

Fig. 1. Conflicts: In the figure, we observe that $\alpha(F') = 3$, and $\alpha(F' - N_G(X')) = 1$. Hence, we get that $\text{CONF}_{F'}(X') = 2$. In other words, the number of conflicts induced by X' on F' is 2.

as the price one has to pay in F' by choosing to include X' in the independent set (Fig. 1).

Observe that $\text{CONF}_{F'}(X')$ is polynomial time computable since the numbers $\alpha(F')$ and $\alpha(F' - Z)$ are polynomial time computable for every $Z \subseteq V(F')$.

Definition 2 *(Chunks).* *Let (G, X, k) be an instance of* IS/PFM. *A set $X' \subseteq X$ is a* chunk *if the following hold:*

- *X' is independent in G,*
- *The size of X' is between 1 and 3, i.e. $1 \le |X'| \le 3$, and*
- *The number of conflicts induced by X' on the pseudoforest F is less than $|X|$, i.e. $\text{CONF}_F(X') < |X|$.*

We let $\mathcal{X}$ be the collection of all chunks of X.

The collection of chunks $\mathcal{X}$ can be seen as all suitable candidate subsets of size at most 3 from X to be included in a maximum independent set I for G. The idea is that I may contain a chunk as a subset, but need not include a subset $X' \subseteq X$ of size at most 3 which is *not* a chunk. This will allow us to discard potential solutions containing non-chunk subsets of X with size at most 3. In order for this intuition to hold, we provide the following lemma, originally by Jansen and Bodlaender [9, Lemma 2] though slightly altered to fit our purposes.

Lemma 1 (⋆). *If there exists an independent set of size k in G, then there exists an independent set I of G such that $|I| \ge k$ and for all subsets $X' \subseteq X \cap I, \text{CONF}_F(X') < |X|$.*

Definition 3 *(Anchor triangle).* *Let (G, X, k) be an instance of* IS/PFM. *Let P be a connected component in F with $V(P) = \{p_1, p_2, p_3\}$ (Fig. 2). Then P is an* anchor triangle *if there exists a set $\{x_1, x_2, x_3\} \subseteq X$ such that:*

- *$N_G(p_1) = \{p_2, p_3, x_1\}$*
- *$N_G(p_2) = \{p_1, p_3, x_2\}$*
- *$N_G(p_3) = \{p_1, p_2, x_3\}$*

An anchor triangle is non-redundant *if there is no other anchor triangle with the same open neighborhood in G.*

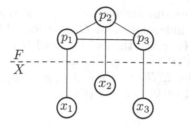

Fig. 2. The connected component $P \subseteq F$ with vertices $V(P) = \{p_1, p_2, p_3\}$ is an anchor triangle for the triple $\{x_1, x_2, x_3\} \subseteq X$.

Definition 4 *(Unnecessary triple).* *A triple $_3X \subseteq X$ is said to be* unnecessary *if there exists an anchor triangle P such that $N_G(P) = {_3}X$.*

The fact that a triple $_3X \subseteq X$ is unnecessary as defined above should intuitively be understood with respect to constructing an independent set. If a triple $_3X$ is unnecessary then there exists a MIS which does not contain all of $_3X$. This intuition is supported by the next lemma.

Lemma 2 ($\star$). *Let (G, X, k) be an instance of IS/PFM. If there exists an independent set of size at least k in G, then there exists an independent set I of G with $|I| \geq k$ containing no unnecessary triple $_3X \subseteq X$.*

2.1 Reduction Rules

We introduce here the reduction rules. Each reduction receives as input an instance (G, X, k) of INDEPENDENT SET/PSEUDOFOREST MODULATOR, and outputs an equivalent instance (G', X', k'). A reduction is *safe* if the input and output instances are equivalent, that is, (G, X, k) is a *yes*-instance if and only if (G', X', k') is a *yes*-instance. Reductions 1, 2 and 4 originates in [9], though Reduction 4 is altered to fit the context of a pseudoforest, which also required some changes to the proof.

Reduction rules will be applied exhaustively starting with lower number rules, until Reduction 4 is no longer applicable. During this process, a lower number rule is always applied before a higher number rule if at any point they are both applicable. Then Reductions 5 and 6 will be applied once each to obtain the final reduced instance. Note that each reduction is computable in polynomial time.

Reduction 1. If there is a vertex $v \in X$ such that $\text{CONF}_F(\{v\}) \geq |X|$, then delete v from the graph G and from the set X. We let $G' := G - v, X' := X - v$ and $k' := k$.

Reduction 2. If there are distinct vertices $u, v \in X$ with $uv \notin E(G)$ for which $\text{CONF}_F(\{u, v\}) \geq |X|$, then add edge uv to G. We let $G' := (V(G), E(G) \cup \{uv\}), X' := X$ and $k' := k$.

Reductions 1 and 2 are safe due to Lemma 1.

Reduction 3. If there are distinct $u, v, w \in X$ such that $\text{CONF}_F(\{u, v, w\}) \geq |X|$, the set $\{u, v, w\}$ is independent in G, and for which there is no anchor triangle P with $N(P) = \{u, v, w\}$, then add an anchor triangle $P' = \{p_u, p_v, p_w\}$ to the graph such that $N(P') = \{u, v, w\}$, and increase k by one. Let $V(G') := V(G) \cup \{p_u, p_v, p_w\}$ and let $E(G') := E(G) \cup \{p_u p_v, p_u p_w, p_v p_w, p_u u, p_v v, p_w w\}$). Further, let $X' := X$ and let $k' := k + 1$ (Fig. 3).

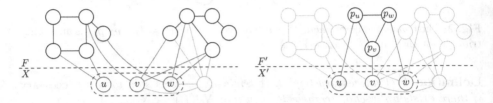

Fig. 3. Reduction 3: Adding an anchor triangle to the independent triple $\{u, v, w\}$ ($k' = k + 1$). This makes $\{u, v, w\}$ an unnecessary triple in the output instance.

Note that Reduction 3 makes the triple $\{u, v, w\} \subseteq X$ unnecessary in the reduced instance as defined in Definition 4.

Lemma 3 ($\star$). *Reduction 3 is safe. Let (G, X, k) be an instance of* IS/PFM *to which Reduction 3 is applicable, and let (G', X', k') be the reduced instance. Then (G, X, k) is a yes-instance if and only if (G', X', k') is a yes-instance.*

Reduction 4. If there exists a connected component P in F which is not a non-redundant anchor triangle, and for every chunk $Y \in \mathcal{X}$ there is no conflict induced by Y on P, i.e. $\text{CONF}_P(Y) = 0$, then remove P from G and reduce k by $\alpha(P)$. We let $G' := G - P, X' := X$ and $k' := k - \alpha(P)$.

To prove that Reduction 4 is safe, we will rely on the following lemma, which states that any pseudotree has a small (at most size three) obstruction in terms of obtaining a maximum independent set.

Lemma 4 ($\star$). *Let P be a pseudotree and let Z be a set of vertices such that $\alpha(P) > \alpha(P - Z)$. Then there exist three (possibly non-distinct) vertices $u, v, w \in Z \cap V(P)$ such that $\alpha(P) > \alpha(P - \{u, v, w\})$.*

The proof of the above lemma is quite technical, and is omitted here in order to preserve the flow of the kernelization algorithm. Taking Lemma 4 as a black box, we are able to make the following observation:

Observation 1. Let $P \subseteq F$ be a connected component in the pseudoforest F and let $X' \subseteq X$ be an independent set such that $\text{CONF}_P(X') > 0$. Then there exists some $X'' \subseteq X'$ with $1 \leq |X''| \leq 3$ such that $\text{CONF}_P(X'') > 0$.

We see that the observation is true, since by Lemma 4 there exist $u, v, w \in N_G(X') \cap V(P)$ such that $\alpha(P) > \alpha(P - \{u, v, w\})$. Then for each element u, v, w, pick an arbitrary neighbor $x_u, x_v, x_w \in X'$ (they need not be distinct) to form the set $X'' := \{x_u, x_v, x_w\}$. See that then $\text{CONF}_P(X'') > 0$. We are now equipped to prove safeness of Reduction 4.

Lemma 5. *Reduction 4 is safe. Let (G, X, k) be an instance of* IS/PFM *to which Reduction 4 is applicable, and let (G', X', k') be the reduced instance. Then (G, X, k) is a yes-instance if and only if (G', X', k') is a yes-instance.*

Proof. Let $P \subseteq F$ be the connected component which triggered the reduction.

For the forward direction of the proof, assume (G, X, k) is a *yes*-instance and let I be an independent set of G with size at least k. Let $I' := I \setminus V(P)$. Clearly I' is an independent set of G'. Now observe that $|I \cap V(P)| \leq \alpha(P)$, and thus $|I'| = |I| - |I \cap V(P)| \geq k - \alpha(P) = k'$. Hence (G', X', k') is a *yes*-instance.

For the backward direction, we assume that (G', X', k') is a *yes*-instance, and has an independent set I' of size at least k'. Because of Lemma 2 we can assume that I' contains no unnecessary triples $_3 X \subseteq X' \cap I'$. We want to show that we can always pick some independent set $I_P \subseteq V(P)$ with $|I_P| = \alpha(P)$ such that $I := I' \cup I_P$ is an independent set with size at least $k' + \alpha(P) = k$. Since I' and $V(P)$ are disjoint in G by construction, it will suffice to show that $\alpha(P - N_G(I')) \geq \alpha(P)$.

Assume for the sake of contradiction that $\alpha(P - N_G(I')) < \alpha(P)$. Since P was a connected component in F, all its neighbors $N_G(P)$ are in X. Thus we have that $\mathrm{CONF}_P(X' \cap I') > 0$. By Observation 1, we further have that there exists some $X'' \subseteq X' \cap I'$ such that $1 \leq |X''| \leq 3$ and $\mathrm{CONF}_P(X'') > 0$.

For any such X'', there are two cases. In the first case, $\mathrm{CONF}_F(X'') < |X|$. Because X'' is also independent and has size at most 3, it is a chunk of X in the input instance. This contradicts the preconditions for Reduction 4, so this case can not happen.

In the second case, $\mathrm{CONF}_F(X'') \geq |X|$. But then one of Reductions 1, 2 or 3 would have previously been applied to X'', yielding it either unfeasible for an independent set or making it an unnecessary triple in the input instance. Because non-redundant anchor triangles are not chosen for removal by Reduction 4, X'' is also an unnecessary triple in the output instance, which contradicts that I' contains no unnecessary triples. This concludes the proof.   □

Notice that Reduction 4 will remove connected components from F. When the reduction is not applicable, we should then be able to give some bound on the number of connected components in F. The next lemma gives such a bound:

Lemma 6 ($\star$). *Let (G, X, k) be an instance of* IS/PFM *which is irreducible with respect to Reductions 1, 2, 3 and 4. Let $\mathcal{C}_F$ denote the set of all connected components $P \subseteq F$. Then $|\mathcal{C}_F| \leq |X|^4 + |X|^3$, i.e. @ the number of connected components in F is at most $|X|^4 + |X|^3$.*

When the above reduction rules have been exhaustively applied, the next two reductions will be executed exactly once each.

Reduction 5. Let $\hat{X} \subseteq V(F)$ be a set such that $\hat{X}$ contains exactly one vertex of each cycle in F. In the reduced graph, let $G' := G, X' := X \cup \hat{X}$, and $k' := k$.

The reduction is safe because neither G nor k was changed. Observe that X is now a feedback vertex set (which is fine, since every feedback vertex set is also a modulator to pseudoforest). This reduction may increase the size of X dramatically. This is why the Reduction is applied only once, such that we can give guarantees for the size of the reduced instance.

Observation 2. Let (G', X', k') be an instance of IS/PFM after Reduction 5 have been applied to (G, X, k). Then $X' \leq |X|^4 + |X|^3 + |X|$.

After Reduction 5 has been applied once, the returned instance (G, X, k) is ready for the final reduction step. Note that since X is now a feedback vertex set, (G, X, k) is now an instance of INDEPENDENT SET/FEEDBACK VERTEX SET as well, and we can for the final reduction apply the kernel of Jansen and Bodlaender.

INDEPENDENT SET/FEEDBACK VERTEX SET (IS/FVS)
Input: A simple undirected graph G, a feedback vertex set $X \subseteq V(G)$, integer k.
Parameter: Size of the feedback vertex set $|X|$.
Question: Does G contain an independence set of size at least k?

Proposition 1 [9, Theorem 2]. INDEPENDENT SET/FEEDBACK VERTEX SET *has a kernel with a cubic number of vertices: There is a polynomial-time algorithm that transforms an instance (G, X, k) into an equivalent instance (G', X', k') such that $|X'| \leq 2|X|$, and $|V(G')| \leq 56|X|^3 + 28|X|^2 + 2|X|$.*

Reduction 6. Let the output instance (G', X', k') be the reduced instance after applying Proposition 1. This reduction is applied once only.

2.2 Bound on Size of Reduced Instances

When no reduction rules can be applied to an instance, we call it *reduced*. In this section we will prove that the number of vertices in a reduced instance (G', X', k') is at most $\mathcal{O}(|X|^{12})$ where $|X|$ is the size of the modulator in the original problem (G, X, k).

Theorem 1. INDEPENDENT SET/PSEUDOFOREST MODULATOR *admits a kernel with $\mathcal{O}(|X|^{12})$ vertices.*

Proof. In order to prove the theorem, we show that there is a polynomial time algorithm that transforms an instance (G, X, k) to an equivalent instance (G', X', k') such that

- $|V(G')| \leq 56(|X|^4 + |X|^3 + |X|)^3 + 28(|X|^4 + |X|^3 + |X|)^2 + 2(|X|^4 + |X|^3 + |X|)$,
- $|X'| \leq 2|X|^4 + 2|X|^3 + 2|X|$, and
- $k' \leq k + |X|^3$.

We will begin with the proof that $k' \leq k + |X|^3$. The only transformation which increases k is Reduction 3, which rise k by 1 each time it is applied. However, this transformation will be done less than $|X|^3$ times, since the rule will be applied at most once for each distinct triple of X.

Next, we focus on the bound $|X'| \leq 2|X|^4 + 2|X|^3 + 2|X|$. The only transformations which increase $|X|$ are Reductions 5 and 6, which are applied only once each. By Observation 2 we then have that $|X'| \leq |X|^4 + |X|^3 + |X|$ after applying Reduction 5, and by Proposition 1 we have that the size is at most doubled after applying Reduction 6. Thus the bound holds.

For the bound on $V(G)$, let us consider the instance of IS/FVS (G'', X'', k'') to which Reduction 6 was applied in order to obtain the final reduced instance (G', X', k'). We have already established that $|X''| \leq |X|^4 + |X|^3 + |X|$. It follows from Proposition 1 that in the reduced instance, $|V(G')| \leq 2|X''| + 28|X''|^2 + 56|X''|^3$, which in terms of $|X|$ yields $|V(G')| \leq 56(|X|^4 + |X|^3 + |X|)^3 + 28(|X|^4 + |X|^3 + |X|)^2 + 2(|X|^4 + |X|^3 + |X|)$.

Finally, observe that each reduction can be done in polynomial time.   □

Corollary 1. VERTEX COVER/PSEUDOFOREST MODULATOR *admits a kernel with* $\mathcal{O}(|X|^{12})$ *vertices.*

3   No Polynomial Kernel for VC/MFM

In this section we show that VERTEX COVER/MOCK FOREST MODULATOR admits no polynomial kernel unless NP $\subseteq$ coNP/poly. Our strategy is to make a reduction from CNF-SAT parameterized by the number of variables to IS/MFM. By the immediate correspondance between VERTEX COVER and INDEPENDENT SET, the result for VERTEX COVER/MOCK FOREST MODULATOR will follow. We define the following problem.

INDEPENDENT SET/MOCK FOREST MODULATOR (IS/MFM)
Input: A simple undirected graph G, a mock forest modulator $X \subseteq V(G)$ such that no two cycles of $G - X$ share a vertex, and an integer k.
Parameter: Size of a mock forest modulator $|X|$.
Question: Does G contain an independent set of size at least k?

Our reduction also shows that there is no polynomial kernel for VERTEX COVER when parameterized by the size of a modulator to cactus graphs as well, under the same condition. Our strategy is a modification of Jansen's proof [8] that VERTEX COVER does not have a polynomial kernel when parameterized by a modulator to outerplanar graphs unless NP $\subseteq$ coNP/poly.

Definition 5 (Polynomial-parameter transformation [1]). *Let* $\mathcal{Q}, \mathcal{Q}' \subseteq \Sigma^* \times \mathbb{N}$ *be parameterized problems. A polynomial-parameter transformation from* $\mathcal{Q}$ *to* $\mathcal{Q}'$ *is an algorithm that, on input* $(x, k) \in \Sigma^* \times \mathbb{N}$, *takes time polynomial in* $|x| + k$, *and outputs an instance* $(x', k') \in \Sigma^* \times \mathbb{N}$ *such that* k' *is polynomially bounded in* k, *and* $(x, k) \in \mathcal{Q}$ *if and only if* $(x', k') \in \mathcal{Q}'$. *For a parameterized problem* $\mathcal{Q} \subseteq \Sigma^* \times \mathbb{N}$, *the unparameterized version of* $\mathcal{Q}$ *is the set* $\hat{\mathcal{Q}} = \{x1^k \mid (x, k) \in \mathcal{Q}\}$ *where 1 is a new symbol that is added to the alphabet.*

Proposition 2 [1]. *Let Q and Q' be parameterized problems and let $\hat{Q}$ and $\hat{Q}'$ be the unparameterized versions of Q and Q' respectively. Suppose $\hat{Q}$ is NP-hard and $\hat{Q}'$ is in NP. If there is a polynomial-parameter transformation form Q to Q', and Q' has a polynomial kernel, then Q also has a polynomial kernel.*

Proposition 3 [3,6]. CNF-SAT *parameterized by the number of variables does not admit a polynomial kernel unless $NP \subseteq coNP/poly$.*

Definition 6 (Clause gadget). *Let $k \geq 1$ be an integer. The* clause gadget *of size k is the graph $\mathcal{G}_k$ consisting of k triangles $T_1, T_2, \ldots, T_k$ and two extra vertices r_0 and l_{k+1} connected as follows: For each triangle T_i, label the three vertices $l_i, r_i,$ and s_i (left vertex, right vertex and spike vertex, respectively). Then for each $i \in \{0\} \cup [k]$, let there be an edge $r_i l_{i+1}$ connecting the right vertex of T_i to the left vertex of T_{i+1}. In this way, $\mathcal{G}_k$ is a "path" of k connected triangles, with two extra degree-1 vertices attached at the ends (Fig. 4).*

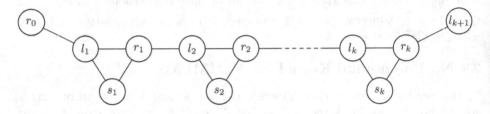

Fig. 4. A clause gadget $\mathcal{G}_k$

Observation 3. For a clause gadget $\mathcal{G}_k$, the independence number $\alpha(\mathcal{G}_k)$ is exactly $k + 2$. This can be obtained by the independent set containing all the spike vertices as well as r_0 and l_{k+1}. We verify that this is also optimal since at most one vertex can be chosen from each triangle T_i, and there are only two non-triangle vertices.

Observation 4. For a clause gadget $\mathcal{G}_k$, every maximum independent set I must contain at least one spike vertex. Removing the spike vertices, what remains of $\mathcal{G}_k$ is an even path with $2k + 2$ vertices, yielding a maximum independent set of size $k + 1$, which is strictly smaller than $\alpha(\mathcal{G}_k)$.

Observation 5. For a clause gadget $\mathcal{G}_k$, let S denote the set of spike vertices. Observe that for each spike vertex $s_i \in S$, there exists a maximum independent set I_i such that s_i is the only spike in I_i, i.e. $I_i \cap S = \{s_i\}$.

Theorem 2. INDEPENDENT SET/MOCK FOREST MODULATOR *does not admit a polynomial kernel unless $NP \subseteq coNP/poly$.*

Proof. Since VERTEX COVER is in NP and CNF-SAT is NP-hard, we have by Propositions 2 and 3, that it is sufficient to show a polynomial-parameter transformation from CNF-SAT parameterized by the number of variables to IS/MFM.

Consider an instance F of CNF-SAT consisting of clauses $C_1, C_2, \ldots, C_m$ over the variables $x_1, x_2, \ldots, x_n$. For a clause C_j, let $h(j)$ denote the number of literals in C_j. We will in polynomial time construct an instance (G, X, k) of IS/MFM such that F is satisfied if and only if (G, X, k) is a *yes*-instance. We construct graph G as follows.

For each variable x_i, we let there be two vertices t_i and f_i in $V(G)$. Let them be connected by an edge $t_i f_i \in E(G)$. Which of t_i and f_i is included in a maximum independent set for G will represent whether the variable x_i is set to true or false.

For each clause C_j, let $\ell_1, \ell_2, \ldots, \ell_{h(j)}$ denote the literals of C_j. Let C_j be a copy of the clause gadget $\mathcal{G}_{h(j)}$, and add it to the graph G. Let the spikes of C_j be denoted $s_1, s_2, \ldots, s_{h(j)}$. We will connect C_j to the rest of G as follows: For each literal $\ell_r \in C_j$, let there be an edge from s_r to f_i if and only if $\ell_r = x_i$. Similarly, let there be an edge from s_r to t_i if and only if $\ell_r = \neg x_i$. By this process, every spike of C_j is connected to exactly one vertex outside of C_j, which is either t_i or f_i for some $i \in [n]$. This concludes the construction of G.

Let the set X consist of the variable gadget vertices, i.e., let $X = \{t_i \mid i \in [n]\} \cup \{f_i \mid i \in [n]\}$. Observe that X is indeed a mock forest modulator for G, since every connected component of $G - X$ is exactly a clause gadget, and thus also a mock forest. Also note that $|X| = 2n$, which is polynomial in the input parameter. Finally, we let $k = n + \sum_{j=1}^{m}(h(j) + 2)$. It remains to show that F is satisfiable if and only if (G, X, k) is a *yes*-instance.

($\Rightarrow$) Assume the formula is satisfiable by the assignment φ. We will now build an independent set I in G which has size at least k. Initially, let $I_X = \emptyset$. For each variable x_i, let t_i be in I_X if $\varphi(x_i)$ is TRUE, and let f_i be in I_X otherwise. In this way, n vertices are added to I_X. Observe that this process preserves independence of I_X.

For each clause C_j, we know that there exists some satisfied literal ℓ_r. In the corresponding clause gadget C_j, observe that $s_r \notin N_G[I_X]$ by the construction of the graph and the choice of I_X. Then by Observation 5, we can choose an independent set I_j for C_j which is disjoint from $N_G[I_X]$.

Finally, let I be the union of I_X and $\bigcup_{j=1}^{m} I_j$. Observe that independence is maintained, since there are no edges between I_X and I_j for all $j \in [m]$, and there are no edges between I_j and $I_{j'}$ for all choices of $j, j' \in [m], j \neq j'$, since there were no edges between C_j and $C_{j'}$. Further, we note that $|I_X| = n$, and $|I_j| = h(j) + 2$ for every $j \in [m]$, and that all the sets are vertex disjoint. Thus we obtain that $|I| = n + \sum_{j=1}^{m}(h(j) + 2)$.

($\Leftarrow$) Assume that there exists an independent set I for G with size $|I| \geq n + \sum_{j=1}^{m}(h(j) + 2)$. We construct an assignment φ which satisfies the SAT formula F. By Observation 3, we know that $|I \setminus X| \leq \sum_{j=1}^{m}(h(j)+2)$. Since $G[X]$ consists exactly of n pairwise joined vertices, we also know that $|I \cap X| \leq n$. Thus, $|I| \leq n + \sum_{j=1}^{m}(h(j)+2)$, and equality holds for all the relations. For each variable x_i it must thus be the case that either $t_i \in I$ and $f_i \notin I$, or vice versa. We let $\varphi(x_i)$ evaluate to TRUE if $t_i \in I$, and to FALSE otherwise.

It remains to show that φ is in fact a satisfying assignment. Consider some clause C_j and its corresponding gadget $\mathcal{C}_j$. Because $|I \cap \mathcal{C}_j| = h(j) + 2$, we have by Observation 4 that there exists a spike vertex $s_r \in \mathcal{C}_j \cap I$. Assume for the sake of contradiction that $\ell_r \in C_j$ is not satisfied by φ. This implies that x_i was assigned a value that would not satisfy ℓ_r. Without loss of generality, (by symmetry) assume $\ell_r = x_i$ and $\varphi(x_i) =$ FALSE. Then $f_i \in I$; however, by the construction of the graph, there is an edge between s_r and f_i. This contradicts that I is independent. Thus $\ell_r \in C_j$ is satisfied by φ and we have concluded the proof.   □

Corollary 2. VERTEX COVER/MOCK FOREST MODULATOR *does not admit a polynomial kernel unless NP $\subseteq$ coNP/poly.*

Finally, let us observe that in the proof of Theorem 2, the graph $G - X$ is a cactus graph. Thus the proof of Theorem 2 can be used to show that INDEPENDENT SET parameterized by the size of a modulator to a cactus graph does not admit a polynomial kernel.

References

1. Bodlaender, H.L., Thomassé, S., Yeo, A.: Kernel bounds for disjoint cycles and disjoint paths. Theoret. Comput. Sci. **412**, 4570–4578 (2011)
2. Cygan, M., Lokshtanov, D., Pilipczuk, M., Pilipczuk, M., Saurabh, S.: On the hardness of losing width. Theor. Comput. Syst. **54**, 73–82 (2014)
3. Dell, H., van Melkebeek, D.: Satisfiability allows no nontrivial sparsification unless the polynomial-time hierarchy collapses. J. ACM **61**, 23 (2014)
4. Fomin, F.V., Lokshtanov, D., Misra, N., Saurabh, S.: Planar F-deletion: approximation, kernelization and optimal FPT algorithms. In: Proceedings of the 53rd Annual Symposium on Foundations of Computer Science (FOCS), pp. 470–479. IEEE (2012)
5. Fomin, F.V., Strømme, T.J.: Vertex cover structural parameterization revisited (2016). arXiv preprint arXiv:1603.00770
6. Fortnow, L., Santhanam, R.: Infeasibility of instance compression and succinct PCPs for NP. J. Comput. Syst. Sci. **77**, 91–106 (2011)
7. Fujito, T.: A unified approximation algorithm for node-deletion problems. Discrete Appl. Math. **86**, 213–231 (1998)
8. Jansen, B.M.P.: The power of data reduction: kernels for fundamental graph problems. Ph.D. thesis, Utrecht University (2013)
9. Jansen, B.M.P., Bodlaender, H.L.: Vertex cover kernelization revisited - upper and lower bounds for a refined parameter. Theor. Comput. Syst. **53**, 263–299 (2013)
10. Majumdar, D., Raman, V., Saurabh, S.: Kernels for structural parameterizations of vertex cover - case of small degree modulators. In: 10th International Symposium on Parameterized and Exact Computation, LIPIcs (IPEC 2015), vol. 43, pp. 331–342. Schloss Dagstuhl-Leibniz-Zentrum fuer Informatik (2015)

On Distance-d Independent Set and Other Problems in Graphs with "few" Minimal Separators

Pedro Montealegre[✉] and Ioan Todinca

Univ. Orléans, INSA Centre Val de Loire, LIFO EA 4022, Orléans, France
{pedro.montealegre,ioan.todinca}@univ-orleans.fr

Abstract. Fomin and Villanger ([14], STACS 2010) proved that MAX-IMUM INDEPENDENT SET, FEEDBACK VERTEX SET, and more generally the problem of finding a maximum induced subgraph of treewith at most a constant t, can be solved in polynomial time on graph classes with polynomially many minimal separators. We extend these results in two directions. Let $\mathcal{G}_{\mathrm{poly}}$ be the class of graphs with at most $\mathrm{poly}(n)$ minimal separators, for some polynomial poly.

We show that the odd powers of a graph G have at most as many minimal separators as G. Consequently, DISTANCE-d INDEPENDENT SET, which consists in finding maximum set of vertices at pairwise distance at least d, is polynomial on $\mathcal{G}_{\mathrm{poly}}$, for any even d. The problem is NP-hard on chordal graphs for any odd $d \geq 3$ [12].

We also provide polynomial algorithms for CONNECTED VERTEX COVER and CONNECTED FEEDBACK VERTEX SET on subclasses of $\mathcal{G}_{\mathrm{poly}}$ including chordal and circular-arc graphs, and we discuss variants of independent domination problems.

1 Introduction

Several natural graph classes are known to have polynomially many minimal separators, w.r.t. the number n of vertices of the graph. It is the case for *chordal* graphs, which have at most n minimal separators [19], *weakly chordal, circular-arc* and *circle* graphs, which have $\mathcal{O}(n^2)$ minimal separators [4,16].

The property of having polynomially many minimal separators has been used in algorithms for decades, initially in an ad-hoc manner, i.e., algorithms were based on minimal separators but also other specific features of particular graph classes (see, e.g., [3,16]). Later, it was observed that minimal separators are sufficient for solving problems like TREEWIDTH or MINIMUM FILL-IN [4,5]. Both problems are related to *minimal triangulations*. Given an arbitrary graph G, a minimal triangulation is a minimal chordal supergraph H of G, on the same vertex set. Bouchitté and Todinca [4] introduced the notion of *potential maximal clique*, that is, a vertex set of G inducing a maximal clique in some minimal triangulation H of G. Their algorithm for treewidth is based on dynamic programming over minimal separators and potential maximal cliques. The same

© Springer-Verlag GmbH Germany 2016
P. Heggernes (Ed.): WG 2016, LNCS 9941, pp. 183–194, 2016.
DOI: 10.1007/978-3-662-53536-3_16

authors proved that the number of potential maximal cliques is polynomially bounded in the number of minimal separators [5].

Fomin and Villanger [14] found a more surprising application of minimal separators and potential maximal cliques, proving that they were sufficient for solving problems like MAXIMUM INDEPENDENT SET, MAXIMUM INDUCED FOREST, and more generally for finding a maximum induced subgraph $G[F]$ of treewidth at most t, where t is a constant.

More formally, let poly be some polynomial. We call $\mathcal{G}_{\mathrm{poly}}$ the family of graphs such that $G \in \mathcal{G}_{\mathrm{poly}}$ if and only if G has at most $\mathrm{poly}(n)$ minimal separators. By [14], the problem of finding a maximum induced subgraph of treewidth at most t can be solved in polynomial time on $\mathcal{G}_{\mathrm{poly}}$. The exponent of the polynomial depends on poly and on t. In [13], Fomin *et al.* further extend the technique to compute large induced subgraphs of bounded treewidth, and satisfying some CMSO property (expressible in counting monadic second-order logic). That allows to capture problems like LONGEST INDUCED PATH. They also point out some limits of the approach. It is asked in [13] whether the techniques can be extended for solving the CONNECTED VERTEX COVER problem, which is equivalent to finding a maximum independent set F such that $G - F$ is connected. More generally, their algorithm computes an induced subgraph $G[F]$ of treewidth at most t satisfying some CMSO property, but is not able to ensure any property relating the induced subgraph to the initial graph.

Here we make some progress in this direction. First, we consider the problem DISTANCE-d INDEPENDENT SET on $\mathcal{G}_{\mathrm{poly}}$, where the goal is to find a maximum independent set F of the input graph G, such that the vertices of F are at pairwise distance at least d in G (in the literature this problem is also known as d-SCATTERED-SET). This is equivalent to finding a maximum independent set in graph G^{d-1}, the $(d-1)$-th power of G. Eto *et al.* [12] already studied the problem on chordal graphs, and proved that it is polynomial for every even d, and NP-hard for any odd $d \geq 3$ (it is even $W[1]$-hard when parameterized by the solution size). Their positive result is based on the observation that for any even d, if G is chordal then so is G^{d-1}. Eto *et al.* [12] ask if DISTANCE-d INDEPENDENT SET is polynomial on chordal bipartite graphs (which are *not* chordal but weakly chordal, see Sect. 2), a subclass of $\mathcal{G}_{\mathrm{poly}}$. We bring a positive answer to their question for even values d, by a result of combinatorial nature: for any graph G and any odd k, the graph G^k has no more minimal separators than G (see Sect. 3). Consequently, DISTANCE-d INDEPENDENT SET is polynomial on $\mathcal{G}_{\mathrm{poly}}$, for any even value d and any polynomial poly, and NP-hard for any odd $d \geq 3$ and any $\mathrm{poly}(n)$ asymptotically larger than n. Such a dichotomy between odd and even values also appears when computing large d-clubs, that are induced subgraphs of diameter at most d [15], and for quite similar reasons.

Second, we consider CONNECTED VERTEX COVER, CONNECTED FEEDBACK VERTEX SET and more generally the problem of finding a maximum induced subgraph $G[F]$ of treewidth at most t, such that $G - F$ is connected. We show (Sect. 4) that the problems are polynomially solvable for subclasses of $\mathcal{G}_{\mathrm{poly}}$, like chordal and circular-arc graphs. This does not settle the complexity of these problems on $\mathcal{G}_{poly}$. As we shall discuss in Sect. 5, when restricted to bipartite graphs in $\mathcal{G}_{poly}$, CONNECTED VERTEX COVER can be reduced from RED-BLUE DOMINATING SET (see [10]). It might be that this latter problem is NP-hard on bipartite graphs of $\mathcal{G}_{poly}$; that was our hope, since the very related problem INDEPENDENT DOMINATING SET is NP-hard on chordal bipartite graphs [8], and on circle graphs [6]. This question is still open, however we will observe that the RED-BLUE DOMINATING SET is polynomial on the two natural classes of bipartite graphs with polynomially many minimal separators: chordal bipartite and circle bipartite graphs.

2 Preliminaries

Let $G = (V, E)$ be a graph. Let $dist_G(u, v)$ denote the distance between vertices u and v (the minimum number of edges of a uv-path). We denote by $N_G^k[v]$ the set of vertices at distance at most k from v. Let also $N_G^k(v) = N_G^k[v] \setminus \{v\}$, and we call these sets the *closed* and *open neighborhoods at distance* k of v, respectively. Similarly, for a set of vertices $U \subseteq V$, we call the sets $N_G^k(U) = \cup_{u \in U} N_G^k(u) \setminus U$ and $N_G^k[U] = \cup_{u \in U} N_G^k[u]$ the open and closed neighborhoods at distance k of U, respectively. For $k = 1$, we simply denote by $N_G(U)$, respectively $N_G[U]$, the open and closed neighborhoods of U; the subscript is omitted if clear from the context.

A *clique* (resp. *independent set*) of G is a set of pairwise adjacent (resp. nonadjacent) vertices. A distance-d independent set is a set of vertices at pairwise distance at least d. Equivalently, it is an independent set of the $(d-1)$-th power G^{d-1} of G. Graph $G^k = (V, E^k)$ is obtained from G by adding an edge between every pair of vertices at distance at most k.

Given a vertex subset C of G, we denote by $G[C]$ the subgraph induced by C. We say that C is a connected component of G if $G[C]$ is connected and C is inclusion-maximal for this property. For $S \subseteq V$, we simply denote $G - S$ the graph $G[V \setminus S]$. We say that S is a a, b-*minimal separator* of G if a and b are in distinct components C and D of $G - S$, and $N(C) = N(D) = S$. We also say that S is a minimal separator if it is an a, b-minimal separator for some pair of vertices a and b.

Proposition 1 ([2]). *Let $G = (V, E)$ be a graph, C be a connected set of vertices, and let D be a component of $G - N[C]$. Then $N(D)$ is an a, b-minimal separator of G, for any $a \in C$ and $b \in D$.*

2.1 Graph Classes

A graph is *chordal* if it has no induced cycle with more than three vertices. A graph G is *weakly chordal* if G and its complement $\overline{G}$ have no induced cycle with more than four vertices.

The classes of *circle* and *circular-arc graphs* are defined by their intersection model. A graph G is a *circle graph* (resp. a *circular-arc graph*) if every vertex of the graph can be associated to a chord (resp. to an arc) of a circle such that two vertices are adjacent in G if and only if the corresponding chords (resp. arcs) intersect. We may assume w.l.o.g. that, in the intersection model, no two chords (resp. no two arcs) share an endpoint. On the circle, we add a *scanpoint* between each two consecutive endpoints of the set of chords (resp. arcs). A *scanline* is a line segment between two scanpoints. Given an intersection model of a circle (resp. circular-arc) graph G, for any minimal separator S of G there is a scanline such that the vertices of S correspond exactly to the chords (resp. arcs) intersecting the scanline, see, e.g., [16].

Chordal graphs have at most n minimal separators [19]; weakly chordal, circle and circular-arc graphs all have $\mathcal{O}(n^2)$ minimal separators [4,16].

Definition 1. *Let* poly *be some polynomial. We call* $\mathcal{G}_{\mathrm{poly}}$ *the family of graphs such that* $G \in \mathcal{G}_{\mathrm{poly}}$ *if and only if G has at most* $\mathrm{poly}(n)$ *minimal separators, where* $n = |V(G)|$.

2.2   Dynamic Programming over Minimal Triangulations

Let $G = (V, E)$ be an arbitrary graph. A chordal supergraph $H = (V, E')$ (i.e., with $E \subseteq E'$), is called a *triangulation* of G. If, moreover, E' is inclusion-minimal among all possible triangulations, we say that H is a *minimal triangulation* of G.

The *treewidth* of a chordal graph is its maximum clique size, minus one. Forests have treewidth 1, and graphs with no edges have treewidth 0. The treewidth $\mathrm{tw}(G)$ of an arbitrary graph G is the minimum treewidth over all (minimal) triangulations H of G.

Cliques of minimal triangulations play a central role in treewidth. A *potential maximal clique* of G is a set of vertices that induces a maximal clique in some minimal triangulation H of G. By [4], if Ω is a potential maximal clique, then for every component C_i of $G - \Omega$, its neighborhood S_i is a minimal separator. Moreover, the sets S_i are exactly the minimal separators of G contained in Ω.

Proposition 2 ([1,5])**.** *For any polynomial* poly, *there is a polynomial-time algorithm enumerating the minimal separators and the potential maximal cliques of graphs on* $\mathcal{G}_{\mathrm{poly}}$.

Minimal separators and potential maximal cliques have been used for computing treewidth and other parameters related to minimal triangulations, on $\mathcal{G}_{\mathrm{poly}}$. Fomin and Villanger [14] extend the techniques to a family of problems:

Proposition 3 ([14])**.** *For any polynomial* poly *and any constant t, there is a polynomial algorithm computing a* MAXIMUM INDUCED SUBGRAPH OF TREEWIDTH AT MOST t *on* $\mathcal{G}_{\mathrm{poly}}$.

Clearly, MAXIMUM INDEPENDENT SET (which is equivalent to MINIMUM VERTEX COVER) and MAXIMUM INDUCED FOREST (which is equivalent to MINIMUM FEEDBACK VERTEX SET) fit into this framework: they consist in finding maximum induced subgraphs $G[F]$ of treewidth at most 0, respectively at most 1. The first ingredient of [14] is the following observation.

Proposition 4 ([14]). *Let $G = (V, E)$ be a graph, $F \subseteq V$, and let H_F be a minimal triangulation of $G[F]$. There exists a minimal triangulation H_G of G such that $H_G[F] = H_F$. We say that H_G respects the minimal triangulation H_F of $G[F]$.*

Note that, for any clique Ω of H_G, we have that $F \cap \Omega$ induces a clique in H_F. In particular, if $\mathrm{tw}(G[F]) \leq t$ and the clique size of H_F is at most $t + 1$, then every maximal clique of H_G intersects F in at most $t + 1$ vertices.

The second ingredient is a dynamic programming scheme that we describe below. Let S be a minimal separator of G, and C be a component of $G - S$ such that $N(C) = S$. The pair (S, C) is called a *block*. Let Ω be a potential maximal clique such that $S \subset \Omega \subseteq S \cup C$. Then (S, C, Ω) is called a *good triple*. In the sequel, W denotes a set of at most $t + 1$ vertices.

Definition 2. *Let (S, C) (resp. (S, Ω, C)) be a block (resp. a good triple) and let $W \subseteq S$ (resp. $W \subseteq \Omega$) be a set of vertices of size at most $t + 1$. We say that a vertex set F is a partial solution compatible with (S, C, W) (resp. with (S, C, Ω, W)) if:*

1. *$G[F]$ is of treewidth at most t,*
2. *$F \subseteq S \cup C$,*
3. *$W = F \cap S$ (resp. $W = F \cap \Omega$),*
4. *there is a minimal triangulation H of G respecting some minimal triangulation of $G[F]$ of treewidth at most t, such that S is a minimal separator (resp. S is a minimal separator and Ω is a maximal clique) of H.*

Observe that the two variants of compatibility differ by parameter Ω and the last two conditions. We denote by $\alpha(S, C, W)$ (resp. $\beta(S, C, \Omega, W)$) the size of a largest partial solution compatible with (S, C, W) (resp. (S, C, Ω, W)). We now show how these quantities can be computed over all blocks and all good triples. The dynamic programming will proceed by increasing size over the blocks (S, C), the size of the block being $|S \cup C|$.

It is based on the following equations (see [13, 14] for details and proofs and Fig. 1 for an illustration).

Base case. It occurs for good triples (S, C, Ω) such that $\Omega = S \cup C$. In this case, for each subset W of Ω of size at most $t + 1$,

$$\beta(S, C, \Omega, W) = |W|. \tag{1}$$

Computing α from β. The following equation allows to compute the α values from β values:

$$\alpha(S, C, W) = \max_{\Omega, W'} \beta(S, C, \Omega, W'), \tag{2}$$

where the maximum is taken over all potential maximal cliques Ω such that (S, C, Ω) is a good triple, and all subsets W' of Ω, of size at most $t + 1$, such that $W = W' \cap S$.

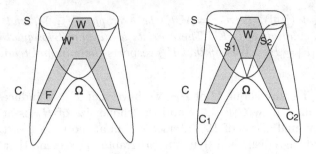

Fig. 1. Computing α form β (left), and β from α (right)

Computing β from α. Let (S, C, Ω) be a good triple, and fix an order $C_1, C_2, \ldots, C_p$ on the connected components of $G[C \setminus \Omega]$. Let $S_i = N_G(C_i)$, for all $1 \leq i \leq p$. By [4], (S_i, C_i) are also blocks of G.

A partial solution F compatible with (S, C, Ω, W) is obtained as a union of partial solutions F_i compatible with $(S_i, C_i, W \cap S_i)$, for each $1 \leq i \leq p$, and the set W.

Denote by $\gamma_i(S, C, \Omega, W)$ the size of the largest partial solution F compatible[1] with (S, C, Ω, W), contained in $\Omega \cup C_1 \cup \cdots \cup C_i$ (hence F is not allowed to intersect the components C_{i+1} to C_p).

We have the following equations.

$$\gamma_1(S, C, \Omega, W) = \alpha(S, C, \Omega, W \cap S_1) + |W| - |W \cap S_1|. \tag{3}$$

For all i, $2 \leq i \leq p$,

$$\gamma_i(S, C, \Omega, W) = \gamma_{i-1}(S, C, \Omega, W) + \alpha(S, C, \Omega, W \cap S_i) - |W \cap S_i|. \tag{4}$$

and finally

$$\beta(S, C, \Omega, W) = \gamma_p(S, C, \Omega, W). \tag{5}$$

For convenience we also consider that $\emptyset$ is a minimal separator, and $(\emptyset, V)$ is a block. Then the size of the optimal global solution is simply $\alpha(\emptyset, V, \emptyset)$. The algorithm can be adapted to output an optimal solution, not only its size.

3 Powers of Graphs with Polynomially Many Minimal Separators

Let us prove that for any odd k, G^k has no more minimal separators than G.

Theorem 1. *Consider a graph G, an odd number $k = 2l + 1$ with $l \geq 0$, and a minimal separator $\overline{S}$ of G^k. Then there exists a minimal separator S of G such that $\overline{S} = N_G^l[S]$.*

---

[1] To be precise, the γ function is not required at this stage, if we only compute largest induced subgraphs of treewidth at most t. However it becomes necessary when we request the solution to satisfy additional properties, as it will happen in Sect. 4.

Proof. The lemma is trivially true if $\overline{S} = \emptyset$. Let $a, b \in V$ such that $\overline{S} \neq \emptyset$ is an a, b-minimal separator in G^k, and call C_a, C_b the components of $G^k - \overline{S}$ that contain a and b, respectively. Let us call $D_a = N_G^l[C_a]$ and $D_b = N_G^l[C_b]$.

Claim 1: $dist_G(D_a, D_b) \geq 2$. Suppose that $dist_G(D_a, D_b) < 2$, and pick $x \in D_a, y \in D_b$ with $dist_G(x, y) \leq 1$ (notice that possibly $x = y$). Let $x_a \in C_a$ and $x_b \in C_b$ be such that there exists an x_a, x-path and a y, x_b-path in G, each one of length at most l, called P_a and P_b, respectively. This implies that there must be a x_a, x_b-path of length at most $2l + 1 = k$ in G, which means that $\{x_a, x_b\} \in E(G^k)$, a contradiction with the fact that $\overline{S}$ separates C_a from C_b in G^k.

Claim 2: $\tilde{S} = \overline{S} \setminus (D_a \cup D_b)$ separates a and b in G. Notice first that $N_G(D_a) \subseteq N_G^{l+1}(C_a) \subseteq \overline{S}$. Suppose that $\tilde{S}$ does not separate a and b, and let P be an a, b-path in G that does not pass through $\tilde{S}$. Let $x_1, \ldots, x_{s-2}$ the internal nodes of P, where $s = |P|$, and consider $i = \max\{j \mid x_j \in D_a \cap P\}$. Since $P \cap \tilde{S} = \emptyset$, necessarily $x_{i+1} \in D_b$, a contradiction with Claim 1.

Claim 3: D_a and D_b are connected subsets of G. This is straightforward from the definition of the sets, $D_a = N_G^l[C_a]$ and $D_b = N_G^l[C_b]$, and the fact that C_a and C_b are connected in G.

Let $\tilde{C}_b$ be the connected component of $G - N_G[D_a]$ that contains b, and denote $S = N_G(\tilde{C}_b)$. Note that $S \subseteq \tilde{S} \subset \overline{S}$. By applying Proposition 1, we have that S is a minimal a, b-separator in G. Call $\tilde{C}_a$ the component of $G - S$ that contains a. Since $S \subseteq \tilde{S}$, we have that $D_b \subseteq \tilde{C}_b$ and $D_a \subseteq \tilde{C}_a$.

Claim 4: $N_G^l[S] = \overline{S}$. We first prove that $N_G^l[S] \subseteq \overline{S}$. By construction, $S \subseteq N_G(D_a)$. Consequently $S \subseteq N_G^{l+1}(C_a) \setminus N_G^l(C_a)$, therefore $N_G^l[S] \subseteq N_G^{2l+1}(C_a) = N_{G^k}(C_a) = \overline{S}$. Conversely, we must show that every vertex x of $\overline{S}$ is in $N_G^l[S]$. By contradiction, let $x \in \overline{S} \setminus N_G^l[S]$. We distinguish two cases : $x \in \tilde{C}_a$, and $x \in \overline{S} \setminus \tilde{C}_a$. In the first case, since $N_{G^k}(C_b) = \overline{S}$, there exists a path from some vertex $y \in C_b$ to x of length at most k, in graph G. Let us call P one of those y, x-paths. Observe that the first $l + 1$ vertices of the path belong to $D_b \subseteq \tilde{C}_b$, and none of the last $l + 1$ vertices of the path belongs to S (otherwise $x \in N_G^l[S]$). Then P is a path that connects $\tilde{C}_b$ with $\tilde{C}_a$ without passing through S, a contradiction with the fact that S separates a and b in graph G.

It remains to prove the last case, when $x \in \overline{S} \setminus \tilde{C}_a$. Since $N_{G^k}(C_a) = \overline{S}$, there exists a node $y \in C_a$ such that there is a y, x-path P of length at most k in G. Since the first $l + 1$ vertices of the path belong to D_a, and the last $l + 1$ vertices of the path do not belong to S, we deduce that P is an y, x-path in G that does not intersect S. The path can be extended (through C_a) into an a, x-path that does not intersect S, a contradiction with the fact that x does not belong to $\tilde{C}_a$. This concludes the proof of our theorem. □

Recall that DISTANCE-d INDEPENDENT SET on G is equivalent to MAXIMUM INDEPENDENT SET on G^{d-1}. Since the latter problem is polynomial on $\mathcal{G}_{\text{poly}}$ by Proposition 3, we deduce:

Theorem 2. *For any even value d, and any polynomial* poly, *problem* DISTANCE-d INDEPENDENT SET *is polynomially solvable on* $\mathcal{G}_{\text{poly}}$.

We remind that for any odd value d, problem DISTANCE-d INDEPENDENT SET is NP-hard on chordal graphs [12], thus on $\mathcal{G}_{poly}$ for any polynomial poly asymptotically larger than n. The construction of [12] also shows that even powers of chordal graphs may contain exponentially many minimal separators.

4   On Connected Vertex Cover and Connected Feedback Vertex Set

Let us consider the problem of finding a maximum induced subgraph $G[F]$ such that $\text{tw}(G[F]) \leq t$ and $G - F$ is connected. One can easily observe that, for $t = 0$ (resp. $t = 1$), this problem is equivalent to CONNECTED VERTEX COVER (resp. CONNECTED FEEDBACK VERTEX SET), in the sense that if F is an optimal solution for the former, than $V(G) - F$ is an optimal solution for the latter.

Our goal is to enrich the dynamic programming scheme described in Subsect. 2.2 in order to ensure the connectivity of $G - F$. One should think of this dynamic programming scheme of Subsect. 2.2 as similar to dynamic programming algorithms for bounded treewidth. The difference is that the bags (here, the potential maximal cliques) are not small but polynomially many, and we parse simultaneously through a set of decompositions. Nevertheless, we can borrow several classical ideas from treewidth-based algorithms.

In general, for checking some property for the solution F, we add a notion of *characteristics* of partial solutions. Then, for a characteristic c, we update the Definition 2 in order to define partial solutions compatible with (S, C, W, c) (resp. (S, C, Ω, W, c)), by requesting the partial solution to be compatible with characteristic c. Parameter c will also appear in the updated version of Eqs. 1 to 5.

As usual in dynamic programming, the characteristics must satisfy several properties: (1) we must be able to compute the characteristic for the base case, (2) the characteristic of a partial solution F obtained from gluing smaller partial solutions F_i must only depend on the characteristics of F_i, and (3) the characteristic of a global solution should indicate whether it is acceptable or not. Moreover, for a polynomial algorithm, we need the set of possible characteristics to be polynomially bounded.

For checking connectivity conditions on $G - F$, we define the characteristics of partial solutions in a natural way. Consider a block (S, C) (resp. a good triple (S, C, Ω)) and a subset W of S (resp. of Ω). Let F be a partial solution compatible with (S, C, W) (resp. (S, C, Ω, W)), see Definition 2. The *characteristic* c of F for (S, C, W) (resp. for (S, C, Ω, W)) is defined as the partition induced on $S \setminus W$ (resp. on $\Omega \setminus W$) by the connected components of $G[S \cup C] - F$. More formally, let $D_1, \ldots, D_q$ denote the connected components of $G[S \cup C] - F$, and let $P_j = D_j \cap S$ (resp. $P_j = D_j \cap \Omega$), for all $1 \leq j \leq q$. Then $c = \{P_1, \ldots, P_q\}$. We decide that if $S \neq \emptyset$, partial solutions F having some component D_j that does not intersect S (resp. Ω) are immediately rejected; indeed, for any extension F' of F, the graph $G - F'$ remains disconnected. Hence we may assume that all sets P_j are non-empty.

We say that a partial solution F is *compatible with* (S, C, W, c) (resp. *with* (S, C, Ω, W, c)) if it satisfies the conditions of Definition 2, and c is the characteristic of F for (S, C, W) (resp. for (S, C, Ω, W)).

We also define functions $\alpha(S, C, W, c)$, $\beta(S, C, \Omega, W, c)$ and $\gamma_i(S, C, \Omega, W, c)$ like in Subsect. 2.2, as the maximum size of partial solutions F compatible with the parameters. For further details refer to [17].

In general, the number of characteristics may be exponential. Nevertheless, there are classes of graphs with the property that each minimal separator S and each potential maximal clique Ω can be partitioned into at most a constant number of cliques. With this constraint, the number of characteristics is polynomial (even constant, for any given triple (S, C, W) or quadruple (S, C, Ω, W)).

This is the case for chordal graphs, where each minimal separator and each potential maximal clique induces a clique in G.

It is also the case for circular-arc graphs. Recall that each minimal separator corresponds to the set of arcs intersecting a pair of scanpoints [16]. Moreover, by [4, 16], each potential maximal clique corresponds to the set of arcs intersecting a triple of scanpoints. Since arcs intersecting a given scanpoint form a clique, we have that each minimal separator can be partitioned into two cliques, and each potential maximal clique can be partitioned into three cliques.

We deduce:

Theorem 3. *On chordal and circular-arc graphs, problems* CONNECTED VERTEX COVER *and* CONNECTED FEEDBACK VERTEX SET *are solvable in polynomial time. More generally, one can compute in polynomial time a maximum vertex subset F such that $G[F]$ is of treewidth at most t and $G - F$ is connected.*

Note that Escoffier *et al.* [11] already observed that CONNECTED VERTEX COVER is polynomial for chordal graphs.

5 Independent Dominating Set and Variants

The INDEPENDENT DOMINATING SET problem consists in finding a *minimum* independent set F of G such that F dominates G. Hence the solution F induces a graph of treewidth 0 and it is natural to ask if similar techniques work in this case. The fact that we have a minimization problem is not a difficulty: the general dynamic programming scheme applies in this case, and for any weighted problem with polynomially bounded weights, including negative ones [13, 14].

INDEPENDENT DOMINATING SET is known to be NP-complete in chordal bipartite graphs [8] and in circle graphs [6]. Therefore, it is NP-hard on $\mathcal{G}_{\text{poly}}$ for some polynomials poly. But, again, we can use our scheme in the case of circular-arc graphs, for this problem or any problem of the type minimum dominating induced subgraph of treewidth at most a constant t.

Let (S, C) be a block an let $F \subseteq S \cup C$ be a partial solution compatible with (S, C, W) for some $W \subseteq S$ of size at most $t + 1$ (in the sense of Definition 2). The natural way for defining the characteristic of F is to specify which vertices of S are dominated by F and which are not (we already know that $F \cap S = W$).

It is thus enough to memorize which vertices of S are dominated by $F \cap C$. In circular-arc graphs, this information can be encoded using a polynomial number of characteristics. Indeed, a minimal separator S corresponds to arcs intersecting a scanline, between two scanpoints p_1 and p_2 of some intersection model of G. Moreover (see [16]), the vertices of component C correspond to the arcs situated on one of the sides of the scanline. Let $s_1^1, s_2^1, \ldots, s_{l_1}^1$ be the arcs of the model containing scanpoint p_1, ordered by increasing intersection with the side of $p_1 p_2$ corresponding to C. Simply observe that if $F \cap C$ dominates vertex s_i^1, it also dominates all vertices s_j^1 with $j > i$. Therefore we only have to store the vertex $s_{min_1}^1$ dominated by $F \cap C$ which has a minimum intersection with the side of the scanline corresponding to component C, and proceed similarly for the arcs of S containing scanpoint p_2. These two vertices of S will define the characteristic of F, and they suffice to identify all vertices of S dominated by $F \cap C$.

These characteristics can be used to compute a minimum dominating induced subgraph of treewidth at most t, for circular-arc graphs, in polynomial time. We will not show, in details, how to do it, since the technique is quite classical. Problem INDEPENDENT DOMINATING SET is already known to be polynomial for this class [7,20]. The algorithm of Vatshelle [20] is more general, based on parameters called *boolean-width* and *MIM-width*, which are small ($\mathcal{O}(\log n)$ for the former, constant for the latter) on circular-arc graphs and also other graph classes. Another problem of similar flavor, combining domination and independence, is RED-BLUE DOMINATING SET. In this problem we are given a bipartite graph $G = (R, B, E)$ with red and blue vertices, and an integer k, and the goal is to find a set of at most k blue vertices dominating all the red ones. RED-BLUE DOMINATING SET can be reduced to CONNECTED VERTEX COVER as follows [10]. Let G' be the graph obtained from $G = (R, B, E)$ by adding a new vertex u adjacent to all vertices of B and then, for each $v \in R \cup \{u\}$, a pendant vertex v' adjacent only to v. Then G has a red-blue dominating set of size at most k if and only if G' has a connected vertex cover of size at most $k + |B| + 1$. Indeed any minimum connected vertex cover of G' must contain u, R, and a subset of B dominating R. It is not hard to prove that this reduction increases the number of minimal separators by at most $\mathcal{O}(n)$.

Therefore, if RED-BLUE DOMINATING SET is NP-hard on (bipartite) $\mathcal{G}_{poly}$ for some *poly*, so is CONNECTED VERTEX COVER. There are two natural, well-studied classes of bipartite graphs with polynomial number of minimal separators, and it turns out that RED-BLUE DOMINATING SET is polynomial for both. One is the class of chordal bipartite graphs (which are actually defined as the bipartite, *weakly* chordal graphs). For this class, RED-BLUE DOMINATING SET is polynomial by [8]. Reference [8] considers the total domination problem for the class, but the approach is based on red-blue domination.

The second natural class is the class of circle bipartite graphs, i.e., bipartite graphs that are also circle graphs. They have an elegant characterization established by de Fraysseix [9]. Let $H = (V, E)$ be a planar multigraph, and partition its edge set into two parts E_R and E_B such that $T = (V, E_R)$ is a spanning tree of H. Let $B(H, E_R) = (E_R, E_B, E')$ be the bipartite graph defined as follows: E_R is the set of red vertices, E_B is the set of blue vertices, and $e_R \in E_R$ is adjacent to $e_B \in E_B$ if the unique cycle obtained from the spanning tree T by adding e_B contains the edge e_R. We say that $B(H, E_R)$ is a fundamental graph of H. By [9], a graph is circle bipartite if and only if it is the fundamental graph $B(H, E_R)$ of a planar multigraph H.

Consider now the TREE AUGMENTATION problem that consists in finding, on input G and a spanning tree T of G, a minimum set of edges $D \subseteq E(G) - E(T)$ such that each edge in $E(T)$ is contained in at least one cycle of $G' = (V, E(T) \cup D)$. In [18] is shown that TREE AUGMENTATION is polynomial when the input graph is planar. Is direct to see that a set $S \subseteq E_B$ is a solution of the TREE AUGMENTATION problem on input $H = (V, E_R \cup E_B)$ and $T = (V, E_R)$, if and only if S is a solution of RED-BLUE DOMINATING SET on input $B(H) = (E_R, E_B, E')$. This observation, together with [9] and [18], imply that RED-BLUE DOMINATING SET is polynomial in circle bipartite graphs.

6 Discussion

We showed how the dynamic programming scheme of [13, 14] can be extended for other optimization problems, on *subclasses* of $\mathcal{G}_{\text{poly}}$. Note that the algorithm of [13] allows to find in polynomial time, on $\mathcal{G}_{\text{poly}}$, a maximum (weight) subgraph $G[F]$ of treewidth at most t, satisfying some property expressible in CMSO. It also handles annotated versions, where the vertices/edges of $G[F]$ must be selected from a prescribed set.

We have seen that DISTANCE-d INDEPENDENT SET can be solved in polynomial time on $\mathcal{G}_{\text{poly}}$ for any even d. This also holds for the more general problem of finding an induced subgraph $G[F]$ whose components are at pairwise distance at least d, and such that each component is isomorphic to a graph in a fixed family. E.g., each component could be an edge, to have a variant of MAXIMUM INDUCED MATCHING where edges should be at pairwise distance at least d. For this we need to solve the corresponding problem on G^{d-1}, using only edges from G, as in [13].

When seeking for maximum (resp. minimum) induced subgraphs $G[F]$ of treewidth at most t such that $G - F$ is connected (resp. F dominates G) on particular subclasses of $\mathcal{G}_{\text{poly}}$, we can add any CMSO condition on $G[F]$. It is not unlikely that the techniques can be extended to other classes than circular-arc graphs (and chordal graphs, for connectivity constraints).

We also believe that the interplay between graphs of bounded MIM-width [20] and $\mathcal{G}_{\text{poly}}$ deserves to be studied. None of the classes contains the other, but several natural graph classes are in their intersection, and they are both somehow related to induced matchings.

We leave as open problems the complexity of CONNECTED VERTEX COVER and CONNECTED FEEDBACK VERTEX SET in weakly chordal graphs, and on $\mathcal{G}_{\text{poly}}$. We have examples showing that, even for weakly chordal graphs, the natural set of characteristics that we used in Sect. 4 is not polynomially bounded.

Acknowledgements. We thank Iyad Kanj for fruitful discussions on the subject.

References

1. Berry, A., Bordat, J.P., Cogis, O.: Generating all the minimal separators of a graph. Int. J. Found. Comput. Sci. **11**(3), 397–403 (2000)
2. Berry, A., Bordat, J.P., Heggernes, P.: Recognizing weakly triangulated graphs by edge separability. Nord. J. Comput. **7**(3), 164–177 (2000)
3. Bodlaender, H., Kloks, T., Kratsch, D.: Treewidth and pathwidth of permutation graphs. In: Lingas, A., Karlsson, R., Carlsson, S. (eds.) Automata Languages and Programming. LNCS, vol. 700, pp. 114–125. Springer, Heidelberg (1993)
4. Bouchitté, V., Todinca, I.: Treewidth and minimum fill-in: grouping the minimal separators. SIAM J. Comput. **31**(1), 212–232 (2001)
5. Bouchitté, V., Todinca, I.: Listing all potential maximal cliques of a graph. Theor. Comput. Sci. **276**(1–2), 17–32 (2002)
6. Bousquet, N., Gonçalves, D., Mertzios, G.B., Paul, C., Sau, I., Thomassé, S.: Parameterized domination in circle graphs. Theor. Comput. Syst. **54**(1), 45–72 (2014)
7. Chang, M.-S.: Efficient algorithms for the domination problems on interval and circular-arc graphs. SIAM J. Comput. **27**(6), 1671–1694 (1998)
8. Damaschke, P., Müller, H., Kratsch, D.: Domination in convex and chordal bipartite graphs. Inf. Process. Lett. **36**(5), 231–236 (1990)
9. de Fraysseix, H.: Local complementation and interlacement graphs. Discrete Math. **33**(1), 29–35 (1981)
10. Dom, M., Lokshtanov, D., Saurabh, S.: Kernelization lower bounds through colors and IDs. ACM Trans. Algorithms **11**(2), 1–20 (2014)
11. Escoffier, B., Gourvès, L., Monnot, J.: Complexity and approximation results for the connected vertex cover problem in graphs and hypergraphs. J. Discrete Algorithms **8**(1), 36–49 (2010)
12. Eto, H., Guo, F., Miyano, E.: Distance-d independent set problems for bipartite and chordal graphs. J. Comb. Optim. **27**(1), 88–99 (2014)
13. Fomin, F.V., Todinca, I., Villanger, Y.: Large induced subgraphs via triangulations and CMSO. SIAM J. Comput. **44**(1), 54–87 (2015)
14. Fomin, F.V., Villanger, Y.: Finding induced subgraphs via minimal triangulations. In: Marion, J.-Y., Schwentick, T. (eds.) 27th International Symposium on Theoretical Aspects of Computer Science, STACS 2010, LIPIcs, March 4–6, 2010, Nancy, France, vol. 5, pp. 383–394. Schloss Dagstuhl - Leibniz-Zentrum fuer Informatik (2010)
15. Golovach, P.A., Heggernes, P., Kratsch, D., Rafiey, A.: Finding clubs in graph classes. Discrete Appl. Math. **174**, 57–65 (2014)
16. Kloks, T., Kratsch, D., Wong, C.K.: Minimum fill-in on circle and circular-arc graphs. J. Algorithms **28**(2), 272–289 (1998)
17. Montealegre, P., Todinca, I.: On distance-d independent set and other problems in graphs with few minimal separators. ArXiv e-prints, July 2016
18. Scott Provan, J., Burk, R.C.: Two-connected augmentation problems in planar graphs. J. Algorithms **32**(2), 87–107 (1999)
19. Donald Rose, J., Tarjan, R.E., Lueker, G.S.: Algorithmic aspects of vertex elimination on graphs. SIAM J. Comput. **5**(2), 266–283 (1976)
20. Vatshelle, M.: New width parameters of graphs. Ph.D. thesis, University of Bergen, Norway (2012)

Parameterized Complexity of the MINCCA Problem on Graphs of Bounded Decomposability

Didem Gözüpek[1], Sibel Özkan[2], Christophe Paul[3], Ignasi Sau[3(⊠)], and Mordechai Shalom[4,5]

[1] Department of Computer Engineering, Gebze Technical University, Kocaeli, Turkey
didem.gozupek@gtu.edu.tr
[2] Department of Mathematics, Gebze Technical University, Kocaeli, Turkey
s.ozkan@gtu.edu.tr
[3] CNRS, LIRMM, Université de Montpellier, Montpellier, France
{paul,sau}@lirmm.fr
[4] TelHai College, Upper Galilee 12210, Israel
cmshalom@telhai.ac.il
[5] Department of Industrial Engineering, Boğaziçi University, Istanbul, Turkey

Abstract. In an edge-colored graph, the cost incurred at a vertex on a path when two incident edges with different colors are traversed is called reload or changeover cost. The *Minimum Changeover Cost Arborescence* (MINCCA) problem consists in finding an arborescence with a given root vertex such that the total changeover cost of the internal vertices is minimized. It has been recently proved by Gözüpek *et al.* [14] that the MINCCA problem is FPT when parameterized by the treewidth and the maximum degree of the input graph. In this article we present the following results for MINCCA:

- the problem is W[1]-hard parameterized by the treedepth of the input graph, even on graphs of average degree at most 8. In particular, it is W[1]-hard parameterized by the treewidth of the input graph, which answers the main open problem of [14];
- it is W[1]-hard on multigraphs parameterized by the tree-cutwidth of the input multigraph;
- it is FPT parameterized by the star tree-cutwidth of the input graph, which is a slightly restricted version of tree-cutwidth. This result strictly generalizes the FPT result given in [14];
- it remains NP-hard on planar graphs even when restricted to instances with at most 6 colors and 0/1 symmetric costs, or when restricted to instances with at most 8 colors, maximum degree bounded by 4, and 0/1 symmetric costs.

Keywords: Minimum Changeover Cost Arborescence · Parameterized complexity · FPT algorithm · Treewidth · Dynamic programming · Planar graph

This work is supported by the bilateral research program of CNRS and TUBITAK under grant no.114E731.

M. Shalom—The work of this author is supported in part by the TUBITAK 2221 Programme.

© Springer-Verlag GmbH Germany 2016
P. Heggernes (Ed.): WG 2016, LNCS 9941, pp. 195–206, 2016.
DOI: 10.1007/978-3-662-53536-3_17

1   Introduction

The cost that occurs at a vertex when two incident edges with different colors are crossed over is referred to as *reload cost* or *changeover cost* in the literature. This cost depends on the colors of the traversed edges. Although the reload cost concept has important applications in numerous areas such as transportation networks, energy distribution networks, and cognitive radio networks, it has received little attention in the literature. In particular, reload/changeover cost problems have been investigated very little from the perspective of parameterized complexity; the only previous work we are aware of is the one in [14].

In heterogeneous networks in telecommunications, transiting from a technology such as 3G (third generation) to another technology such as wireless local area network (WLAN) has an overhead in terms of delay, power consumption etc., depending on the particular setting. This cost has gained increasing importance due to the recently popular concept of vertical handover [6], which is a technique that allows a mobile user to stay connected to the Internet (without a connection loss) by switching to a different wireless network when necessary. Likewise, switching between different service providers even if they have the same technology has a non-negligible cost. Recently, cognitive radio networks (CRN) have gained increasing attention in the communication networks research community. Unlike other wireless technologies, CRNs are envisioned to operate in a wide range of frequencies. Therefore, switching from one frequency band to another frequency band in a CRN has a significant cost in terms of delay and power consumption [2,13]. This concept has applications in other areas as well. For instance, the cost of transferring cargo from one mode of transportation to another has a significant cost that outweighs even the cost of transporting the cargo from one place to another using a single mode of transportation [19]. In energy distribution networks, transferring energy from one type of carrier to another has an important cost corresponding to reload costs [8].

The reload cost concept was introduced in [19], where the considered problem is to find a spanning tree having minimum diameter with respect to reload cost. In particular, they proved that the problem cannot be approximated within a factor better than 3 even on graphs with maximum degree 5, in addition to providing a polynomial-time algorithm for graphs with maximum degree 3. The work in [8] extended these inapproximability results by proving that the problem is inapproximable within a factor better than 2 even on graphs with maximum degree 4. When reload costs satisfy the triangle inequality, they showed that the problem is inapproximable within any factor better than 5/3.

The work in [10] focused on the minimum reload cost cycle cover problem, which is to find a set of vertex-disjoint cycles spanning all vertices with minimum total reload cost. They showed an inapproximability result for the case when there are 2 colors, the reload costs are symmetric and satisfy the triangle inequality. They also presented some integer programming formulations and computational results.

The authors in [12] study the problems of finding a path, trail or walk connecting two given vertices with minimum total reload cost. They present several polynomial and NP-hard cases for (a)symmetric reload costs and reload costs with(out) triangle inequality. Furthermore, they show that the problem is polynomial for walks, as previously mentioned by [19], and re-proved later for directed graphs by [1].

The work in [9] introduced the *Minimum Changeover Cost Arborescence* (MINCCA) problem. Given a root vertex, MINCCA problem is to find an arborescence with minimum total changeover cost starting from the root vertex. They proved that even on graphs with bounded degree and reload costs adhering to the triangle inequality, MINCCA on directed graphs is inapproximable within $\beta \log \log(n)$ for $\beta > 0$ when there are two colors, and within $n^{1/3-\epsilon}$ for any $\epsilon > 0$ when there are three colors. The work in [15] investigated several special cases of the problem such as bounded cost values, bounded degree, and bounded number of colors. In addition, [15] presented inapproximability results as well as a polynomial-time algorithm and an approximation algorithm for the considered special cases.

In this paper, we study the MINCCA problem from the perspective of parameterized complexity; see [3,5,7,17]. Unlike the classical complexity theory, parameterized complexity theory takes into account not only the total input size n, but also other aspects of the problem encoded in a parameter k. It mainly aims to find an exact resolution of NP-complete problems. A problem is called *fixed parameter tractable* (FPT) if it can be solved in time $f(k) \cdot p(n)$, where $f(k)$ is a function depending solely on k and $p(n)$ is a polynomial in n. An algorithm constituting such a solution is called an FPT algorithm for the problem. Analogously to NP-completeness in classical complexity, the theory of W[1]-hardness can be used to show that a problem is unlikely to be FPT, i.e., for every algorithm the parameter has to appear in the exponent of n. The parameterized complexity of reload cost problems is largely unexplored in the literature. To the best of our knowledge, [14] is the only work that focuses on this issue by studying the MINCCA problem on bounded treewidth graphs. In particular, [14] showed that the MINCCA problem is in XP when parameterized by the treewidth of the input graph and it is FPT when parameterized by the treewidth *and* the maximum degree of the input graph. In this paper, we prove that the MINCCA problem is W[1]-hard parameterized by the treedepth of the input graph, even on graphs of average degree at most 8. In particular, it is W[1]-hard parameterized by the treewidth of the input graph, which answers the main open issue pointed out by [14]. Furthermore, we prove that it is W[1]-hard on multigraphs parameterized by the tree-cutwidth of the input multigraph. On the positive side, we present an FPT algorithm parameterized by the star tree-cutwidth of the input graph, which is a slightly restricted version of tree-cutwidth that we introduce here. This algorithm strictly generalizes the FPT algorithm given in [14]. We also prove that the problem is NP-hard on planar graphs, which are also graphs of bounded decomposability, even when restricted to instances with at most 6 colors and 0/1 symmetric costs. In addition, we prove that it remains NP-hard on planar graphs even when restricted to instances with at most 8 colors, maximum degree bounded by 4, and 0/1 symmetric costs.

The rest of this paper is organized as follows. In Sect. 2 we introduce some basic definitions and preliminaries as well as a formal definition of the MinCCA problem. We present our hardness results in Sect. 3. Finally, Sect. 4 concludes the paper. Due to space limitations, the proofs of the results marked with '[⋆]', our algorithmic results with respect to star tree-cutwidth, as well as several figures, can be found in the full version of the article, which is permanently available at [arXiv:1605.00532].

2   Preliminaries

We say that two partial functions f and f' *agree* if they have the same value everywhere they are both defined, and we denote it by $f \sim f'$. For a set A and an element x, we use $A + x$ (resp., $A - x$) as a shorthand for $A \cup \{x\}$ (resp., $A \setminus \{x\}$). We denote by $[i, k]$ the set of all integers between i and k inclusive, and $[k] = [1, k]$.

Graphs, Digraphs, Trees, and Forests. Given an undirected (multi)graph $G = (V(G), E(G))$ and a subset $U \subseteq V(G)$ of the vertices of G, $\delta_G(U) :=$ $\{uu' \in E(G) \mid u \in U, u' \notin U\}$ is the *cut* of G determined by U, i.e., the set of edges of G that have exactly one end in U. In particular, $\delta_G(v)$ denotes the set of edges incident to v in G, and $d_G(v) := |\delta_G(v)|$ is the *degree* of v in G. The *minimum and maximum degrees* of G are defined as $\delta(G) := \min\{d_G(v) \mid v \in V(G)\}$ and $\Delta(G) := \max\{d_G(v) \mid v \in V(G)\}$ respectively. We denote by $N_G(U)$ (resp., $N_G[U]$) the *open* (resp., *closed*) *neighborhood* of U in G. $N_G(U)$ is the set of vertices of $V(G) \setminus U$ that are adjacent to a vertex of U, and $N_G[U] := N_G(U) \cup U$. When there is no ambiguity about the graph G we omit it from the subscripts. For a subset of vertices $U \subseteq V(G)$, $G[U]$ denotes the subgraph of G *induced* by U.

A digraph T is a *rooted tree* or *arborescence* if its underlying graph is a tree and it contains a *root* vertex denoted by root(T) with a directed path from every other vertex to it. Every other vertex $v \neq $ root(T) has a parent in T, and v is a *child* of its parent.

A rooted forest is the disjoint union of rooted trees, that is, each connected component of it has a root, which will be called a *sink* of the forest.

Tree Decompositions, Treewidth, and Treedepth. A *tree decomposition* of a graph $G = (V(G), E(G))$ is a tree $\mathcal{T}$, where $V(\mathcal{T}) = \{B_1, B_2, \ldots\}$ is a set of subsets (called *bags*) of $V(G)$ such that the following three conditions are met:

1. $\bigcup V(\mathcal{T}) = V(G)$.
2. For every edge $uv \in E(G)$, $u, v \in B_i$ for some bag $B_i \in V(\mathcal{T})$.
3. For every $B_i, B_j, B_k \in V(\mathcal{T})$ such that B_k is on the path $P_{\mathcal{T}}(B_i, B_j)$, $B_i \cap B_j \subseteq B_k$.

The *width* $\omega(\mathcal{T})$ of a tree decomposition $\mathcal{T}$ is defined as the size of its largest bag minus 1, i.e., $\omega(\mathcal{T}) = \max\{|B| \mid B \in V(\mathcal{T})\} - 1$. The *treewidth* of a graph G, denoted as $\mathbf{tw}(G)$, is defined as the minimum width among all tree decompositions of G. When the treewidth of the input graph is bounded, many efficient algorithms are known for problems that are in general NP-hard. In fact, most problems are known to be FPT when parameterized by the treewidth of the input graph. Hence, what we prove in this paper, i.e., the MINCCA problem is W[1]-hard when parameterized by treewidth, is an interesting result.

The *treedepth* $\mathbf{td}(G)$ of a graph G is the smallest natural number k such that each vertex of G can be labeled with an element from $\{1,\dots,k\}$ so that every path in G joining two vertices with the same label contains a vertex having a larger label. Intuitively, where the treewidth parameter measures how far a graph is from being a tree, treedepth measures how far a graph is from being a star. The treewidth of a graph is at most one less than its treedepth; therefore, a W[1]-hardness result for treedepth implies a W[1]-hardness for treewidth.

Tree-Cutwidth. We now explain the concept of tree-cutwidth and follow the notation in [11]. A *tree-cut decomposition* of a graph G is a pair $(T, \mathcal{X})$ where T is a rooted tree and $\mathcal{X}$ is a near-partition of $V(G)$ (that is, empty sets are allowed) where each set X_t of the partition is associated with a node t of T. That is, $\mathcal{X} = \{X_t \subseteq V(G) : t \in V(T)\}$. The set X_t is termed the *bag* associated with the *node* t. For a node t of T we denote by Y_t the union of all the bags associated with t and its descendants, and $G_t = G[Y_t]$. $\mathsf{cut}(t) = \delta(Y_t)$ is the set of all edges with exactly one endpoint in Y_t.

The *adhesion* $\mathsf{adh}(t)$ of t is $|\mathsf{cut}(t)|$. The *torso* of t is the graph H_t obtained from G as follows. Let $t_1,\dots,t_\ell$ be the children of t, $Y_i = Y_{t_i}$ for $i \in [\ell]$ and $Y_0 = V(G) \setminus (X_t \cup_{i=1}^\ell Y_i)$. We first contract each set Y_i to a single vertex y_i, by possibly creating parallel edges. We then remove every vertex y_i of degree 1 (with its incident edge), and finally *suppress* every vertex y_i of degree 2 having 2 neighbors, by connecting its two neighbors with an edge and removing y_i. The *torso size* $\mathsf{tor}(t)$ of t is the number of vertices in H_t. The *width* of a tree-cut decomposition $(T, \mathcal{X})$ of G is $\max_{t \in V(T)}\{\mathsf{adh}(t), \mathsf{tor}(t)\}$. The *tree-cutwidth* of G, or $\mathbf{tcw}(G)$ in short, is the minimum width of $(T, \mathcal{X})$ over all tree-cut decompositions $(T, \mathcal{X})$ of G.

Figure 1 shows the relationship between the graph parameters that we consider in this article. As depicted in Fig. 1, tree-cutwidth provides an intermediate measurement which allows either to push the boundary of fixed parameter tractability or strengthen W[1]-hardness result (cf. [11,16,20]). Furthermore, Fig. 1 also shows that treedepth and tree-cutwidth are unrelated.

Reload and Changeover Costs. We follow the notation and terminology of [19] where the concept of reload cost was defined. We consider edge colored graphs G, where the colors are taken from a finite set X and $\chi : E(G) \to X$ is the *coloring function*. Given a coloring function χ, we denote by E_x^χ, or simply by E_x the set of edges of E colored x, and $G_x = (V(G), E(G)_x)$ is the subgraph of G having the same vertex set as G, but only the edges colored x. The costs are given by a non-negative function $cc : X^2 \to \mathbb{N}_0$ satisfying

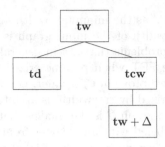

Fig. 1. Relationships between several graph parameters. A being a child of B means that every graph class with bounded A has also bounded B [11]

1. $cc(x_1, x_2) = cc(x_2, x_1)$ for every $x_1, x_2 \in X$.
2. $cc(x, x) = 0$ for every $x \in X$.

The cost of traversing two incident edges e_1, e_2 is $cc(e_1, e_2) := cc(\chi(e_1), \chi(e_2))$. The changeover cost of a path $P = (e_1 - e_2 - \ldots - e_\ell)$ of length ℓ is $cc(P) := \sum_{i=2}^{\ell} cc(e_{i-1}, e_i)$. Note that $cc(P) = 0$ whenever $\ell \leq 1$.

We extend this definition to trees as follows: Given a directed tree T rooted at r, (resp., an undirected tree T and a vertex $r \in V(T)$), for every outgoing edge e of r (resp., incident to r) we define $prev(e) = e$, and for every other edge $prev(e)$ is the edge preceding e on the path from r to e. The changeover cost of T with respect to r is $cc(T, r) := \sum_{e \in E(T)} cc(prev(e), e)$. When there is no ambiguity about the vertex r, we denote $cc(T, r)$ by $cc(T)$.

Statement of the Problem. The MINCCA problem aims to find a spanning tree rooted at r with minimum changeover cost [9]. Formally,

---

MINCCA
Input: A graph $G = (V, E)$ with an edge coloring function $\chi : E \to X$, a vertex $r \in V$ and a changeover cost function $cc : X^2 \to \mathbb{N}_0$.
Output: A spanning tree T of G minimizing $cc(T, r)$.

---

3   Hardness Results

In this section we prove several hardness results for the MINCCA problem. Our main result is in Subsect. 3.1, where we prove that the problem is W[1]-hard parameterized by the treedepth of the input graph. We also prove that the problem is W[1]-hard on *multigraphs* parameterized by the tree-cutwidth of the input graph. Both results hold even if the input graph has bounded average degree. Finally, in Subsect. 3.2 we prove that the problem remains NP-hard on planar graphs.

3.1 W[1]-hardness with Parameters Treedepth and Tree-Cutwidth

We need to define the following parameterized problem.

MULTICOLORED k-CLIQUE
Input: A graph G, a coloring function $c : V(G) \to \{1, \ldots, k\}$, and a positive integer k.
Parameter: k.
Question: Does G contain a clique on k vertices with one vertex from each color class?

MULTICOLORED k-CLIQUE is known to be W[1]-hard on general graphs, even in the special case where all color classes have the same number of vertices [18], and therefore we may make this assumption as well.

Theorem 1. *The* MINCCA *problem is* W[1]-*hard parameterized by the treedepth of the input graph, even on graphs with average degree at most 8.*

Proof. We reduce from MULTICOLORED k-CLIQUE, where we may assume that k is odd. Indeed, given an instance (G, c, k) of MULTICOLORED k-CLIQUE, we can trivially reduce the problem to itself as follows. If k is odd, we do nothing. Otherwise, we output $(G', c', k + 1)$, where G' is obtained from G by adding a universal vertex v, and $c' : V(G') \to \{1, \ldots, k + 1\}$ is such that its restriction to G equals c, and $c(v) = k + 1$.

Given an instance (G, c, k) of MULTICOLORED k-CLIQUE with k odd, we proceed to construct an instance (H, X, χ, r, cc) of MINCCA. Let $V(G) = V_1 \uplus V_2 \uplus \cdots \uplus V_k$, where the vertices of V_i are colored i for $1 \le i \le k$. Let W be an arbitrary Eulerian circuit of the complete graph K_k, which exists since k is odd. If $V(K_k) = \{v_1, \ldots, v_k\}$, we can clearly assume without loss of generality[1] that W starts by visiting, in this order, vertices $v_1, v_2, \ldots, v_k, v_1$, and that the last edge of W is $\{v_3, v_1\}$. For every edge $\{v_i, v_j\}$ of W, we add to H a vertex $s_{i,j}$. These vertices are called the *selector vertices* of H. For every two consecutive edges $\{v_i, v_j\}, \{v_j, v_\ell\}$ of W, we add to H a vertex $v_j^{i,\ell}$ and we make it adjacent to both $s_{i,j}$ and $s_{j,\ell}$. We also add to H a new vertex $v_1^{0,2}$ adjacent to $s_{1,2}$, a new vertex $v_1^{3,0}$ adjacent to $s_{3,1}$, and a new vertex r adjacent to $v_1^{0,2}$, which will be the *root* of H. Note that the graph constructed so far is a simple path P on $2\binom{k}{2} + 2$ vertices. We say that the vertices of the form $v_j^{i,\ell}$ are *occurrences* of vertex $v_j \in V(K_k)$. For $2 \le j \le k$, we add an edge between the root r and the first occurrence of vertex v_j in P (note that the edge between r and the first occurrence of v_1 already exists).

The first k selector vertices, namely $s_{1,2}, s_{2,3}, \ldots, s_{k-1,k}, s_{k,1}$ will play a special role that will become clear later. To this end, for $1 \le i \le k$, we add an edge

---

[1] This assumption is not crucial for the construction, but helps in making it conceptually and notationally easier.

between the selector vertex $s_{i,i \pmod{k}+1}$ and each of the occurrences of v_i that appear after $s_{i,i \pmod{k}+1}$ in P. These edges will be called the *jumping edges* of H.

Let us denote by F the graph constructed so far. Finally, in order to construct H, we replace each vertex of the form $v_j^{i,\ell}$ in F with a whole copy of the vertex set V_j of G and make each of these new vertices adjacent to all the neighbors of $v_j^{i,\ell}$ in F. This completes the construction of H. Note that $\mathbf{td}(H) \le \binom{k}{2} + 1$, as the removal of the $\binom{k}{2}$ selector vertices from H results in a star centered at r and isolated vertices.

We now proceed to describe the color palette X, the coloring function χ, and the cost function cc, which altogether will encode the edges of G and will ensure the desired properties of the reduction. For simplicity, we associate a distinct color with each edge of H, and thus, with slight abuse of notation, it is enough to describe the cost function cc for every ordered pair of incident edges of H. We will use just three different costs: 0, 1, and B, where B can be set as any real number strictly greater than $\binom{k}{2}$. For each ordered pair of incident edges e_1, e_2 of H, we define

$$
cc(e_1, e_2) = \begin{cases}
0, & \text{if } e_1 = \{\hat{x}, s_{i,j}\} \text{ and } e_2 = \{s_{i,j}, \hat{y}\} \text{ is a jumping edge such that} \\
& \hat{x}, \hat{y} \text{ are copies of vertices } x, y \in V_i, \text{ respectively, with } x \ne y, \text{ or} \\
& \text{if } e_1 = \{r, \hat{x}\} \text{ and } e_2 = \{\hat{x}, s_{1,2}\}, \text{ where } \hat{x} \text{ is a copy of a vertex} \\
& x \in V_1, \text{ or} \\
& \text{if } e_1 \text{ and } e_2 \text{ are the two edges that connect a vertex in a copy} \\
& \text{of a color class } V_i \text{ to a selector vertex.} \\[1.5ex]
1, & \text{if } e_1 = \{\hat{x}, s_{i,j}\} \text{ and } e_2 = \{s_{i,j}, \hat{y}\}, \text{ where } \hat{x} \text{ is a copy of a vertex} \\
& x \in V_i \text{ and } \hat{y} \text{ is a copy of a vertex } y \in V_j \text{ such that } \{x, y\} \in E(G). \\[1.5ex]
B, & \text{otherwise.}
\end{cases}
$$

This completes the construction of (H, X, χ, r, cc), which can be clearly performed in polynomial time.

Claim 1 [⋆]. *The average degree of H is bounded by 8.*

We now claim that H contains and arborescence T rooted at r with cost at most $\binom{k}{2}$ if and only if G contains a multicolored k-clique[2]. Note that the simple path P described above naturally defines a partial left-to-right ordering among the vertices of H, and hence any arborescence rooted at r contains *forward* and *backward* edges defined in an unambiguous way. Note also that all costs that involve a backward edge are equal to B, and therefore no such edge can be contained in an arborescence of cost at most $\binom{k}{2}$.

---

[2] If the costs associated with colors are restricted to be strictly positive, we can just replace cost 0 with cost ε, for an arbitrarily small positive real number ε, and ask for an arborescence in H of cost strictly smaller than $\binom{k}{2} + 1$.

Suppose first that G contains a multicolored k-clique with vertices $v_1, v_2, \ldots, v_k$, where $v_i \in V_i$ for $1 \le i \le k$. Then we define the edges of the spanning tree T of H as follows. Tree T contains the edges of a left-to-right path Q that starts at the root r, contains all $\binom{k}{2}$ selector vertices and connects them, in each occurrence of a set V_i, to the copy of vertex v_i defined by the k-clique. Since in Q the selector vertices connect copies of pairwise adjacent vertices of G, the cost incurred so far by T is exactly $\binom{k}{2}$. For $1 \le i \le k$, we add to Q the edges from r to all vertices in the first occurrence of V_i that are not contained in Q. Note that the addition of these edges to T incurs no additional cost. Finally, we will use the jumping edges to reach the uncovered vertices of H. Namely, for $1 \le i \le k$, we add to T an edge between the selector vertex $s_{i,i \pmod{k}+1}$ and all occurrences of the vertices in V_i distinct from v_i that appear after $s_{i,i \pmod{k}+1}$. Note that since the jumping edges in T contain copies of vertices distinct from the the ones in the k-clique, these edges incur no additional cost either. Therefore, $cc(T, r) = \binom{k}{2}$, as we wanted to prove.

Conversely, suppose now that H has an arborescence T rooted at r with cost at most $\binom{k}{2}$. Clearly, all costs incurred by the edges in T are either 0 or 1. For a selector vertex $s_{i,j}$, we call the edges joining $s_{i,j}$ to the vertices in the occurrence of V_i right before $s_{i,j}$ (resp., in the occurrence of V_j right after $s_{i,j}$) the *left* (resp., *right*) edges of this selector vertex.

Claim 2 [⋆]. *Tree T contains exactly one left edge and exactly one right edge of each selector vertex of H.*

By Claim 2, tree T contains a path Q' that chooses exactly one vertex from each occurrence of a color class of G. We shall now prove that, thanks to the jumping edges, these choices are *coherent*, which will allow us to extract the desired multicolored k-clique in G.

Claim 3. *For every $1 \le i \le k$, the vertices in the copies of color class V_i contained in Q' all correspond to the same vertex of G, denoted by v_i.*

Proof. Assume for contradiction that for some index i, the vertices in the copies of color class V_i contained in Q' correspond to at least two distinct vertices v_i and v'_i of G, in such a way that v_i is the selected vertex in the first occurrence of V_i, and v'_i occurs later, say in the jth occurrence of V_i. Therefore, the copy of v_i in the jth occurrence of V_i does not belong to path Q', so for this vertex to be contained in T, by construction it is necessarily an endpoint of a jumping edge e starting at the selector vertex $s_{i,i \pmod{k}+1}$. But then the cost incurred in T by the edges e' and e, where e' is the edge joining the copy of v_i in the first occurrence of V_i to the selector vertex $s_{i,i \pmod{k}+1}$, equals B, contradicting the hypothesis that $cc(T, r) \le \binom{k}{2}$.   □

Finally, we claim that the vertices $v_1, v_2, \ldots, v_k$ defined by Claim 3 induce a multicolored k-clique in G. Indeed, assume for contradiction that there exist two such vertices v_i and v_j such that $\{v_i, v_j\} \notin E(G)$. Then the cost in T incurred by the two edges connecting the copies of v_i and v_j to the selector vertex $s_{i,j}$

(by Claim 2, these two edges indeed belong to T) would be equal to B, contracting again the hypothesis that $cc(T, r) \leq \binom{k}{2}$. This concludes the proof of the theorem.   □

In the next theorem we prove that the MINCCA problem is W[1]-hard on multigraphs parameterized by the tree-cutwidth of the input graph. Note that this result does not imply Theorem 1, which applies to graphs without multiple edges.

Theorem 2 [⋆]. *The* MINCCA *problem is* W[1]-*hard on multigraphs parameterized by the tree-cutwidth of the input multigraph.*

3.2   NP-hardness on Planar Graphs

In this subsection we prove that the MINCCA problem remains NP-hard on planar graphs. In order to prove this result, we need to introduce the PLANAR MONOTONE 3-SAT problem. An instance of 3-SAT is called *monotone* if each clause is monotone, that is, each clause consists only of positive variables or only of negative variables. We call a clause with only positive (resp., negative) variables a *positive* (resp., *negative*) clause. Given an instance ϕ of 3-SAT, we define the bipartite graph G_ϕ that has one vertex per each variable and each clause, and has an edge between a variable-vertex and a clause-vertex if and only if the variable appears (positively or negatively) in the clause. A *monotone rectilinear representation* of a monotone 3-SAT instance ϕ is a *planar* drawing of G_ϕ such that all variable-vertices lie on a path, all positive clause-vertices lie *above* the path, and all negative clause-vertices lie *below* the path. In the PLANAR MONOTONE 3-SAT problem, we are given a monotone rectilinear representation of a planar monotone 3-SAT instance ϕ, and the objective is to determine whether ϕ is satisfiable. Berg and Khosravi [4] proved that the PLANAR MONOTONE 3-SAT problem is NP-complete.

Theorem 3. *The* MINCCA *problem is* NP-*hard on planar graphs even when restricted to instances with at most 6 colors and 0/1 symmetric costs.*

Proof. We reduce from the PLANAR MONOTONE 3-SAT problem. Given a monotone rectilinear representation of a planar monotone 3-sat instance ϕ, we build an instance (H, X, χ, r, f) of MINCCA as follows. We denote the variable-vertices of G_ϕ as $\{x_1, \ldots, x_n\}$ and the clause-vertices of G_ϕ as $\{C_1, \ldots, C_m\}$. Without loss of generality, we assume that the variable-vertices appear in the order $x_1, \ldots, x_n$ on the path P of G_ϕ that links the variable-vertices. For every variable-vertex x_i of G_ϕ, we add to H a gadget consisting of four vertices $x_i^\ell, x_i^r, x_i^+, x_i^-$ and five edges $\{x_i^\ell, x_i^+\}$, $\{x_i^+, x_i^r\}$, $\{x_i^r, x_i^-\}$, $\{x_i^-, x_i^\ell\}$, $\{x_i^+, x_i^-\}$. We add to H a new vertex r, which we set as the root, and we add the edge $\{r, x_1^\ell\}$. For every $i \in \{1, \ldots, n-1\}$, we add to H the edge $\{x_i^r, x_{i+1}^\ell\}$. We add to H all clause-vertices $C_1, \ldots, C_m$. For every $i \in \{1, \ldots, n\}$, we add an edge between vertex x_i^+ and each clause-vertex of G_ϕ in which variable x_i appears positively, and an edge between vertex x_i^- and each clause-vertex of G_ϕ in which variable x_i appears negatively. This completes the construction of H. Since G_ϕ is planar and all positive (resp., negative) clause-vertices appear above (resp., below) the path P, it is easy to see that the graph H is planar as well.

We define the color palette as $X = \{1,2,3,4,5,6\}$. Let us now describe the edge-coloring function χ. For every clause-vertex C_j, we color arbitrarily its three incident edges with the colors $\{4,5,6\}$, so that each edge incident to C_j gets a different color. For every $i \in \{1,\ldots,n\}$, we define $\chi(\{x_i^\ell, x_i^+\}) = \chi(\{x_i^r, x_i^-\}) = 1$, $\chi(\{x_i^+, x_i^r\}) = \chi(\{x_i^-, x_i^\ell\}) = 2$, and $\chi(\{x_i^+, x_i^-\}) = 3$. We set $\chi(\{r, x_1^\ell\}) = 4$ and for every $i \in \{1,\ldots,n-1\}$, $\chi(\{x_i^r, x_{i+1}^\ell\}) = 4$. Finally, we define the cost function cc to be symmetric and, for every $i \in \{1,2,3,4,5,6\}$, we set $cc(i,i) = 0$. We define $cc(1,2) = 1$ and $cc(1,3) = cc(2,3) = 0$. For every $i \in \{4,5,6\}$, we set $cc(1,i) = cc(2,i) = 0$ and $cc(3,i) = 1$. Finally, for every $i,j \in \{4,5,6\}$ with $i \neq j$ we set $cc(i,j) = 1$. The following claim concludes the proof.

Claim 4 [⋆]. *H contains an arborescence T rooted at r with cost 0 if and only if the formula ϕ is satisfiable.* □

Note that the above proof actually implies that MINCCA cannot be approximated to any positive ratio on planar graphs in polynomial time, since an optimal solution has cost 0. We do not know whether such a strong inapproximability result holds even if we do not allow to use costs 0 among different colors.

In the next theorem we present a modification of the previous reduction showing that the MINCCA problem remains hard even if the maximum degree of the input planar graph is bounded.

Theorem 4 [⋆]. *The MINCCA problem is NP-hard on planar graphs even when restricted to instances with at most 8 colors, maximum degree bounded by 4, and 0/1 symmetric costs.*

4   Conclusions and Further Research

In this article we proved several hardness results for the MINCCA problem. In particular, we proved that the problem is W[1]-hard parameterized by treewidth on general graphs, and that it is NP-hard on planar graphs, but we do not know whether it is W[1]-hard parameterized by treewidth (or treedepth) on planar graphs.

On the other hand, we provided an FPT algorithm for a restricted version of tree-cutwidth, and we proved that the problem is W[1]-hard on multigraphs parameterized by tree-cutwidth. While we were not able to prove this W[1]-hardness result on graphs without multiple edges, we believe that it is indeed the case. It would be natural to consider other structural parameters such as the size of a vertex cover or a feedback vertex set.

Finally, it would be interesting to try to generalize our techniques to prove hardness results or to provide efficient algorithms for other reload cost problems that have been studied in the literature [6,8,10,19].

Acknowledgment. We would like to thank the anonymous referees for helpful comments that improved the presentation of the manuscript

References

1. Amaldi, E., Galbiati, G., Maffioli, F.: On minimum reload cost paths, tours, and flows. Networks **57**(3), 254–260 (2011)
2. Arkoulis, S., Anifantis, E., Karyotis, V., Papavassiliou, S., Mitrou, N.: On the optimal, fair and channel-aware cognitive radio network reconfiguration. Comput. Netw. **57**(8), 1739–1757 (2013)
3. Cygan, M., Fomin, F.V., Kowalik, L., Lokshtanov, D., Marx, D., Pilipczuk, M., Pilipczuk, M., Saurabh, S.: Parameterized Algorithms. Springer, Switzerland (2015)
4. de Berg, M., Khosravi, A.: Optimal binary space partitions for segments in the plane. Int. J. Comput. Geom. Appl. **22**(3), 187–206 (2012)
5. Downey, R.G., Fellows, M.R.: Fundamentals of Parameterized Complexity. Texts in Computer Science. Springer, London (2013)
6. Çelenlioğlu, M. R., Gözüpek, D., Mantar, H. A.: A survey on the energy efficiency of vertical handover mechanisms. In: Proceedings of the International Conference on Wireless and Mobile Networks (WiMoN) (2013)
7. Flum, J., Grohe, M.: Parameterized Complexity Theory. Texts in Theoretical Computer Science. Springer, Heidelberg (2006)
8. Galbiati, G.: The complexity of a minimum reload cost diameter problem. Discrete Appl. Math. **156**(18), 3494–3497 (2008)
9. Galbiati, G., Gualandi, S., Maffioli, F.: On minimum changeover cost arborescences. In: Pardalos, P.M., Rebennack, S. (eds.) SEA 2011. LNCS, vol. 6630, pp. 112–123. Springer, Heidelberg (2011)
10. Galbiati, G., Gualandi, S., Maffioli, F.: On minimum reload cost cycle cover. Discrete Appl. Math. **164**, 112–120 (2014)
11. Ganian, R., Kim, E.J., Szeider, S.: Algorithmic applications of tree-cut width. In: Italiano, G.F., Pighizzini, G., Sannella, D.T. (eds.) MFCS 2015. LNCS, vol. 9235, pp. 348–360. Springer, Heidelberg (2015)
12. Gourvès, L., Lyra, A., Martinhon, C., Monnot, J.: The minimum reload s-t path, trail and walk problems. Discrete Appl. Math. **158**(13), 1404–1417 (2010)
13. Gozupek, D., Buhari, S., Alagoz, F.: A spectrum switching delay-aware scheduling algorithm for centralized cognitive radio networks. IEEE Trans. Mobile Comput. **12**(7), 1270–1280 (2013)
14. Gözüpek, D., Shachnai, H., Shalom, M., Zaks, S.: Constructing minimum changeover cost arborescenses in bounded treewidth graphs. Theorerical Comput. Sci. **621**, 22–36 (2016)
15. Gözüpek, D., Shalom, M., Voloshin, A., Zaks, S.: On the complexity of constructing minimum changeover cost arborescences. Theorerical Comput. Sci. **540**, 40–52 (2014)
16. Kim, E., Oum, S., Paul, C., Sau, I., Thilikos, D.M.: An FPT 2-approximation for tree-cut decomposition. In: Sanità, L., et al. (eds.) WAOA 2015. LNCS, vol. 9499, pp. 35–46. Springer, Heidelberg (2015). doi:10.1007/978-3-319-28684-6_4
17. Niedermeier, R.: Invitation to Fixed-Parameter Algorithms, vol. 31. Oxford University Press, Oxford (2006)
18. Pietrzak, K.: On the parameterized complexity of the fixed alphabet shortest common supersequence and longest common subsequence problems. J. Comput. Syst. Sci. **67**(4), 757–771 (2003)
19. Wirth, H.-C., Steffan, J.: Reload cost problems: minimum diameter spanning tree. Discrete Appl. Math. **113**(1), 73–85 (2001)
20. Wollan, P.: The structure of graphs not admitting a fixed immersion. J. Comb. Theor. Ser. B **110**, 47–66 (2015)

On Edge Intersection Graphs of Paths with 2 Bends

Martin Pergel[1] and Paweł Rzążewski[2,3]([⊠])

[1] Department of Software and Computer Science Education, Charles University,
Praha, Czech Republic
perm@kam.mff.cuni.cz
[2] Institute of Computer Science and Control,
Hungarian Academy of Sciences (MTA SZTAKI), Budapest, Hungary
p.rzazewski@mini.pw.edu.pl
[3] Faculty of Mathematics and Information Science,
Warsaw University of Technology, Warszawa, Poland

Abstract. An EPG-representation of a graph G is a collection of paths in a grid, each corresponding to a single vertex of G, so that two vertices are adjacent if and only if their corresponding paths share infinitely many points. In this paper we focus on graphs admitting EPG-representations by paths with at most 2 bends. We show hardness of the recognition problem for this class of graphs, along with some subclasses.

We also initiate the study of graphs representable by unaligned polylines, and by polylines, whose every segment is parallel to one of prescribed slopes. We show hardness of recognition and explore the trade-off between the number of bends and the number of slopes.

1 Introduction

The concept of *edge intersection graphs of paths in a grid (EPG-graphs)* was introduced by Golumbic et al. [7]. By an EPG-representation of a graph G we mean a mapping from vertices of G to paths in a grid, such that two vertices are adjacent if and only if their corresponding paths share a grid edge. As each graph can be represented in this way [7], it makes sense to consider representations with some restricted set of shapes. A usual parameterization is by bounding the number k of times each path is allowed to change the direction. Graphs with such a representation are called *k-bend graphs*. So far, the case of 1-bend graphs received most attention [4,7].

Since 0-bend graphs are just *interval graphs*, they can be recognized in polynomial time [1]. The recognition of 1-bend graphs is NP-complete [8], even if the representation is restricted to any prescribed set of 1-bend objects [4]. However, the problem becomes trivially solvable when k is at least the maximum degree of the input graph [8]. Thus it is unclear whether k-bend graphs are hard to recognize for all $k \geq 2$.

M. Pergel—Partially supported by a Czech research grant GAČR GA14-10799S.
P. Rzążewski—Supported by ERC Starting Grant PARAMTIGHT (No. 280152).

P. Heggernes (Ed.): WG 2016, LNCS 9941, pp. 207–219, 2016.
DOI: 10.1007/978-3-662-53536-3_18

It is worth mentioning the closely related notion of B_k-*VPG-graphs*. These graphs are defined as intersection graphs of axis-aligned paths with at most k bends. So, unlike in the EPG-representation, paths that share a finite number of points define adjacent vertices. Chaplick *et al.* [5] showed it is NP-complete to recognize B_k-VPG-graphs, for all $k \geq 0$.

In this paper we explore the problem of recognition of subclasses of EPG-graphs. Namely, we show that it is NP-complete to recognize 2-bend graphs. We also consider some restrictions, where we permit just some types of the curves in an EPG-representation (similarly to [4]). One of these restrictions, i.e., *monotonic EPG-representations*, where each path ascends in rows and columns, was already considered by Golumbic *et al.* [7]. Our hardness proof even shows that between monotonic 2-bend graphs and 2-bend graphs, no polynomially recognizable class can be found.

The class of 2-bend graphs can be perceived as a generalization of quite well-studied class of 1-bend graphs. We also consider some generalizations of the concept of EPG-representations. We do not require individual segments to be axis-aligned, but we permit them to use any slope. We call such graphs *unaligned EPG-graphs* and study the number of bends in this setting. After this generalization, we may ask about particular restrictions. These restrictions are represented by restricting number of slopes that segments may use or even by using just prescribed shapes (in a flavor similar to [4]).

For unaligned EPG-graphs, we show that it is NP-hard to determine whether a graph is an unaligned 2-bend graph (hardness of the recognition for 1-bend graphs follows from [4]).

Having introduced unaligned EPG-graphs, we observe that there is a trade-off between the number of bends and the number of slopes used in a representation. We also show that representing an unaligned 2-bend graph on n vertices, may require using $\Omega(\sqrt{n})$ slopes. This result follows from our hardness reduction.

2   Preliminaries

For an EPG-representation of a graph G, by P_v we shall denote the path representing a vertex v. Often we shall identify the vertex v with P_v. For example, if we say that two paths are adjacent, we mean that they share infinitely many points (note that if two paths *intersect*, one common point is enough).

A central notion in the study of EPG-graphs is the *bend number*. The bend number of a graph G, denoted by $b(G)$, is the minimum k, such that G has an EPG-representation, in which every paths changes it direction at most k times. W.l.o.g. we can assume that every path in a k-bend EPG-representation bends exactly k times [4].

Each 2-bend path will be classified as *vertical* or *horizontal*, if its middle segment is resp. vertical or horizontal. This middle segment will be called the *body* of the path, while the remaining two segments will be referenced as its *legs*.

For a set X of shapes of polylines (i.e., piecewise-linear curves), by X-graphs we shall denote the class of graphs admitting an EPG-representation, in which the shape of every path is in X (similar notation was used in [4]). So for example monotonic 2-bend graphs are exactly $\{\llcorner, \ulcorner\}$-graphs.

Golumbic *et al.* [7] analyzed the structure of cliques in 1-bend graphs and proved that in 1-bend graphs each clique C is either an *edge clique* or a *claw-clique*. A maximal edge clique consists of vertices whose representing paths share a common grid edge. A *claw* is a set of three distinct grid edges sharing a single endpoint and a maximal claw-clique consists of all paths containing two out of three edges of a given claw. Since we can safely assume that each 1-bend representation of a graph with n vertices can be embedded in a $2n \times 2n$ grid, we obtain that the number of maximal cliques in a 1-bend graph is at most $O(n^2)$, i.e., is polynomial in n. This is no longer the case with 2-bend graphs.

Let n be an integer and let K_{2n}^- be the *cocktail-party graph*, i.e., a complete graph on $2n$ vertices with a perfect matching removed. It is clear that K_{2n}^- has $2^n = 2^{|V(K_{2n}^-)|/2}$ maximal cliques. Figure 1 (left) shows that K_{2n}^- is a 2-bend graph.

Proposition 1. *2-bend graphs can have exponentially many maximal cliques.*

The restricted structure of cliques in 1-bend graphs follows from the fact that the 1-bend paths representing pairwise adjacent vertices must all share at least one grid point. It is easy to observe that cliques in 2-bend graphs do not have such a simple structure. One could be inclined by Fig. 1 (left) that every maximal clique is contained in the union of two edge-cliques or claw-cliques (a similar situation appears in unit disk graphs and is the main ingredient of a polynomial algorithm for CLIQUE in these graphs [3]). However, Fig. 1 (right) shows it is not true.

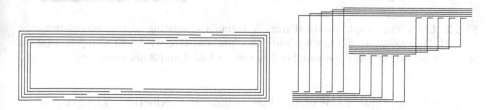

Fig. 1. Left: K_{10}^- as 2-bend graph. **Right:** A clique is not contained in two edge-cliques.

3 Aligned 2-Bend Graphs

The main results of this section is the following complexity result.

Theorem 1. *It is NP-complete to decide if a given graph is a 2-bend graph.*

Proof. The NP-membership is obvious. As a polynomial certificate we use a list of coordinates denoting start- and end-points of straight-line segments. Such a representation has polynomial size w.r.t. the given graph.

For the NP-hardness we use a polynomial reduction from PURE-NAE-3-SAT. The instance of this problem is a set of clauses, each containing three variables. We ask for the existence of a truth assignment, such that each clause contains at least one true variable and at least one false variable (we say that such a clause is *satisfied*). The problem is NP-complete and equivalent to 2-coloring of 3-uniform hypergraphs [9].

For a given formula φ, we shall construct a graph G, which is a 2-bend graph iff the formula is satisfiable. We start by replicating φ 21 times (each time over a distinct copy of the set of variables), obtaining an equivalent formula φ'. The reason of this operation will be made clear in a while.

We start the construction of G with two special vertices a and b. Then for each variable i of φ, we add a vertex v_i adjacent to both a and b. For each occurence of i in a clause z of φ', we add another vertex $o_{i,z}$, adjacent to a, b, and v_i. Finally, for each clause $z = (i, j, k)$ we add mutually non-adjacent vertices c_z, d_z, e_z, and f_z, with the following neighbors: $N(c_z) = \{o_{i,z}, o_{j,z}, o_{k,z}\}$; $N(d_z) = \{o_{i,z}, o_{j,z}\}$; $N(e_z) = \{o_{i,z}, o_{k,z}\}$; and $N(f_z) = \{o_{j,z}, o_{k,z}\}$ (see Fig. 2 (left)).

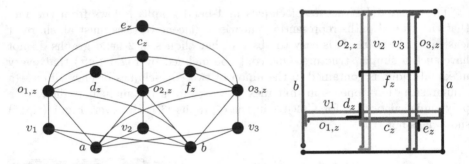

Fig. 2. Left: The graph obtained from a formula consisting of a single clause $z = (1, 2, 3)$. For clarity we did not replicate the formula. **Right:** An EPG-representation of the graph on the left. The variable 1 is false, while 2 and 3 are true.

Now let us explain the main ideas behind the reduction. The purpose of vertices a and b is to cover the legs of each P_{v_i} and $P_{o_{i,z}}$, keeping just their bodies exposed for possible intersections with clause-vertices. This assumption may fail, as some P_{v_i} or $P_{o_{i,z}}$ can be positioned over an end of a segment of P_a or P_b, or on an intersection point of P_a and P_b. However, each end can be used at most once and each intersection point at most twice (see Fig. 3). As P_a and P_b have (together) 12 ends of segments and at most 4 intersection points, we have at most 20 special situations. But since we replicated φ 21 times, we are sure that for at least one copy of φ our assumption holds (this type of trick we call the "quantitative trick" and we use it to cope with some obstructions which may appear only a constant number of times). Let us focus on this "clean" copy of φ in φ'. One leg of each $P_{o_{i,j}}$ is adjacent to a and the other one is adjacent to b. Also, at least one of them has to be adjacent to P_{v_i}, since otherwise clause-vertices would be adjacent to v_i. Thus the body of $P_{o_{i,j}}$ is exposed for representing clause-related vertices. Moreover, the orientation of the body (and this of whole path) is the same as the orientation of P_{v_i}, so all variable-occurences are "synchronized". The orientation of the paths will decide on truth assignment (horizontal means false, vertical means true).

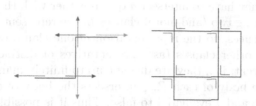

Fig. 3. Left: 6 pairwise non-adjacent segments may exit a 2-bend path without having to bend inside it. **Right:** At most 8 pairwise non-adjacent 2-bend paths may be adjacent to both P_a and P_b and contain their intersection point.

First we show irrepresentabililty of the graph for an unsatisfiable formula. Let $z = (i, j, k)$ be an unsatisfied clause. We will show that it cannot be represented. Observe that it is impossible to have a 2-bend path adjacent to three parallel, pairwise non-collinear segments, while it is possible for two parallel and one perpendicular segments (see Fig. 4 left).

The situation with three parallel segments corresponds to all-true or all-false clause. So, if no pair of middle segments of $P_{o_{i,z}}$, $P_{o_{j,z}}$, $P_{o_{k,z}}$ (and thus P_{v_i}, P_{v_j}, P_{v_k}) is collinear, we cannot represent c_z.

However, it might still happen that the bodies of, say, $P_{o_{i,z}}$ and $P_{o_{j,z}}$ are lying on the same line. But this pair of segments cannot be adjacent to more than one 2-bend path (see Fig. 4 (middle)). So if we represent c_z, then we cannot represent d_z, e_z, or f_z. This shows irrepresentability of an unsatisfied clause.

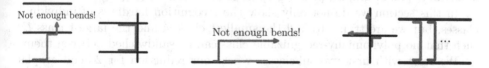

Fig. 4. Left: It is impossible to intersect three parallel, pairwise non-collinear segments with a 2-bend path, while two parallel and one perpendicular segments can be intersected. **Middle:** Two collinear segments cannot be adjacent to two mutually non-adjacent 2-bend paths. **Right:** This is possible for two mutually intersecting segments or two non-collinear parallel segments.

For a representable formula, we build a canonical representation shown in Fig. 2 (right) (for a clause with one false and two true literals, we rotate everything except of a and b by 90 ∘). Figure 2 shows one clause in one of (21) replicated copies and one occurence of each variable. The full construction with all

21 copies would consist of 21 copies of all items present in the picture, except for P_a and P_b. Note that there are no edges between vertices belonging to different copies of φ. Further occurences, e.g., of v_2 in the same formula can be represented next to $o_{2,z}$ intersecting v_2 in the bottom (or top) horizontal leg (where it simultaneously intersects a (or b, respectively and it has to avoid legs of other possible occurences). Anyway, their truth assignments are "synchronized" in all possible cases as they have to intersect a or b together with the vertex representative v_2. Considering two (and more) clauses in the representation, each clause has its own occurences, so the representation of one clause does not influence representations of other clauses (as representatives of distinct occurences are not mutually adjacent, i.e., they are disjoint up to finitely many points). In this representation, the body of each $P_{o_{i,z}}$ intersects the body of each $P_{o_{j,z}}$, for all i evaluated to true and j evaluated to false. Thus it is possible to represent all clause-vertices, just as depicted. □

3.1   Subclasses of Aligned 2-Bend Graphs

Here we focus on the recognition of particular subclasses of 2-bend graphs. Note that as there are many classes (whose recognition is often NP-hard), it is important to ask whether even some polynomially recognizable class can exist "in between". This concept is called *sandwiching*. Formally, having two classes of graphs $\mathcal{A} \subseteq \mathcal{B}$, a class $\mathcal{C}$ is *sandwiched between $\mathcal{A}$ and $\mathcal{B}$* if $\mathcal{A} \subseteq \mathcal{C} \subseteq \mathcal{B}$. For optimization problems, it holds that if an algorithm works for class $\mathcal{B}$, it works also for the class $\mathcal{A}$. Also a hardness result for $\mathcal{A}$ carries over to $\mathcal{B}$. However, the recognition problem behaves in a different way. As a trivial example we may pick a class $\mathcal{A}$ containing only complete graphs (this class is polynomially recognizable), for class $\mathcal{B}$ we may take class of all graphs (which is also polynomially recognizable) and between them we can find, e.g., classes of 2-bend graphs, whose recognition is NP-complete, as shown in Theorem 1. Similarly, between two NP-hard classes, a polynomially-recognizable class can be sandwiched (consider e.g. 3-colorable planar graphs, planar graphs, and 4-colorable graphs).

In this section we do not only show the recognition hardness of individual classes, but we are trying to find the smallest class $\mathcal{A}$ and the largest class $\mathcal{B}$, such that no polynomially-recognizable class can be sandwiched between them.

We start with first two subclasses where our reduction for 2-bend graphs can be applied directly. One of them is a class of monotonic 2-bend graphs (i.e., $\{\llcorner\neg,\neg\lrcorner\}$-graphs) and the other is the class of $\{\llcorner\neg,\urcorner\neg\}$-graphs.

We observe that in the proof of Theorem 1 we produce a monotonic 2-bend graph from each satisfiable formula. As a non-satisfiable formula cannot be represented by any 2-bend graph, if there was a polynomially-recognizable class between monotonic 2-bend graphs and 2-bend graphs, we would be able to distinguish satisfiable formulae from non-satisfiable ones, showing P=NP.

It is very simple to redraw the representation used in the proof of Theorem 1, using only $\llcorner\neg$ and $\urcorner\neg$ -shapes.

Corollary 1. *It is NP-complete to recognize monotonic 2-bend graphs and $\{\llcorner\lrcorner,\urcorner\lrcorner\}$-graphs. Moreover, between 2-bend graphs and any of these classes, or even their intersection, no polynomially recognizable class can be sandwiched (unless P=NP).*

Now we shall modify the construction a bit to show a cascade of further results. Note that there are four possible patterns of horizontal paths ($\sqcap$, $\sqcup$, $\urcorner$, $\urcorner$) and another four for vertical paths. As we want to show that it is NP-complete to recognize graphs of any class $X \in \{\sqcap,\sqcup,\urcorner,\urcorner\} \times \{\sqsubset,\sqsupset,\urcorner,\llcorner\}$, we need to start with exploring the symmetries, to classify possible classes X.

So consider a pair or shapes, one of which is horizontal and the other one is vertical. If both legs of each shape bend in the same direction, we obtain the class $\{\sqcap,\sqsupset\}$, which is equivalent to each $\{\sqcup,\sqsupset\}$, $\{\sqcap,\sqsubset\}$, and $\{\sqcup,\sqsubset\}$ (consider a rotation of flipping of an EPG-representation). If both legs of one shape bend in the same direction, and the legs of the other shape bend in opposite directions, we get the class $\{\sqcap,\urcorner\}$ (again, up to symmetry). Finally, if the legs of both shapes bend in opposite directions, we get two possibilities, i.e., $\{\llcorner\lrcorner,\urcorner\}$ (monotonic 2-bend graphs) and $\{\llcorner\lrcorner,\urcorner\lrcorner\}$. Although for the latter two classes we have already shown NP-hardness, now we show yet one construction that works for all four cases. Such a general construction is important from the point of view of sandwiching.

The new construction, in fact, is just a simplified version of the one in the proof of Theorem 1. Again, for a formula φ, we replicate it to obtain φ' (using "quantitative trick") and introduce variable-vertices v_i and occurence-vertices $o_{i,z}$. The difference is that now each clause $z = (i, j, k)$ is represented by just one vertex c_z, adjacent to $o_{i,z}, o_{j,z}$, and $o_{k,z}$ (so we omit vertices d_z, e_z, and f_z). For a formula φ, let us call such constructed graph $G(\varphi')$.

Using this construction we can show that it is NP-complete to recognize X-graphs for each of the pairs X of permitted shapes, one of which is vertical and the other horizontal.

Lemma 1. *It is NP-complete to recognize X-graphs, for any $X \in \{\sqcap,\sqcup,\urcorner,\urcorner\} \times \{\sqsubset,\sqsupset,\urcorner,\llcorner\}$.*

Note that the lemma above shows that, both, an intersection and a union of the mentioned subclasses (as well as anything sandwiched between them) is NP-hard to get recognized. Also, note that it does not show that all classes representable by a given subset of 2-bend shapes (which includes at least one vertical and at least one horizontal shape) are NP-complete to get recognized. It still may happen that there exists such a set X of patterns, that X-graphs can be polynomially recognized. However, we know that if such a class exists, it must not contain even the intersection of $\{\llcorner\lrcorner,\urcorner\}$-graphs and $\{\llcorner\lrcorner,\urcorner\lrcorner\}$-graphs.

Finally, let us try to explore limits of the original hardness reduction for 2-bend graphs (Theorem 1). We know that it works for 2-bend graphs, for $\{\llcorner\lrcorner,\urcorner\}$-graphs, and for $\{\llcorner\lrcorner,\urcorner\lrcorner\}$-graphs (and where the inclusion-relation applies, then also for everything in between). However, we may show that the reduction works also for all triples of 2-bend shapes, in which at least one shape is vertical, at least one is horizontal, and they are not symmetric to the triple $\{\sqsupset,\urcorner,\sqcup\}$, i.e., w.l.o.g., two vertical shapes, one having its legs in the same direction, the other having legs in mutually opposite directions, and the legs of the horizontal one go in the same direction and yet in the direction "towards the common angle" of the other two gadgets. It is easy to observe that the "simplified" construction can be represented, so we need to show, for a particular satisfied clause $z = (i, j, k)$, how to represent vertices d_z, e_z, and f_z. Suppose w.l.o.g. i, j are evaluated true and k is evaluated false. The path P_{c_z} passes through the intersection point of $P_{o_{i,z}}$ and $P_{o_{k,z}}$, and through the intersection point of $P_{o_{j,z}}$ and $P_{o_{k,z}}$. In order to represent d_z (adjacent to $o_{i,z}$ and $o_{j,z}$) we need to use the same intersection-point, i.e., we need the angle obtained from c_z rotated by 180∘. The case analysis shows that this is possible.

As a corollary of the previous statement, the reduction works for all such 4-tuples of 2-bend shapes, where at least one shape is vertical and at least one horizontal (non-trivial situation arises only when extending $\{\neg, \neg, \sqcup\}$). Note also that the reduction works for any k-tuple of 2-bend shapes for $k \geq 5$ (as there are just 4 vertical and 4 horizontal shapes, we are sure that at least one will be horizontal and at least one will be vertical).

Summing up the results from this section, we obtain the following.

Theorem 2. *It is NP-complete to recognize X-graphs, where X is:*

(i) *any of $\{\sqcap, \sqcup, \rightharpoondown, \rightharpoonup\} \times \{\sqsubset, \sqsupset, \neg, \llcorner\}$,*

(ii) *any triple of 2-bend shapes containing at least one vertical and one horizontal shape, and is not symmetric to $\{\sqsupset, \neg, \sqcup\}$.*

(iii) *any 4-tuple of 2-bend shapes, containing at least one horizontal and one vertical shape.*

(iv) *any k-tuple of 2-bend shapes for $k \geq 5$.*

Moreover, one cannot sandwich any polynomially recognizable class between:

(a) *the intersection of $\{\sqcap, \sqcup, \rightharpoondown, \rightharpoonup\} \times \{\sqsubset, \sqsupset, \neg, \llcorner\}$*

(b) *intersection of classes given in (ii),*

and the class of 2-bend graphs.

4   More Slopes

In this section we relax the definition of an EPG-representation. By an *unaligned EPG-representation* of a graph G we mean a mapping from vertices of G to a set of polylines (piecewise linear curves), such that two vertices are adjacent iff their corresponding polylines share infinitely many points. Again, we are interested in keeping the number of bends (or equivalently, segments in a polyline) small.

Here we show hardness of the recognition of unaligned 2-bend graphs and conclude the section with discussion of a trade-off between the number of slopes used and the number of bends.

Theorem 3. *It is NP-hard to recognize unaligned 2-bend graphs.*

Proof. This time we reduce from 3-COLORING. For a graph G we shall construct a graph H, which is an unaligned 2-bend graph iff G is 3-colorable.

The reduction uses ideas similar to the reduction for aligned 2-bend graphs. This time we use 12 service vertices and again we want our gadgets to avoid being represented over the ends of segments of these service vertices, and over their mutual intersection points. So we use the "quantitative trick" again. This time we may have no more than 1 260 special places ($12 \cdot 2 \cdot 3$ ends of segments, $\binom{12}{2} \cdot 9$ possible intersection points, each of which can be used at most twice). Thus we take 1 261 disjoint copies of the graph G, obtaining the graph G' (clearly G' is 3-colorable iff G is 3-colorable).

The main idea of the reduction is that one service vertex of H, named a, simulates the 3-coloring of G'. The individual segments of P_a correspond to three color classes. Each vertex v of G' will be represented by several vertices of H. One of them, called v_2, will have the property that one of the legs of P_{v_2} lies on a segment of P_a (thus defining the color of v in a 3-coloring of G'), and the remaining two segments of P_{v_2} will be fully covered by some other paths, non-adjacent to edge-representatives. An edge uv of G' will be represented by a pair of mutually non-adjacent vertices of H. Both of them will be made adjacent to a and the representatives of both u and v. The main idea is that we cannot construct edge-representatives, if v_2 and u_2 are adjacent to the same segment of a (and thus v and u get the same color). This part of H is illustrated in Fig. 5.

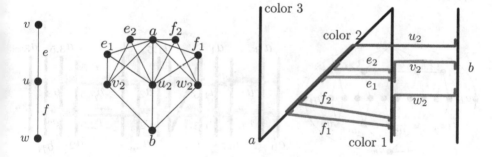

Fig. 5. Left: The graph G. **Middle:** The main part of H. For clarity, just the main vertex-representants are depicted. Also the replication ("quantitative trick") was not performed. **Right:** An unaligned 2-bend representation of H. Note that we are unable to represent the edge vw (having fixed representations of v_2 and u_2).

Formally, the graph H has 12 service vertices $a_0, a_{0.5}, a_1, a_{1.5}, a_2, a_{2.5}, a_3, a_{3.5},$ $a, b, a_B,$ and b_B. For each vertex v of G', we add to H vertices $v_1, v_{1.5}, v_2, v_{2.5}, v_3,$ and v_b (we will call them v-vertices). The vertex v_b is adjacent to all other v-vertices. Furthermore, $v_{1.5}$ is adjacent to $v_1, v_2,$ and $v_{2.5}$ is adjacent to v_2, v_3. Finally, each v-vertex is adjacent to two service vertices: v_1 to $a_0, a_1, v_{1.5}$ to $a_{0.5}, a_{1.5}, v_2$ to $a, b, v_{2.5}$ to $a_{2.5}, a_{3.5}, v_3$ to a_2, a_3. For each edge $e = uv$ we add a pair of mutually non-adjacent vertices e_1, e_2, both adjacent to $a, u_2,$ and v_2.

Suppose we have an unaligned 2-bend representation of H. First, by the "quantitative trick", we know that at least for one copy of G, for any vertex v, all vertices v_i ($i \in \{1, 1.5, 2, 2.5, 3, b\}$) are represented by 2-bend paths having both legs covered by the segments of the appropriate pair of service vertices. Let us focus on this copy of G.

We observe that the body of P_{v_2} (for any v) is covered by (at least) P_{v_b}. This follows from the fact that P_{v_b} can intersect the other v-vertices only by its body (as one leg lies on P_{a_B}, and the second on P_{b_B}). Thus the bodies of $P_{v_1}, P_{v_{1.5}}, \ldots, P_{v_3}, P_{v_b}$ must form an interval representation of $H[\{v_1, v_{1.5}, \ldots, v_3, v_b\}]$ and in no such representation the body of P_{v_2} can exceed the body of P_{v_b}. Therefore the body of P_{v_2} is fully covered by (at least) the body of P_{v_b}.

Now, we are in a desired situation. Consider an edge $e = uv$. For each P_{u_2} and P_{v_2}, only the leg lying on P_a, can be made adjacent to both P_{e_1} and P_{e_2}, as using any other segment would cause some unwanted adjacency. If these legs are on distinct segments of P_a, obviously we can represent both e_1 and e_2. Conversely, if they are on the same segment of P_a, we can represent at most one of them (similarly to Fig. 4 (left)). This shows irrepresentability for a non-3-colorable G.

On the other hand, if G has a 3-coloring, we use it for distributing segments of P_{v_2} of each vertex v over the segments of P_a. Note that we may create a representation, where the bodies of P_{v_2}, for all v, are parallel. Then other v-vertices may be represented in the way shown in Fig. 6. For any edge e, paths P_{e_1} and P_{e_2} connect two non-collinear segments, which can be easily done.   □

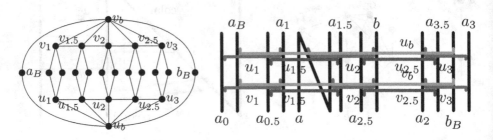

Fig. 6. Left: A graph H for G being an edge uv (replication is omitted). Unlabeled vertices between a_B and b_B are, respectively: $a_0, a_1, a_{0.5}, a_{1.5}, a, b, a_{2.5}, a_{3.5}, a_2, a_3$. **Right:** Unaligned 2-bend representation of H.

4.1   Slopes and Bends

Defining unaligned bend graphs permits us to introduce a new measure of complexity of a representation, i.e., the number of slopes used. There is an obvious trade-off between the number of bends and the number of slopes. Before we explore this relation a little more, let us try to minimize the number of different slopes used by the unaligned 2-bend representation.

Proposition 2. *In order to represent all unaligned 2-bend graphs on n vertices, we need $\Omega(\sqrt{n})$ slopes.*

Proof. The proof follows from the construction in the proof of Theorem 3. Let $G \sim K_{m,m,1}$ be a complete bipartite graph with biparition classes X, Y, both of size m, and one extra vertex z adjacent to all other ones.

We replicate G 1261 times, obtaining G', and construct H in the way described in the proof of Theorem 3. Since G has $2m + 1$ vertices and $\Theta(m^2)$ edges, H has $n = \Theta(m^2)$ vertices.

As G is 3-colorable, H has an unaligned 2-bend representation. As always, we will focus on the "clean" copy of G. Consider the path P_a, and let p, q, r denote its three segments. By the properties of H, w.l.o.g. one leg of every P_{x_2} for $x \in X$ lies on p, while one leg of every P_{y_2} for $y \in Y$ lies on r.

Now consider the paths P_{e_1} (for $e = xy$, $x \in X$, $y \in Y$). There are m^2 such paths. We observe that every slope ℓ can be used by the bodies of at most $2m$ paths P_{e_1}. To see this, we use a sweeping line, parallel to ℓ. As each path P_{e_1} connects a pair of segments of a different pair (P_{x_2}, P_{y_2}), the sweeping line must leave at least one of the segments before meeting a new one. As there are in total $2m$ segments of P_{x_2} or P_{y_2} on P_a, at most $2m$ paths P_{e_1} can have their bodies parallel to ℓ. Thus we need at least $\left\lceil \frac{m^2}{2m} \right\rceil = \Theta(m) = \Theta(\sqrt{n})$ different slopes to represent the bodies of paths P_{e_1}. □

To see a trade-off between the number of bends and the number of slopes, observe that the for $G \sim K_{m,m,1}$, the graph H can be easily represented by 3-bend paths, using only 2 slopes (P_a is represented by a ⊐-shape with segments of P_{v_2} on three different segments of it).

4.2 *d*-bend Number

Let us conclude the section with some generalization of the bend number. Fix a set D of d pairwise non-parallel lines (slopes) containing the origin point. We say that an unaligned EPG-representation is a *EPG(D)-representation* if every segment of each polyline is parallel to some line in D.

The *d-bend number* $b_d(G)$ of a graph G is the minimum k for which there exists a set D of d slopes, such that G has an $EPG(D)$-representation in which every path bends at most k times. We also define $b_\infty(G) := \min_{d \in \mathbb{N}} b_d(G)$, which corresponds to unaligned EPG-representations.

Observe that the 2-bend number is just the classical bend number. It is also straightforward to observe that if $d_1 < d_2$, then $b_{d_1}(G) \geq b_{d_2}(G)$ for all graphs G. Moreover, if there exists $d \in \mathbb{N}$ such that $b_d(G) = 0$, then $b_{d'}(G) = 0$ for all $d' \in \mathbb{N}$ (as this means that G is an interval graph).

As we have seen in Proposition 2, introducing more slopes may help us reduce the number of bends needed to represent a given graph. Here we show two more examples of this. Consider a *wheel graph* W_n on $n+1$ vertices ($n \geq 3$). It follows from the work of Golumbic *et al.* [7] that W_n is not a 1-bend graph (using 2 slopes only) and one can easily find a representation using 2 bends. On the other hand, for $d \geq 3$, we can represent W_n using 1-bend paths (see Fig. 7 (left)). Thus $b_2(W_n) = 2$ and $b_d(W_n) = 1$ for all $d \geq 3$.

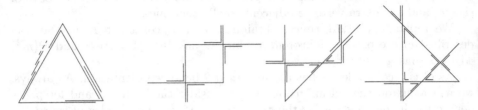

Fig. 7. Left: Representation of a wheel using 1-bend paths. **Right:** Representations of $K_{2,s}$ with 1-bend paths.

Another examples of graphs with bend number depending on the number of slopes are complete bipartite graphs. Consider e.g. a graph $K_{2,s}$. When only 2 slopes are available, then $K_{2,s}$ has a 1-bend representation only for $s \leq 4$. Introducing a third slope allows us to represent $K_{2,5}$ and $K_{2,6}$. Fourth slope allows representing $K_{2,7}$ and $K_{2,8}$. By analyzing the possible intersection points of two 1-bend paths, we observe that $K_{2,s}$ for any $s \geq 9$ does not have a 1-bend representation for any number of slopes. On the other hand, every $K_{2,s}$ is a 2-bend graph on 2 slopes (see Fig. 7 (right) and Fig. 4 (right)).

5   Conclusions and Open Problems

Although all non-trivial classes of EPG-graphs are considered hard for recognition, not much is known. It is an open problem whether the recognition problem remains NP-hard for k-bend graphs (for $k \geq 3$).

Problem 1. Is the recognition of k-bend graphs NP-complete for every fixed $k \geq 1$?
For unaligned bend graphs and aligned bend graphs, using more than 2 slopes, naturally arises the question on inclusions between different classes. Also the complexity of the recognition problem is unknown (for more than 1 bend, when we restrict the number of slopes). Note that none of our reductions can be easily used. The unaligned version increases the number of slopes, while in the aligned version a new slope introduces a new "truth value", but in a way that does not seem to be suitable for a reduction from any form of coloring.

As mentioned before, the CLIQUE problem is polynomially solvable in 1-bend graphs. On the other hand, the problem is shown to be NP-complete in *2-interval* graphs [6]. Since every 2-interval graph is a 3-bend graph and also a 2-bend graph with 3 slopes, we know that the problem is NP-complete is these classes as well. The complexity for 2-bend graphs remains open.

Problem 2. What is the complexity of the CLIQUEproblem is 2-bend graphs?

It is not hard to observe that for any two sets D, D' with $|D| = |D'| = 3$, one can transform an $EPG(D)$-representation of any graph G to its $EPG(D')$-representation. However, it is not clear if the same holds for sets with at least 4 direction of slopes. It is worth mentioning that there are infinitely many classes of intersection graphs of segments, each of which is parallel one of 4 slopes [2]. *Problem 3. Is the minimum number of bends (per path) in an $EPG(D)$-representation of a graph G always equal to $b_d(G)$, for any set D of $d > 3$ slopes?*

Our generalization rises yet further questions. Especially, we may put individual vertices into points with integral coordinates. Now, we may ask, how large grid is necessary and sufficient to represent any graph with n vertices and prescribed number of permitted slopes, or even, with prescribed slopes.

References

1. Booth, K., Lueker, G.: Testing for the consecutive ones property, interval graphs, and planarity using PQ-tree algorithms. J. Comput. Syst. Sci. **13**, 335–389 (1976)
2. Černý, J., Král, D., Nyklová, H., Pangrác, O.: On intersection graphs of segments with prescribed slopes. In: Mutzel, P., Jünger, M., Leipert, S. (eds.) GD 2001. LNCS, vol. 2265, pp. 261–271. Springer, Heidelberg (2002). doi:10.1007/3-540-45848-4_21
3. Clark, B., Colbourn, C., Johnson, D.: Unit disk graphs. Disc. Math. **86**, 165–177 (1990)
4. Cameron, K., Chaplick, S., Hoang, C.T.: Edge intersection graphs of L-shaped paths in grids. Disc. Appl. Math. (2015, in press, available online)
5. Chaplick, S., Jelínek, V., Kratochvíl, J., Vyskočil, T.: Bend-bounded path intersection graphs: sausages, noodles, and waffles on a grill. In: Golumbic, M.C., Stern, M., Levy, A., Morgenstern, G. (eds.) WG 2012. LNCS, vol. 7551, pp. 274–285. Springer, Heidelberg (2012)
6. Francis, M., Gonçalves, D., Ochem, P.: The maximum clique problem in multiple interval graphs. Algoritmica **71**, 812–836 (2015)
7. Golumbic, M., Lipshteyn, M., Stern, M.: Edge intersection graphs of single bend paths on a grid. Networks **54**, 130–138 (2009)
8. Heldt, D., Knauer, K., Ueckerdt, T.: Edge-intersection graphs of grid paths: the bend-number. Disc. Appl. Math. **167**, 144–162 (2014)
9. Lovász, L: Coverings and coloring of hypergraphs. In: Proceedings of the 4th SEIC-CGTC, pp. 3–12 (1973)

Almost Induced Matching: Linear Kernels and Parameterized Algorithms

Mingyu Xiao[1,2(✉)] and Shaowei Kou[1]

[1] School of Computer Science and Engineering, UESTC, Chengdu 611731, China
myxiao@gmail.com, kou_sw@163.com
[2] Center for Information in Medicine, UESTC, Chengdu 611731, China

Abstract. The ALMOST INDUCED MATCHING problem asks whether we can delete at most k vertices from a graph such that the remaining graph is an induced matching, i.e., a graph with each vertex of degree 1. This paper studies parameterized algorithms for this problem by taking the size of deletion set k as the parameter. By using the techniques of finding maximal 3-path packings and an extended crown decomposition, we obtain the first linear vertex kernel for this problem, improving the previous quadratic kernel. We also present an $O^*(1.7485^k)$-time and polynomial-space algorithm, which is the best known parameterized algorithm for this problem.

Keywords: Maximum induced matching · Linear kernel · Parameterized Algorithms · Graph algorithms

1 Introduction

An induced subgraph is called an induced matching if each vertex is a degree-1 vertex in the subgraph. The MAXIMUM INDUCED MATCHING problem, to find an induced matching of maximum size, is an important problem in algorithmic graph theory. Golumbic and Lewenstein [7] demonstrated applications of induced matchings in secure communication channels, VLSI design and network flow problems. Some other applications of induced matchings can be found in [4,8, 17,19,21].

A maximum induced matching can be found in polynomial time in many graph classes, such as trees [7], chordal graphs [2], circular arc graphs [8] and interval graphs [7]. However, MAXIMUM INDUCED MATCHING is NP-hard even in planar 3-regular graphs or planar bipartite graphs with degree-2 vertices in one part and degree-3 vertices in the other part [4,11,19]. The NP-hardness of this problem in Hamiltonian graphs, claw-free graphs, chair-free graphs, line graphs and regular graphs is proved by Kobler and Rotics [12].

This work is supported by the National Natural Science Foundation of China, under grant 61370071, and the Fundamental Research Funds for the Central Universities, under grant ZYGX2015J057.

© Springer-Verlag GmbH Germany 2016
P. Heggernes (Ed.): WG 2016, LNCS 9941, pp. 220–232, 2016.
DOI: 10.1007/978-3-662-53536-3_19

In terms of exact algorithms, Gupta, Raman and Saurabh showed that MAX-IMUM INDUCED MATCHING can be solved in $O^*(1.6957^n)$ time [9]. Xiao and Tan gave an $O^*(1.4391^n)$-time algorithm [21] and further improved the result to $O^*(1.3752^n)$ [22]. Basavaraju et al. showed that all maximal induced matchings in a triangle-free graph can be listed in $O^*(1.4423^n)$ time [1].

In terms of parameterized complexity, the size k' of the induced matching is one of the frequently considered parameters. However, the problem parameter-ized by k' is W[1]-hard in general graphs [16]. Moser and Sidkar showed that it becomes fixed-parameter tractable (FPT) when the graph is a planar graph by giving a linear-size problem kernel [15]. The kernel size was improved to $40k'$ by Kanj et al. [10]. In this paper, we study parameterized algorithms for MAXIMUM INDUCED MATCHING by considering another parameter. We take the number k of vertices not in the induced matching as our parameter. Formally, our problem is defined as follows.

ALMOST INDUCED MATCHING
Instance: A graph $G = (V, E)$ and an integer parameter k.
Question: Is there a vertex subset $S \subseteq V$ of size at most k whose deletion makes the graph an induced matching?

When the size k of the deletion set is taken as the parameter, the problem becomes FPT. Moser and Thilikos gave a kernel of $O(k^3)$ vertices [16], and Mathieson and Szeider improved the kernel size to $O(k^2)$ [14]. In fact, these two papers consider more general problems: to delete at most k vertices to make the remaining graph a regular graph with a constant degree or a graph such that every vertex has a specified degree bounded by a constant. In this paper, we will give the first linear-vertex kernel for ALMOST INDUCED MATCHING. We first find a proper 3-path packing in the graph and partition the vertex set of the graph according to the 3-path packing. Then we use a technique, called "double bi-crown decomposition," to reduce the graph size. Finally, we can get a problem kernel of $8k$ vertices. We also design a parameterized algorithm for ALMOST INDUCED MATCHING, which runs in $O^*(1.7485^k)$ time and polynomial space. Due to space limitation, the proofs of some lemmas are omitted, which can be found in the full version of this paper.

2   Preliminaries

In this paper, we use $G = (V, E)$ to denote a simple and undirected graph with $n = |V|$ vertices and $m = |E|$ edges. A singleton $\{v\}$ may be simply denoted by v. The vertex set and edge set of a graph G' are denoted by $V(G')$ and $E(G')$, respectively. For a vertex subset X, the subgraph induced by X is denoted by $G[X]$, and $G[V \setminus X]$ is also written as $G \setminus X$. A vertex in a subgraph or a vertex subset X is called an X-vertex. Let $N(X)$ denote the set of *open neighbors* of a vertex subset X, i.e., the vertices in $V \setminus X$ adjacent to some vertex in X, and

$N[X] = N(X) \cup X$ denote the set of *closed neighbors* of X. The *degree* of a vertex v in G, denoted by $d(v)$, is defined to be $|N(v)|$, and a vertex in $N(v)$ is called a *neighbor* of v. Two vertex-disjoint subgraphs X_1 and X_2 are *adjacent* if there is an edge with one endpoint in X_1 and the other in X_2. A graph is called an *induced matching* if each vertex in the graph is a degree-1 vertex. A vertex subset S is called an *AIM-deletion set* of G if $G \setminus S$ is an induced matching.

A 3-*path*, denoted by P_3, is a simple path with three different vertices and two edges. Given a graph $G = (V, E)$, a P_3-*packing* $\mathcal{P} = \{L_1, L_2, ..., L_t\}$ of size t is a collection of t vertex-disjoint 3-paths, i.e., each element $L_i \in \mathcal{P}$ is a 3-path and $V(L_{i_1}) \cap V(L_{i_2}) = \emptyset$ for any two different 3-paths $L_{i_1}, L_{i_2} \in \mathcal{P}$. A P_3-packing is *maximal* if it is not properly contained in any strictly larger P_3-packing in G. The set of vertices in 3-paths in $\mathcal{P}$ is denoted by $V(\mathcal{P})$.

For a maximal P_3-packing $\mathcal{P}$, each connected component of the graph induced by $Q = V \setminus V(\mathcal{P})$ is either a single vertex or a single edge [13], which follows immediately from the maximality of $\mathcal{P}$. Let Q_0 be the set of degree-0 vertices in $G[Q]$ and Q_1 be the set of degree-1 vertices in $G[Q]$. A component of two vertices in $G[Q]$ is called a Q_1-*edge*. For each $L_i \in \mathcal{P}$, Let $Q(L_i)$ denote the set of Q-vertices that are in the components of $G[Q]$ adjacent to L_i.

3   Kernelization

In this section, we show that ALMOST INDUCED MATCHING allows a kernel of $8k$ vertices. The main idea of our algorithm is as follows. We first partition the vertex set of G into two parts P and Q, where the size of P is at most $3k$ and Q induces a graph of maximum degree at most 1. Then we use a technique to bound the number of components in the induced graph $G[Q]$ and then bound the size of Q. We will find a maximal P_3-packing $\mathcal{P}$ in G and let $P = V(\mathcal{P})$. Each 3-path must contain at least one vertex in each AIM-deletion set. So the instance is a **yes**-instance if and only if $|\mathcal{P}| \leq k$. Initially we let $\mathcal{P}$ be an arbitrary maximal P_3-packing in G. Then we use two rules to update $\mathcal{P}$. In fact, these two rules contain several rules used in [3,20] to design kernelization algorithms for the P_3-packing problem.

Rule 1. *If there is a 3-path $L_i \in \mathcal{P}$ with $V_i = Q(L_i) \cup V(L_i)$ such that $G[V_i]$ contains at least two vertex-disjoint 3-paths, then replace L_i by these 3-paths in $\mathcal{P}$.*

Some cases of this rule are illustrated in Fig. 1. Each execution of this rule can in implemented in linear time since there are constant number of different configurations. Each execution will increase the number of 3-paths in $\mathcal{P}$ by at least 1, where the size of 3-path packing $\mathcal{P}$ is a lower bound for k.

Rule 2. *If there is a 3-path $L_i \in \mathcal{P}$ with two vertices adjacent to two different Q_0-vertices respectively, then use the operations in Fig. 2 to replace L_i with another 3-path in $\mathcal{P}$.*

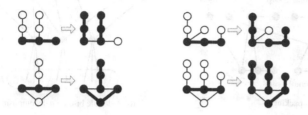

Fig. 1. Some cases of Rule 1: the 3-paths with dark edges mean the 3-paths in $\mathcal{P}$

Fig. 2. The operations of Rule 2: the 3-paths with dark edges mean the 3-paths in $\mathcal{P}$

It is not hard to see that each execution of Rule 2 will increase the number of Q_1-edges by at least 1. We will apply Rule 2 only when Rule 1 is not applicable on the graph. Note that in a graph where Rule 1 is not applicable, after each application of Rule 2, the P_3-packing $\mathcal{P}$ is still maximal.

A maximal P_3-packing $\mathcal{P}$ is *proper* if neither Rule 1 nor Rule 2 is applicable. We have the following properties for proper P_3-packings.

Lemma 1. *For any initially maximal P_3-packing $\mathcal{P}$ in G, we can apply Rules 1 and 2 for less than $2n$ times to change $\mathcal{P}$ to a proper P_3-packing.*

Lemma 2. *Let $\mathcal{P}$ be a proper P_3-packing in G and L_i be an arbitrary 3-path in $\mathcal{P}$. It holds that:*

(a) *If more than one Q_0-vertex is adjacent to L_i, then all these Q_0-vertices are adjacent to the same and unique vertex in L_i;*

(b) *If more than one Q_1-edge is adjacent to L_i, then all these Q_1-edges are adjacent to the same and unique vertex in L_i;*

(c) *If more than one vertex in L_i is adjacent to Q_0-vertices, then all these vertices in L_i are adjacent to the same and unique Q_0-vertex;*

(d) *If more than one vertex in L_i is adjacent to Q_1-edges, then all these vertices in L_i are adjacent to the same and unique Q_1-edge.*

It is easy to verify Lemma 2: (a) and (c) hold because Rule 2 is not applicable, and (b) and (d) hold because Rule 1 is not applicable.

For a proper P_3-packing $\mathcal{P}$, a 3-path $L_i \in \mathcal{P}$ is *crucial* if exactly one vertex in L_i is adjacent to some Q_1-edge, a 3-path $L_i \in \mathcal{P}$ is *normal* if it is not crucial, a Q_1-edge is *crucial* if it is adjacent to only crucial 3-paths in $\mathcal{P}$, and a Q_1-edge is

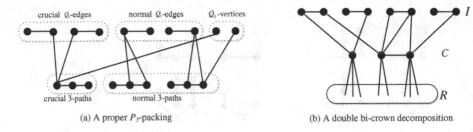

(a) A proper P_3-packing   (b) A double bi-crown decomposition

Fig. 3. Two decompositions

normal if it is adjacent to at least one normal 3-path in $\mathcal{P}$. A vertex in a crucial (resp., normal) Q_1-edge is also called a *crucial* (resp., *normal*) Q_1-*vertex*. See Fig. 3(a) for an illustration of these notations. By Lemma 2, we know that

Lemma 3. *Given a proper P_3-packing $\mathcal{P}$, the number of normal Q_1-edges is at most the number of normal 3-paths in $\mathcal{P}$.*

Next we are going to bound the number of vertices in $Q = V \setminus V(\mathcal{P})$ based on a proper P_3-packing $\mathcal{P}$. We can bound the number of Q_0-vertices directly by using the following lemma.

Lemma 4. *If there is an AIM-deletion set of size at most k in G, then there are at most k components of a single vertex in $G \setminus V(\mathcal{P})$ for any proper P_3-packing $\mathcal{P}$ in G.*

Proof. Assume that $S^* \subseteq V$ is a minimum AIM-deletion set of G. Recall that Q_0 is the set of vertices in components of a single vertex in $G \setminus V(\mathcal{P})$. We only need to prove that $|S^*| \geq |Q_0|$.

Let $X_1 = Q_0 \cap S^*$ and $X_2 = Q_0 \setminus X_1$. In $G \setminus S^*$, each vertex $v_i \in X_2$ has exactly one neighbor u_i since each vertex in $G \setminus S^*$ has degree 1, and u_i is a vertex in $V(\mathcal{P})$ since v_i is a Q_0-vertex. Let $L_i \in \mathcal{P}$ denote the 3-path containing u_i. Since u_i is also a degree-1 vertex in $G \setminus S^*$, we know that at least one vertex in L_i is in S^*, which is a neighbor of u_i. For any two different vertices $v_i, v_j \in X_2$, the two neighbors u_i and u_j of them in $G \setminus S^*$ are different, otherwise $G \setminus S^*$ would have a vertex of degree 2 and it would not be an induced matching. Furthermore, u_i and u_j are in two different 3-paths L_i and L_j in $\mathcal{P}$ by Lemma 2(a) and (c). Thus, for each vertex $v_i \in X_2$, there is a corresponding vertex in a 3-path L_i that is in S^*, and for any two vertices in X_2, the two corresponding vertices in S^* are different. We know that $|S^* \setminus Q_0| \geq |X_2|$, which implies $|S^*| = |S^* \setminus Q_0| + |X_1| \geq |X_2| + |X_1| = |Q_0|$. □

Lemma 4 is not enough to get a kernel for our problem. We still need to bound the number of Q_1-vertices. Lemma 3 gives a good bound on the number of normal Q_1-edges. We only need to bound the number of crucial Q_1-edges. To do this, we use a technique called "double bi-crown decomposition". Let A and B be two disjoint vertex subsets. An edge set M between A and B is called a

2-*full matching from* A *to* B if each vertex in A is in exactly two edges in M and each vertex in B is in at most one edge in M.

Definition 1. *A double bi-crown decomposition of a graph G is a decomposition* (I, C, R) *of the vertex set of G such that*

1. *there is no edge between I and R;*
2. *each connected component of the induced subgraph $G[I]$ has exactly two vertices;*
3. *there is a 2-full matching M from C to I.*

Figure 3(b) illustrates a double bi-crown decomposition. Note that our double bi-crown decomposition is different from the double crown decomposition defined in [18], which requires $G[I]$ to be an independent set. We have the following lemma for double bi-crown decompositions.

Lemma 5. *Let (I, C, R) be a double bi-crown decomposition of a graph G. For any minimum AIM-deletion set K of the induced subgraph $G[R]$, $S = K \cup C$ is a minimum AIM-deletion set of G.*

Based on Lemma 5, given a double bi-crown decomposition (I, C, R) of the graph we can reduce the instance by including vertex set C to the deletion set and deleting it from the graph. We will show that when the graph satisfies some properties the graph always allows a double bi-crown decomposition and it can be computed in polynomial time.

Lemma 6. *Let $G = (V, E)$ be a graph with each component containing more than two vertices and (A, B, D) be a decomposition of V such that (i) no vertex in A is adjacent to a vertex in D; (ii) each connected component in the induced subgraph $G[A]$ has exactly two vertices. If $|A| \geq 4|B|$, then the graph allows a double bi-crown decomposition (I, C, R) with $\emptyset \neq I \subseteq A$ and $\emptyset \neq C \subseteq B$, and (I, C, R) can be computed in $O(\sqrt{n}m)$ time.*

We construct an algorithm to prove this lemma. The main idea of the algorithm is as follows. Let $A' \subseteq A$ be the set of vertices in A that are adjacent to some vertices in B. Our algorithm first finds a maximum edge set M between A' and B with the constraints that each vertex in A' appears in at most one edge in M and each vertex in B appears in at most two edges in M. The edge set M can be computed in $O(\sqrt{n}m)$ time by finding a certain maximum flow between A' and B: We add a vertex s adjacent to each vertex in B via an edge with capacity 2, and add a vertex t adjacent to each vertex in A' via an edge with capacity 1. The capacity of each edge between A' and B is 1. We only need to find a maximum flow from s to t. By using the algorithm in [6] we can solve the maximum flow problem in $O(\sqrt{n}m)$ time. Next, our algorithm computes sets I and C based on M. A vertex in A' is M-*unsaturated* if it does not appear in any edge in M. A path with all edges between A' and B is called an M-*alternating path* if it alternates between edges not in M and edges in M. We will let I' be the set of vertices in A' that are reachable from an M-unsaturated vertex via

---

Input: A graph $G = (V, E)$ with each component containing more than two vertices and a partition (A, B, D) of the vertex set V satisfying the conditions in Lemma 6.
Output: Two sets $I \subseteq A$ and $C \subseteq B$ such that $(I, C, V \setminus (I \cup C))$ is a double bi-crown decomposition.

1. Let $A' = N(B) \cap A$.
2. Compute a maximum edge set M between A' and B with the constraints that each vertex in A' appears in at most one edge in M and each vertex in B appears in at most two edges in M.
3. Let I' be $\emptyset$ if there is no M-unsaturated vertex, and the set of vertices in A' connected with at least one M-unsaturated vertex via an M-alternating path (together with all M-unsaturated vertices) otherwise.
   Let $I = I' \cup (N(I') \cap A)$.
   Let C be $\emptyset$ if there is no M-unsaturated vertex, and the set of vertices in B connected with at least one M-unsaturated vertex via an M-alternating path otherwise.
4. Return (I, C).

Fig. 4. Algorithm $\mathtt{decomp}(G, A, B, D)$

an M-alternating path (I' also includes all M-unsaturated vertices), I be the set of vertices in components of $G[A]$ containing vertices in I', and C be the set of vertices in B that are reachable from an M-unsaturated vertex via an M-alternating path. We will prove that $(I, C, R = V \setminus (I \cup C))$ is a double bi-crown decomposition. The detailed steps of the algorithm is presented in Fig. 4.

Lemma 7. *Algorithm* $\mathtt{decomp}(G, A, B, D)$ *runs in* $O(\sqrt{n}m)$ *time and returns two vertex subsets* $I \subseteq A$ *and* $C \subseteq B$ *such that*

(a) $(I, C, V \setminus (I \cup C))$ *is a double bi-crown decomposition of* G;
(b) $|A \setminus I| \leq 4|B \setminus C|$.

Lemma 7 implies Lemma 6. Our algorithm will first compute a vertex partition (A, B, D) satisfying the condition in Lemma 6 based on a proper P_3-packing. We will let A be the set of crucial Q_1-vertices and $B = N(A)$. Then we use $\mathtt{decomp}(G, A, B, D)$ to find a double bi-crown decomposition. The whole algorithm is described in Fig. 5.

Based on the above analysis, we can get the following result, the full proof of which can be found in the full version of this paper.

Lemma 8. *Algorithm* $\mathtt{kernel}(G, k)$ *runs in* $O(n(n + m))$ *time and returns an equivalent instance with at most* $8k$ *vertices.*

4   A Parameterized Algorithm

In this section we will design a parameterized algorithm for ALMOST INDUCED MATCHING. Our algorithm is a branch-and-reduce algorithm that runs in $O^*(1.7485^k)$ time and polynomial space. In branch-and-reduce algorithms, the

Input: An undirected graph $G = (V, E)$ and a non-negative integer k.
Output: a graph G' and an integer k' such that $k' \leq k$ and G' has at most $8k$ vertices, or 'no' to indicate that the instance is a **no**-instance.

1. Delete any connected component of two vertices from the graph.
2. Let V_0 be the set of single vertices in G. Let $G = G \setminus V_0$ and $k = k - |V_0|$.
3. Find an arbitrary maximal P_3-packing $\mathcal{P}$ in G
4. Iteratively apply Rules 1 and 2 to update $\mathcal{P}$ until none of them can be applied anymore.
5. **If** $|\mathcal{P}| > k$, **then** return 'no' and halt.
6. let $P = V(\mathcal{P})$ and $Q = V \setminus P$.
7. **If** $|Q_0| > k$, **then** return 'no' and halt.
8. Let A be the set of crucial Q_1-vertices and $B = N(A)$.
9. Let $(I, C) = \texttt{decomp}(G, A, B, V \setminus (A \cup B))$.
10. Return $G' = G \setminus (I \cup C)$ and $k' = k - |C|$.

Fig. 5. Algorithm $\texttt{kernel}(G, k)$

exponential part of the running time is determined by the branching operations in the algorithm. In a branching operation, the algorithm solves the current instance I by solving several smaller instances. We will use the parameter k as the measure of the instance and use $T(k)$ to denote the maximum size of the search tree generated by the algorithm running on any instance with parameter at most k. Clearly, when $k \leq 0$ we can solve the instance in linear time. If a branching operation generates l branches and the measure k in the i-th instance decreases by at least c_i, then this operation creates a recurrence relation

$$T(k) \leq T(k - c_1) + T(k - c_2) + \cdots + T(k - c_l) + 1.$$

The largest root of the function $f(x) = 1 - \sum_{i=1}^{l} x^{-c_i}$ is called the *branching factor* of the recurrence. Let γ be the maximum branching factor among all branching factors in the algorithm. The running time of the algorithm is bounded by $O^*(\gamma^k)$. More details about the analysis and how to solve recurrences can be found in the monograph [5].

4.1 Branching Rules

For any vertex v, it is either deleted (included to the deletion set) or remained in the induced matching. We get a simple branching rule

Branching rule (B1): *Branch on v to generate $|N[v]|$ branches by either (i) deleting v from the graph and including it to the deletion set, or (ii) for each neighbor u of v, deleting $N[\{u, v\}]$ from the graph and including $N(\{u, v\})$ to the deletion set.*

This branching rule may not always be effective. To obtain a better running time bound, we also use different branching rules for some special graph structures.

A vertex v is *dominated* by a neighbor u of it if v is adjacent to all neighbors of u. The following property of dominated vertices has been proved and used in [9, 21].

Lemma 9. *Let v be a vertex dominated by u. If there is a maximum induced matching M of G such that $v \in V(M)$, then there is a maximum induced matching M' of G such that $v, u \in V(M)$.*

Based on this lemma, we design the following branching rule.

Branching rule (B2): *Branch on a vertex v dominated by another vertex u to generate two instances by either (i) deleting v from the graph and including it to the deletion set, or (ii) deleting $N[\{u, v\}]$ from the graph and including $N(\{u, v\})$ to the deletion set.*

4.2   The Algorithm

We will use $\mathtt{aim}(G, k)$ to denote our parameterized algorithm. The algorithm contains 7 steps. When we execute one step, we assume that all previous steps are not applicable anymore on the current graph. We will analyze each step after describing it.

Step 1 (Trivial cases). *If $k < 0$ or the graph is an empty graph, then return the result directly. If the graph has a component of maximum degree 2, then solve this component directly in linear time.*

After Step 1, each component of the graph contains at least three vertices. A degree-1 vertex v is called a *tail* if its neighbor u is a degree-2 vertex. We have the following property for tails.

Lemma 10. *If a graph G has a tail v, then there is a maximum induced matching of G containing the unique edge incident on v.*

Proof. Let u be the degree-2 neighbor of the tail v, and w be the other neighbor of u. If an edge e incident on w is in a maximum induced matching M, then the edge vu cannot be in M. For this case, we can replace e with vu in M to get a maximum induced matching containing vu. If no edge incident on w is in a maximum induced matching M, then by the maximality of M we know that vu is in M. So there is always a maximum induced matching containing vu.   □

Step 2 (Tails). *If there is a degree-1 vertex v with a degree-2 neighbor u, then return $\mathtt{aim}(G \setminus N[\{v, u\}], k - 1)$.*

Step 3 (Dominated vertices of degree ≥ 3). *If there is a vertex v of degree ≥ 3 dominated by u, then branch on v with Rule (B2) to generate two branches*

$$\mathtt{aim}(G \setminus \{v\}, k - 1) \quad and \quad \mathtt{aim}(G \setminus N[\{v, u\}], k - |N(\{v, u\})|).$$

Lemma 9 guarantees the correctness of this step. Note that $|N(\{v, u\})| = d(v)-1$. This step generates a recurrence

$$T(k) \le T(k-1) + T(k - (d(v) - 1)) + 1, \tag{1}$$

where $d(v) \ge 3$. For the worst case that $d(v) = 3$, the branching factor of it is 1.6181.

After Step 2 the remaining graph may still have degree-1 vertices with a neighbor of degree ≥ 3. These vertices will be handled in Step 3. So after Step 3, the graph has no vertices of degree ≤ 1.

Next, we consider degree-2 vertices. A path $u_0 u_1 u_2 u_3 u_4$ of five vertices is called a *chain* if the first vertex u_0 is of degree ≥ 3 and the three middle vertices are of degree 2, where we allow $u_4 = u_0$. A path $u_0 u_1 u_2 u_3$ of four vertices is called a *short chain* if the first vertex u_0 and last vertex u_3 are of degree ≥ 3 and the two middle vertices are of degree 2, where we allow $u_3 = u_0$. A chain or a short chain can be found in linear time if it exists.

Step 4 (Chains). *If there is a chain $u_0 u_1 u_2 u_3 u_4$, then branch on u_1 with Rule (B1). In the branch where u_1 is deleted and included to the deletion set, we get a tail u_2 and then further deal with the tail as what we do in Step 2. Then we get the following three branches*

$$\mathrm{aim}(G \setminus \{u_1, u_2, u_3, u_4\}, k-2), \quad \mathrm{aim}(G \setminus N[\{u_0, u_1\}], k - |N(\{u_0, u_1\})|)$$
$$and \quad \mathrm{aim}(G \setminus N[\{u_1, u_2\}], k - |N(\{u_1, u_2\})|).$$

The corresponding recurrence is

$$T(k) \le T(k-2) + T(k - d(u_0)) + T(k-2) + 1,$$

where $d(u_0) \ge 3$. For the worst case that $d(u_0) = 3$, the branching factor of it is 1.6181.

Step 5 (Short chains). *If there is a short chain $u_0 u_1 u_2 u_3$, then branch on u_1 with Rule (B1). In the branch where u_1 is deleted and included to the deletion set, we get a dominated vertex u_3 and then further branch on u_3 with Rule (B2). We get the following four branches*

$$\mathrm{aim}(G \setminus \{u_1, u_2, u_3\}, k-3), \mathrm{aim}(G \setminus N[\{u_2, u_3\}], k - |N(\{u_2, u_3\})|),$$
$$\mathrm{aim}(G \setminus N[\{u_0, u_1\}], k - |N(\{u_0, u_1\})|), and \ \mathrm{aim}(G \setminus N[\{u_1, u_2\}], k - |N(\{u_1, u_2\})|).$$

The corresponding recurrence is

$$T(k) \le T(k-3) + T(k - d(u_3)) + T(k - d(u_0)) + T(k-2) + 1,$$

where $d(u_0), d(u_3) \ge 3$. For the worst case that $d(u_0) = d(u_3) = 3$, the branching factor of it is 1.6717.

After Step 5, each degree-2 vertex in the graph has two neighbors of degree ≥ 3. Furthermore, the two neighbors of it are not adjacent to each other, since otherwise they would be dominated vertices of degree ≥ 3 and Step 3 would be applied.

Step 6 (Vertices of degree $\neq 3$). *If there is a vertex v of $d(v) \neq 3$, then branch on v with Rule (B1) to generate $d(v) + 1$ branches*

$\text{aim}(G \setminus \{v\}, k - 1)$ *and* $\text{aim}(G \setminus N[\{v, u\}], k - |N(\{v, u\})|)$ *for each $u \in N(v)$.*

We analyze the corresponding recurrence by considering the degree or v is 2 or not.

Assume that $d(v) = 2$. Let u_1 and u_2 denote the two neighbors of v. Then u_1 and u_2 are nonadjacent vertices of degree ≥ 3. The branching operation will generate three branches. Since u_1 and u_2 are not adjacent, we can see that $|N(v, u_i)| = d(u_i) - 1 + 1 = d(u_i) \geq 3$ $(i = 1, 2)$. This leads to a recurrence

$$T(k) \leq T(k - 1) + T(k - d(u_1)) + T(k - d(u_2)) + 1,$$

where $d(u_1), d(u_2) \geq 3$. For the worst case that $d(u_1) = d(u_2) = 3$, the branching factor of it is 1.6957.

Assume that $d(v) \geq 4$. Since v a vertex of degree ≥ 3 and then it can not be dominated by any neighbor of it now, we know that each neighbor u of it is adjacent to a vertex not in $N[v]$ and then $|N(v, u)| \geq d(v)$. So we get a recurrence

$$T(k) \leq T(k - 1) + d(v) \cdot T(k - d(v)) + 1,$$

where $d(v) \geq 4$. For the worst case that $d(v) = 4$, the branching factor of 1.7485.

After Step 6, if the graph is not an empty graph, then the graph can only be a 3-regular graph.

Step 7 (3-regular graphs). *Pick up an arbitrary vertex v and branch on it with Rule (B1).*

We do not analyze the branching factor for this step, because this step will not exponentially increase the running time bound of the algorithm. Note that any proper subgraph of a connected 3-regular graph is not a 3-regular graph. For each connected component of a 3-regular graph, Step 7 can be applied for at most one time and all other branching operations have a branching factor of at most 1.6957. Then each connected component of a 3-regular graph can be solved in $O^*(1.6957^k)$ time. Before getting a connected component of a 3-regular graph, the algorithm can always branch with branching factors of at most 1.7485. Therefore, we have the following result.

Theorem 1. ALMOST INDUCED MATCHING *can be solved in $O^*(1.7485^k)$ time and polynomial space.*

5   Concluding Remarks

In this paper, we give the first linear kernel and a fast parameterized algorithm for ALMOST INDUCED MATCHING. ALMOST INDUCED MATCHING is a

special case of REGULAR INDUCED SUBGRAPH and DEGREE-SPECIFIED VER-
TEX DELETION, which are going to check whether we can delete k vertices from
a graph to make the remaining graph a d-regular graph or a graph such that
every vertex has a specified degree bounded by d. There is a quadratic kernel for
DEGREE-SPECIFIED VERTEX DELETION for each constant $d \geq 0$ [14]. Whether
DEGREE-SPECIFIED VERTEX DELETION (even REGULAR INDUCED SUBGRAPH)
allows a linear kernel for each fixed $d \geq 2$ is still unknown. The techniques in this
paper cannot be used to solve this general problem directly. The main reason is
that we do not find a decomposition similar to Definition 1 and a lemma similar
to Lemma 5 for general d.

References

1. Basavaraju, M., Heggernes, P., Saei, R., Villanger, Y.: Maximal induced matchings in triangle-free graphs. J. Graph Theor. (2015). doi:10.1002/jgt.21994
2. Cameron, K.: Induced matchings. Discrete Appl. Math. **24**(1–3), 97–102 (1989)
3. Chen, J., Fernau, H., Shaw, P., Wang, J., Yang, Z.: Kernels for packing and covering problems. In: Snoeyink, J., Lu, P., Su, K., Wang, L. (eds.) AAIM 2012 and FAW 2012. LNCS, vol. 7285, pp. 199–211. Springer, Heidelberg (2012)
4. Duckworth, W., Manlove, D., Zito, M.: On the approximability of the maximum induced matching problem. J. Discrete Algorithms **3**(1), 70–91 (2005)
5. Fomin, F.V., Kratsch, D.: Exact Exponential Algorithms. Springer, Heidelberg (2010)
6. Goldberg, A.V., Kaplan, H., Hed, S., Tarjan, R.E.: Minimum cost flows in graphs with unit capacites. In: STACS 2015, LIPIcs 30, Dagstuhl, Germany, pp. 406–419 (2015)
7. Golumbic, M.C., Lewenstein, M.: New results on induced matchings. Discrete Appl. Math. **101**(1–3), 157–165 (2000)
8. Golumbic, M.C., Laskar, R.: Irredundancy in circular arc graphs. Discrete Appl. Math. **44**(1–3), 79–89 (1993)
9. Gupta, S., Raman, V., Saurabh, S.: Maximum r-regular induced subgraph problem: fast exponential algorithms and combinatorial bounds. SIAM J. Discrete Math. **26**(4), 1758–1780 (2012)
10. Kanj, I.A., Pelsmajer, M.J., Schaefer, M., Xia, G.: On the induced matching problem. J. Comput. Syst. Sci. **77**(6), 1058–1070 (2011)
11. Ko, C.W., Shepherd, F.B.: Bipartite domination and simultaneous matroid covers. SIAM J. Discrete Math. **16**(4), 517–523 (2003)
12. Kobler, D., Rotics, U.: Finding maximum induced matchings in subclasses of claw-free and P5-free graphs, and in graphs with matching and induced matching of equal maximum size. Algorithmica **37**(4), 327–346 (2003)
13. Koutis, I.: Faster algebraic algorithms for path and packing problems. In: Aceto, L., Damgård, I., Goldberg, L.A., Halldórsson, M.M., Ingólfsdóttir, A., Walukiewicz, I. (eds.) ICALP 2008, Part I. LNCS, vol. 5125, pp. 575–586. Springer, Heidelberg (2008)
14. Mathieson, L., Szeider, S.: Editing graphs to satisfy degree constraints: a parameterized approach. J. Comput. Syst. Sci. **78**(1), 179–191 (2012)
15. Moser, H., Sikdar, S.: The parameterized complexity of the induced matching problem. Discrete Appl. Math. **157**(4), 715–727 (2009)

16. Moser, H., Thilikos, D.M.: Parameterized complexity of finding regular induced subgraphs. J. Discrete Algorithms **7**(2), 181–190 (2009)
17. Orlovich, Y.L., Finke, G., Gordon, V.S., Zverovich, I.E.: Approximability results for the maximum and minimum maximal induced matching problems. Discrete Optimaization **5**(3), 584–593 (2008)
18. Prieto, E., Sloper, C.: Looking at the stars. Theor. Comput. Sci. **351**(3), 437–445 (2006)
19. Stockmeyer, L.J., Vazirani, V.V.: NP-completness of some generalizations of the maximum matching problem. Inf. Proc. Lett. **15**(1), 14–19 (1982)
20. Wang, J., Ning, D., Feng, Q., Chen, J.: An improved kernelization for P_2-packing. Inf. Proc. Lett. **110**(5), 188–192 (2010)
21. Xiao, M., Tan, H.: An improved exact algorithm for maximum induced matching. In: Jain, R., Jain, S., Stephan, F. (eds.) TAMC 2015. LNCS, vol. 9076, pp. 272–283. Springer, Heidelberg (2015)
22. Xiao, M., Tan, H.: Exact Algorithms for Maximum Induced Matching (2016, to appear)

Parameterized Vertex Deletion Problems for Hereditary Graph Classes with a Block Property

Édouard Bonnet, Nick Brettell[(⊠)], O-joung Kwon, and Dániel Marx

Institute for Computer Science and Control,
Hungarian Academy of Sciences (MTA SZTAKI), Budapest, Hungary
edouard.bonnet@dauphine.fr, nbrettell@gmail.com, ojoungkwon@gmail.com,
dmarx@cs.bme.hu

Abstract. For a class of graphs $\mathcal{P}$, the BOUNDED $\mathcal{P}$-BLOCK VERTEX DELETION problem asks, given a graph G on n vertices and positive integers k and d, whether there is a set S of at most k vertices such that each block of $G - S$ has at most d vertices and is in $\mathcal{P}$. We show that when $\mathcal{P}$ satisfies a natural hereditary property and is recognizable in polynomial time, BOUNDED $\mathcal{P}$-BLOCK VERTEX DELETION can be solved in time $2^{\mathcal{O}(k \log d)} n^{\mathcal{O}(1)}$, and this running time cannot be improved to $2^{o(k \log d)} n^{\mathcal{O}(1)}$, in general, unless the Exponential Time Hypothesis fails. On the other hand, if $\mathcal{P}$ consists of only complete graphs, or only K_1, K_2, and cycle graphs, then BOUNDED $\mathcal{P}$-BLOCK VERTEX DELETION admits a $c^k n^{\mathcal{O}(1)}$-time algorithm for some constant c independent of d. We also show that BOUNDED $\mathcal{P}$-BLOCK VERTEX DELETION admits a kernel with $\mathcal{O}(k^2 d^7)$ vertices.

1 Introduction

Vertex deletion problems are formulated as follows: given a graph G and a class of graphs $\mathcal{G}$, is there a set of at most k vertices whose deletion transforms G into a graph in $\mathcal{G}$? A graph class $\mathcal{G}$ is *hereditary* if whenever G is in $\mathcal{G}$, every induced subgraph H of G is also in $\mathcal{G}$. Lewis and Yannakakis [14] proved that for every non-trivial hereditary graph class decidable in polynomial time, the vertex deletion problem for this class is NP-complete. On the other hand, a class is hereditary if and only if it can be characterized by a set of forbidden induced subgraphs $\mathcal{F}$, and Cai [2] showed that if $\mathcal{F}$ is finite, with each graph in $\mathcal{F}$ having at most c vertices, then there is an $\mathcal{O}(c^k n^{c+1})$-time algorithm for the corresponding vertex deletion problem.

A *block* of a graph is a maximal connected subgraph B such that B has no cut vertices. Every maximal 2-connected subgraph is a block, but a block may just consist of one or two vertices. We consider vertex deletion problems for hereditary graph classes where all blocks of a graph in the class satisfy a certain

---

All authors are supported by ERC Starting Grant PARAMTIGHT (No. 280152).

P. Heggernes (Ed.): WG 2016, LNCS 9941, pp. 233–244, 2016.
DOI: 10.1007/978-3-662-53536-3_20

common property. It is natural to describe such a class by the set of permissible blocks $\mathcal{P}$. For ease of notation, we do not require that $\mathcal{P}$ is itself hereditary, but the resulting class, where graphs consist of blocks in $\mathcal{P}$, should be. To achieve this, we say that a class of graphs $\mathcal{P}$ is *block-hereditary* if, whenever G is in $\mathcal{P}$ and H is an induced subgraph of G, every block of H is isomorphic to a graph in $\mathcal{P}$. For a block-hereditary class of graphs $\mathcal{P}$, we define $\Phi_{\mathcal{P}}$ as the class of all graphs whose blocks are in $\mathcal{P}$. Several well-known graph classes can be defined in this way. For instance, a *forest* is a graph in the class $\Phi_{\{K_1,K_2\}}$, a *cactus graph* is a graph in the class $\Phi_{\mathcal{C}}$ where $\mathcal{C}$ consists of K_1, K_2 and all cycles, and a *complete-block graph*[1] is a graph in $\Phi_{\mathcal{K}}$ where $\mathcal{K}$ consists of all complete graphs. We note that $\mathcal{C}$ is not a hereditary class, but it is block-hereditary; this is what motivates our use of the term.

Let $\mathcal{P}$ be a block-hereditary class such that $\Phi_{\mathcal{P}}$ is a non-trivial hereditary class. The result of Lewis and Yannakakis [14] implies that the vertex deletion problem for $\Phi_{\mathcal{P}}$ is NP-complete. We define the following parameterized problem for a fixed block-hereditary class of graphs $\mathcal{P}$.

$\mathcal{P}$-BLOCK VERTEX DELETION   **Parameter:** k
Input: A graph G and a non-negative integer k.
Question: Is there a set $S \subseteq V(G)$ with $|S| \leqslant k$ such that each block of $G - S$ is in $\mathcal{P}$?

This problem generalizes the well-studied parameterized problems VERTEX COVER, when $\mathcal{P} = \{K_1\}$, and FEEDBACK VERTEX SET, when $\mathcal{P} = \{K_1, K_2\}$. Moreover, if $\Phi_{\mathcal{P}}$ can be characterized by a finite set of forbidden induced subgraphs, then Cai's approach [2] can be used to obtain a fixed-parameter tractable (FPT) algorithm that runs in time $2^{\mathcal{O}(k)} n^{\mathcal{O}(1)}$.

In this paper, we are primarily interested in the variant of this problem where, additionally, the number of vertices in each block is at most d. The value d is a parameter given in the input.

BOUNDED $\mathcal{P}$-BLOCK VERTEX DELETION   **Parameter:** d, k
Input: A graph G, a positive integer d, and a non-negative integer k.
Question: Is there a set $S \subseteq V(G)$ with $|S| \leqslant k$ such that each block of $G - S$ has at most d vertices and is in $\mathcal{P}$?

We also consider this problem when parameterized only by k. When $d = 1$, this problem is equivalent to VERTEX COVER. This implies that the BOUNDED $\mathcal{P}$-BLOCK VERTEX DELETION problem is para-NP-hard when parameterized only by d.

The BOUNDED $\mathcal{P}$-BLOCK VERTEX DELETION problem is also equivalent to VERTEX COVER when $\mathcal{P}$ is a class of edgeless graphs. Since VERTEX COVER is well studied, we assume that $d \geqslant 2$, and focus on classes that contain a graph

---

[1] A *block graph* is the usual name in the literature for a graph where each block is a complete subgraph. However, since we are dealing here with both blocks and block graphs, to avoid confusion we instead use the term *complete-block graph* and call the corresponding vertex deletion problem COMPLETE BLOCK VERTEX DELETION.

with at least one edge. We call such a class *non-degenerate*. When $\mathcal{P}$ is the class of all graphs, we refer to BOUNDED $\mathcal{P}$-BLOCK VERTEX DELETION as BOUNDED BLOCK VD.

Related Work. The analogue of BOUNDED BLOCK VD for connected components, rather than blocks, is known as COMPONENT ORDER CONNECTIVITY. For this problem, the question is whether a given graph G has a set of vertices S of size at most k such that each connected component of $G - S$ has at most d vertices. Drange et al. [3] showed that COMPONENT ORDER CONNECTIVITY is $W[1]$-hard when parameterized by k or by d, but FPT when parameterized by $k + d$, with an algorithm running in $2^{\mathcal{O}(k \log d)}n$ time.

Clearly, the vertex deletion problem for either cactus graphs, or complete-block graphs, is a specialization of $\mathcal{P}$-BLOCK VERTEX DELETION. A graph is a cactus graph if and only if it does not contain a subdivision of the *diamond* [4], the graph obtained by removing an edge from the complete graph on four vertices. For this reason, the problem for cactus graphs is known as DIAMOND HITTING SET. For complete-block graphs, we call it COMPLETE BLOCK VERTEX DELETION. General results imply that there is a $c^k n^{\mathcal{O}(1)}$-time algorithm for DIAMOND HITTING SET [6,9,11], but an exact value for c is not forthcoming from these approaches. However, Kolay et al. [13] obtained a $12^k n^{\mathcal{O}(1)}$-time randomized algorithm. For the variant where each cycle must additionally be odd (that is, $\mathcal{P}$ consists of K_1, K_2, and all odd cycles), there is a $50^k n^{\mathcal{O}(1)}$-time deterministic algorithm due to Misra et al. [15]. For COMPLETE BLOCK VERTEX DELETION, Kim and Kwon [10] showed that there is an algorithm that runs in $10^k n^{\mathcal{O}(1)}$ time, and there is a kernel with $\mathcal{O}(k^6)$ vertices. Agrawal et al. [1] improved this running time to $4^k n^{\mathcal{O}(1)}$, and also obtained a kernel with $\mathcal{O}(k^4)$ vertices.

When considering a minor-closed class, rather than a hereditary class, the vertex deletion problem is known as $\mathcal{F}$-MINOR-FREE DELETION. When $\mathcal{F}$ is a set of connected graphs containing at least one planar graph, Fomin et al. [6] showed there is a deterministic FPT algorithm for this problem running in time $2^{\mathcal{O}(k)} \cdot \mathcal{O}(n \log^2 n)$. One can observe that the class of all graphs whose blocks have size at most d is closed under taking minors. Thus, $\mathcal{P}$-BLOCK VERTEX DELETION has a single-exponential FPT algorithm and a polynomial kernel, when $\mathcal{P}$ contains all connected graphs with no cut vertices and at most d vertices. However, it does not tell us anything about the parameterized complexity of BOUNDED $\mathcal{P}$-BLOCK VERTEX DELETION, which we consider in this paper.

Our Contribution. The main contribution of this paper is the following:

Theorem 1.1. *Let $\mathcal{P}$ be a non-degenerate block-hereditary class of graphs that is recognizable in polynomial time. Then, BOUNDED $\mathcal{P}$-BLOCK VERTEX DELETION*

(i) *can be solved in $2^{\mathcal{O}(k \log d)}n^{\mathcal{O}(1)}$ time, and*
(ii) *admits a kernel with $\mathcal{O}(k^2 d^7)$ vertices.*

Theorem 1.1 (i) can be viewed as a generalization of the single-exponential FPT algorithm for FEEDBACK VERTEX SET [12]. The running time is essentially optimal when $\Phi_{\mathcal{P}}$ is the class of all graphs, unless the Exponential Time

Hypothesis (ETH) [8] fails. One may expect that if the permissible blocks in $\mathcal{P}$ have a simpler structure, then the problem becomes easier. However, we obtain the same lower bound when $\Phi_{\mathcal{P}}$ contains all split graphs.

Theorem 1.2. *Let $\mathcal{P}$ be a block-hereditary class. If $\Phi_{\mathcal{P}}$ contains all split graphs, then* BOUNDED $\mathcal{P}$-BLOCK VERTEX DELETION *is not solvable in* $2^{o(k \log d)} n^{\mathcal{O}(1)}$ *time, unless the ETH fails.*

Formally, there is no function $f(x) = o(x)$ such that there is a $2^{f(k \log d)} n^{\mathcal{O}(1)}$-time algorithm for BOUNDED $\mathcal{P}$-BLOCK VERTEX DELETION, unless the ETH fails. This theorem can be proved by a reduction from the $k \times k$ CLIQUE problem, similar to [3, Theorem 14].

On the other hand, the running time can be improved to $c^k n^{\mathcal{O}(1)}$ for some c, independent of d, when $\mathcal{P}$ consists of all complete graphs, or when $\mathcal{P}$ consists of K_1, K_2, and all cycles. We refer to these problems as BOUNDED COMPLETE BLOCK VD and BOUNDED CACTUS GRAPH VD respectively.

Theorem 1.3. BOUNDED COMPLETE BLOCK VD *can be solved in time* $\mathcal{O}^*(10^k)$, *and admits a kernel with* $\mathcal{O}(k^2 d^3)$ *vertices.*

Theorem 1.4. BOUNDED CACTUS GRAPH VD *can be solved in time* $\mathcal{O}^*(26^k)$, *and admits a kernel with* $\mathcal{O}(k^2 d^4)$ *vertices.*

The proofs of Theorems 1.3 and 1.4 use the well-known technique of iterative compression [16], but are omitted due to space constraints. Note that, when $d = |V(G)|$, these become $\mathcal{O}^*(c^k)$-time algorithms for COMPLETE BLOCK VERTEX DELETION and DIAMOND HITTING SET respectively.

To obtain the general FPT algorithm in Theorem 1.1 (i), we start by searching for one of two types of obstruction of bounded size: a biconnected subgraph with at least $d + 1$ and at most $2d - 2$ vertices, or a biconnected subgraph with at most d vertices that is not in $\mathcal{P}$. Once we find such an obstruction, we branch by removing one of the vertices. The key observation is that if there are no such obstructions, then the graph can be decomposed into small pieces, which we call "clusters", that are biconnected induced subgraphs in $\mathcal{P}$. Then, it remains only to detect long cycles that are not fully contained in a cluster (Lemmas 3.1 and 3.3), which we can do by reducing the problem to SUBSET FEEDBACK VERTEX SET (Proposition 3.2).

Theorem 1.1 (ii) is a generalization of the $4k^2$ kernel for FEEDBACK VERTEX SET given by Thomassé in [18]. In order to obtain this kernel, we first develop a $(2d + 6)$-approximation algorithm for the unparameterized version of the problem, using the 8-approximation algorithm for SUBSET FEEDBACK VERTEX SET [5]. The rest of the kernelization algorithm is similar to the one for FEEDBACK VERTEX SET; however, a more general result than Gallai's A-path theorem is needed. To this end, we develop a 'packing and covering'-type result for (A, d)-trees, which are trees with at least d vertices whose leaves are in A (Proposition 4.4). Note that Gallai's A-path theorem can be seen as a packing and covering result for $(A, 2)$-trees. Using this, we can efficiently find either a flower structure, or a small deleting set, which helps to reduce the instance to a smaller instance using the α-expansion lemma.

2   Preliminaries

All graphs considered in this paper are undirected, and have no loops and no parallel edges. Let G be a graph. We denote by $N_G(v)$ the set of neighbors of a vertex v in G, and let $N_G(S) := \bigcup_{v \in S} N_G(v) \setminus S$ for any set of vertices S. For $X \subseteq V(G)$, the *deletion* of X from G is the graph obtained by removing X and all edges incident to a vertex in X, and is denoted $G - X$. For $x \in V(G)$, we simply use $G - x$ to refer to $G - \{x\}$. Let $\mathcal{F}$ be a set of graphs; then G is $\mathcal{F}$-free if it has no induced subgraph isomorphic to a graph in $\mathcal{F}$. For $n \geqslant 1$, the complete graph on n vertices is denoted K_n.

A vertex v of G is a *cut vertex* if the deletion of v from G increases the number of connected components. We say G is *biconnected* if it is connected and has no cut vertices. A *block* of G is a maximal biconnected subgraph of G. The graph G is *2-connected* if it is biconnected and $|V(G)| \geqslant 3$. In this paper we are frequently dealing with blocks, so the notion of being biconnected is often more natural than that of being 2-connected. The *block tree* of G is a bipartite graph $B(G)$ with bipartition $(\mathcal{B}, X)$, where $\mathcal{B}$ is the set of blocks of G, X is the set of cut vertices of G, and a block $B \in \mathcal{B}$ and a cut vertex $x \in X$ are adjacent in $B(G)$ if and only if B contains x.

Parameterized Complexity. A parameterized problem $Q \subseteq \Sigma^* \times N$ is *fixed-parameter tractable (FPT)* if there is an algorithm that decides whether (x, k) belongs to Q in time $f(k) \cdot |x|^{\mathcal{O}(1)}$ for some computable function f. Such an algorithm is called an *FPT algorithm*. A parameterized problem is said to admit a *polynomial kernel* if there is a polynomial time algorithm in $|x| + k$, called a *kernelization algorithm*, that reduces an input instance into an instance with size bounded by a polynomial function in k, while preserving the YES or NO answer.

3   Clustering

Agrawal et al. [1] described an efficient FPT algorithm for COMPLETE BLOCK VERTEX DELETION using a two stage approach. Firstly, small forbidden induced subgraphs are eliminated using a branching algorithm. More specifically, for each diamond or cycle of length four, at least one vertex must be removed in a solution, which can be done in $\mathcal{O}^*(4^k)$ time. The resulting graph has the following structural property: any two distinct maximal cliques have at most one vertex in common. Thus, in the second stage, it remains only to eliminate all cycles not fully contained in a maximal clique, so the problem can be reduced to an instance of WEIGHTED FEEDBACK VERTEX SET. We generalize this process and refer to it as "clustering", where the "clusters", in the case of COMPLETE BLOCK VERTEX DELETION, are the maximal cliques. We use this to obtain an algorithm for BOUNDED $\mathcal{P}$-BLOCK VERTEX DELETION in Sect. 3.2.

3.1 $\mathcal{P}$-clusters

Let $\mathcal{P}$ be a block-hereditary class of graphs. We may assume that $\mathcal{P}$ contains only biconnected graphs; otherwise there is some block-hereditary $\mathcal{P}'$ such that

$\mathcal{P}' \subset \mathcal{P}$ and $\Phi_{\mathcal{P}'} = \Phi_{\mathcal{P}}$. Let G be a graph. A $\mathcal{P}$-*cluster* of G is a maximal induced subgraph H of G with the property that H is isomorphic either to K_1 or a graph in $\mathcal{P}$. We say that G is $\mathcal{P}$-*clusterable* if for any distinct $\mathcal{P}$-clusters H_1 and H_2 of G, we have $|V(H_1) \cap V(H_2)| \leqslant 1$. For a $\mathcal{P}$-clusterable graph, if $v \in V(G)$ is contained in at least two distinct $\mathcal{P}$-clusters, then v is called an *external* vertex.

The following property of $\mathcal{P}$-clusters is essential. We say that $X \subseteq V(G)$ *hits* a cycle C if $X \cap V(C) \neq \emptyset$, and a cycle C is *contained* in a $\mathcal{P}$-cluster of G if $V(C) \subseteq V(H)$ for some $\mathcal{P}$-cluster H of G.

Lemma 3.1. *Let $\mathcal{P}$ be a non-degenerate block-hereditary class of graphs, let G be a graph, and let $S \subseteq V(G)$. Then $G - S \in \Phi_{\mathcal{P}}$ if and only if S hits every cycle not contained in a $\mathcal{P}$-cluster of G.*

The next proposition follows from the fact that $\mathcal{P}$-BLOCK VERTEX DELETION can be reduced to SUBSET FEEDBACK VERTEX SET if the input graph is $\mathcal{P}$-clusterable, and SUBSET FEEDBACK VERTEX SET can be solved in time $\mathcal{O}^*(4^k)$ [19]. We omit the proof, but the idea of the reduction is illustrated in Fig. 1.

SUBSET FEEDBACK VERTEX SET	Parameter: k
Input: A graph G, a set $X \subseteq V(G)$, and a non-negative integer k.	
Question: Is there a set $S \subseteq V(G)$ with $|S| \leqslant k$ such that no cycle in $G - S$ contains a vertex of X?	

Proposition 3.2. *Let $\mathcal{P}$ be a non-degenerate block-hereditary class of graphs recognizable in polynomial time. Given a $\mathcal{P}$-clusterable graph G together with the set of $\mathcal{P}$-clusters of G, and a non-negative integer k, there is an $\mathcal{O}^*(4^k)$-time algorithm that determines whether there is a set $S \subseteq V(G)$ with $|S| \leqslant k$ such that $G - S \in \Phi_{\mathcal{P}}$.*

By Proposition 3.2, the $\mathcal{P}$-BLOCK VERTEX DELETION problem admits an efficient FPT algorithm provided we can reduce the input to $\mathcal{P}$-clusterable graphs. In the next section, we show that this is possible for any finite block-hereditary $\mathcal{P}$ where the permissible blocks in $\mathcal{P}$ have at most d vertices. In particular, we use this to show there is an $\mathcal{O}^*(2^{\mathcal{O}(k \log d)})$-time algorithm for BOUNDED $\mathcal{P}$-BLOCK VERTEX DELETION.

3.2 An FPT Algorithm for Bounded $\mathcal{P}$-Block Vertex Deletion

In this section we describe an FPT algorithm for BOUNDED $\mathcal{P}$-BLOCK VERTEX DELETION using the clustering approach. For positive integers x and y, let $\mathcal{B}_{x,y}$ be the class of all biconnected graphs with at least x vertices and at most y vertices. When $x > y$, $\mathcal{B}_{x,y} = \emptyset$.

Lemma 3.3. *Let $\mathcal{P}$ be a non-degenerate block-hereditary class, and let $d \geqslant 2$ be an integer. If a graph G is $\mathcal{B}_{d+1,2d-2}$-free and $(\mathcal{B}_{2,d} \setminus \mathcal{P})$-free, then G is $(\mathcal{P} \cap \mathcal{B}_{2,d})$-clusterable.*

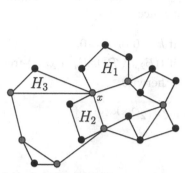
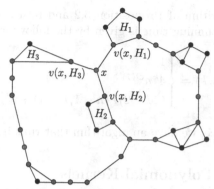

(a) A $\mathcal{P}$-clusterable graph G, where the red (light gray) vertices are shared by multiple $\mathcal{P}$-clusters.

(b) Graph G' of SUBSET FEEDBACK VERTEX SET instance (G', X, k), where the vertices of X are red (or light gray).

Fig. 1. Example construction of a SUBSET FEEDBACK VERTEX SET instance (G', X, k) from a $\mathcal{P}$-clusterable graph G. (Color figure online)

Proposition 3.4. *Let $d \geqslant 2$ be an integer, and let $\mathcal{P}$ be a non-degenerate block-hereditary class recognizable in polynomial time. There is a polynomial-time algorithm that, given a graph G, either*

(i) *outputs an induced subgraph of G in $\mathcal{B}_{2,d} \setminus \mathcal{P}$, or*
(ii) *outputs an induced subgraph of G in $\mathcal{B}_{d+1,2d-2}$, or*
(iii) *correctly answers that G is $((\mathcal{B}_{2,d} \setminus \mathcal{P}) \cup \mathcal{B}_{d+1,2d-2})$-free.*

Lemma 3.5. *Let $d \geqslant 2$ be an integer, and let $\mathcal{P}$ be a non-degenerate block-hereditary class recognizable in polynomial time. Then there is a polynomial-time algorithm that, given a $((\mathcal{B}_{2,d} \setminus \mathcal{P}) \cup \mathcal{B}_{d+1,2d-2})$-free graph G, outputs the set of $(\mathcal{P} \cap \mathcal{B}_{2,d})$-clusters of G.*

Theorem 3.6. *Let $\mathcal{P}$ be a non-degenerate block-hereditary class of graphs recognizable in polynomial time. Then BOUNDED $\mathcal{P}$-BLOCK VERTEX DELETION can be solved in time $2^{\mathcal{O}(k \log d)} n^{\mathcal{O}(1)}$.*

Proof. We describe a branching algorithm for BOUNDED $\mathcal{P}$-BLOCK VERTEX DELETION on the instance (G, d, k). If G contains an induced subgraph in $(\mathcal{B}_{2,d} \setminus \mathcal{P}) \cup \mathcal{B}_{d+1,2d-2}$, then any solution S contains at least one vertex of this induced subgraph. We first run the algorithm of Proposition 3.4, and if it outputs such an induced subgraph J, then we branch on each vertex $v \in V(J)$, recursively applying the algorithm on $(G - v, d, k - 1)$. Since $|V(J)| \leqslant 2d - 2$, there are at most $2d - 2$ branches. If one of these branches has a solution S', then $S' \cup \{v\}$ is a solution for G. Otherwise, if every branch returns NO, we return that (G, d, k) is a NO-instance. On the other hand, if there is no such induced subgraph, then G is $(\mathcal{P} \cap \mathcal{B}_{2,d})$-clusterable, by Lemma 3.3, and we can find the set of all $(\mathcal{P} \cap \mathcal{B}_{2,d})$-clusters in polynomial time, by Lemma 3.5. We can now run the $\mathcal{O}^*(4^k)$-time algorithm of Proposition 3.2 and return the result. Thus, an upper bound for the running time is given by the following recurrence:

$$T(n, k) = \begin{cases} 1 & \text{if } k = 0 \text{ or } n = 0, \\ 4^k n^{\mathcal{O}(1)} & \text{if } ((\mathcal{B}_{2,d} \setminus \mathcal{P}) \cup \mathcal{B}_{d+1,2d-2})\text{-free}, \\ (2d - 2)T(n - 1, k - 1) + n^{\mathcal{O}(1)} & \text{otherwise.} \end{cases}$$

Hence, we have an algorithm that runs in time $\mathcal{O}^*(2^{\mathcal{O}(k \log d)})$. $\qquad \square$

4   Polynomial Kernels

In this section, we prove the following:

Theorem 4.1. *Let $\mathcal{P}$ be a non-degenerate block-hereditary class of graphs recognizable in polynomial time. Then* BOUNDED $\mathcal{P}$-BLOCK VERTEX DELETION *admits a kernel with $\mathcal{O}(k^2 d^7)$ vertices.*

We fix a block-hereditary class of graphs $\mathcal{P}$ recognizable in polynomial time. The block tree of a graph can be computed in time $\mathcal{O}(|V(G)| + |E(G)|)$ [7]. Thus, one can test whether a given graph is in $\Phi_{\mathcal{P} \cap \mathcal{B}_{2,d}}$ in polynomial time.

Before describing the algorithm, we observe that there is a $(2d + 6)$-approximation algorithm for the (unparameterized) minimization version of the BOUNDED $\mathcal{P}$-BLOCK VERTEX DELETION problem. We first run the algorithm of Proposition 3.4. When we find an induced subgraph in $(\mathcal{B}_{2,d} \setminus \mathcal{P}) \cup \mathcal{B}_{d+1,2d-2}$, instead of branching on the removal of one of the vertices, we remove all the vertices of the subgraph, then rerun the algorithm. Hence, we can reduce to a $(\mathcal{P} \cap \mathcal{B}_{2,d})$-clusterable graph by removing at most $(2d - 2) \cdot \text{OPT}$ vertices. Moreover, we can obtain the set of all $(\mathcal{P} \cap \mathcal{B}_{2,d})$-clusters using the algorithm in Lemma 3.5. Arguments in the proof of Proposition 3.2 and the known 8-approximation algorithm for SUBSET FEEDBACK VERTEX SET [5] imply that there is a $(2d + 6)$-approximation algorithm for BOUNDED $\mathcal{P}$-BLOCK VERTEX DELETION.

We start with the straightforward reduction rules. Let (G, d, k) be an instance of BOUNDED $\mathcal{P}$-BLOCK VERTEX DELETION.

Reduction Rule 1 (Component rule). If G has a connected component $H \in \Phi_{\mathcal{P} \cap \mathcal{B}_{2,d}}$, then remove H.

Reduction Rule 2 (Cut vertex rule). Let v be a cut vertex of G such that $G - v$ contains a connected component H where $G[V(H) \cup \{v\}]$ is a block in $\mathcal{P} \cap \mathcal{B}_{2,d}$. Then remove H from G.

Now, we introduce a so-called bypassing rule. We first run the $(2d + 6)$-approximation algorithm, and if it outputs a solution of more than $(2d + 6)k$ vertices, then we have a NO-instance. Thus, we may assume that the algorithm outputs a solution of size at most $(2d + 6)k$. Let us fix such a set U.

Reduction Rule 3 (Bypassing rule). Let $v_1, v_2, \ldots, v_t$ be a sequence of cut vertices of $G - U$ with $2 \leqslant t \leqslant d + 1$, and let $B_1, \ldots, B_{t-1}$ be blocks of $G - U$ such that

(1) for each $i \in \{1, \ldots, t-1\}$, B_i is the unique block of $G - U$ containing v_i and v_{i+1} and no other cut vertices of $G - U$;
(2) G has no edges between $(\bigcup_{1 \leqslant i \leqslant t-1} V(B_i)) \setminus \{v_1, v_t\}$ and U; and
(3) $|\bigcup_{1 \leqslant i \leqslant t-1} V(B_i))| \geqslant d + 1$.

If $\bigcup_{1 \leqslant i \leqslant t-1} V(B_i) \setminus \{v_1, \ldots, v_t\} = \emptyset$, then contract $v_1 v_2$; otherwise, choose a vertex in $\bigcup_{1 \leqslant i \leqslant t-1} V(B_i)$ that is not a cut vertex of $G - U$, and remove it.

See Fig. 2 for an example application of Reduction Rule 3. Note that this rule can be applied in polynomial time using the block tree of $G - U$.

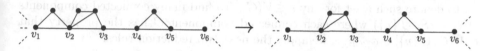

Fig. 2. An example application of Reduction Rule 3 when $d = 9$

Lemma 4.2. *Reduction Rule 3 is safe.*

We show that after applying Reduction Rules 1 to 3, if the reduced graph is still large, then there is a vertex of large degree. This follows from the fact that the block tree of $G - U$ has no path of $2d + 2$ vertices where the internal vertices have degree 2 in $G - U$.

Lemma 4.3. *Let (G, d, k) be an instance reduced under Reduction Rules 1 to 3. If (G, d, k) is a YES-instance and $|V(G)| \geqslant 4d(2d + 3)(d + 3)k\ell$, for some integer ℓ, then G contains a vertex of degree at least $\ell + 1$.*

Now, we discuss a "sunflower structure" that allows us to find a vertex that can be safely removed. A similar technique was used in [1, 10, 17]; there, Gallai's A-path Theorem is used to find many obstructions whose pairwise intersections are exactly one vertex; here, we use different objects to achieve the same thing.

Let $A \subseteq V(G)$ and let $d \geqslant 2$. An (A, d)-*tree* in G is a tree subgraph of G on at least d vertices whose leaves are contained in A. Let v be a vertex of G. If there is an $(N_G(v), d)$-tree T in $G - v$, then $G[V(T) \cup \{v\}]$ is a 2-connected graph with at least $d + 1$ vertices. This implies that if there are $k + 1$ pairwise vertex-disjoint $(N_G(v), d)$-trees in $G - v$, then we can safely remove v, as any solution should contain v.

We prove that if G does not have any set of $k + 1$ pairwise vertex-disjoint (A, d)-trees, then there exists $S \subseteq V(G)$ where the size of S is bounded by a function of k and d, and every connected component of $G - S$ has fewer than d vertices of A. Note that $G - S$ may still have some (A, d)-trees, as a path of length $d - 1$ between two vertices in A is also an (A, d)-tree.

Proposition 4.4. *Let G be a graph, let k and d be positive integers, and let $A \subseteq V(G)$. There is an algorithm that, in time $\mathcal{O}(d|V(G)|^3)$, finds either:*

(i) *k pairwise vertex-disjoint (A, d)-trees in G, or*
(ii) *a vertex subset $S \subseteq V(G)$ of size at most $2(2k - 1)(d^2 - d + 1)$ such that each connected component of $G - S$ contains fewer than d vertices of A.*

Reduction Rule 4 (Sunflower rule 1). Let v be a vertex of G. If there are $k+1$ pairwise vertex-disjoint $(N_G(v), d)$-trees in $G - v$, then remove v and reduce k by 1.

After exhaustively applying Reduction Rule 4, we may assume, by Proposition 4.4, that for each $v \in V(G)$, there exists $S_v \subseteq V(G - v)$ with $|S_v| \leq 2(2k + 1)(d^2 - d + 1)$ such that v has at most $d - 1$ neighbors in each connected component of $G - (S_v \cup \{v\})$. In the remainder of this section, we use S_v to denote such a set for any $v \in V(G)$. To find many connected components of $G - (S_v \cup \{v\})$ where each connected component C has the property that $G[V(C) \cup \{v\}] \in \phi_{\mathcal{P} \cap \mathcal{B}_{2,d}}$, we apply the next two reduction rules.

Reduction Rule 5 (Disjoint obstructions rule). If there are $k+1$ connected components of $G - (S_v \cup \{v\})$ such that each connected component is not in $\phi_{\mathcal{P} \cap \mathcal{B}_{2,d}}$, then conclude that (G, d, k) is a No-instance.

Reduction Rule 6 (Sunflower rule 2). If there are $k + 1$ connected components of $G - (S_v \cup \{v\})$ where each connected component C is in $\phi_{\mathcal{P} \cap \mathcal{B}_{2,d}}$ but $G[V(C) \cup \{v\}] \notin \phi_{\mathcal{P} \cap \mathcal{B}_{2,d}}$, then remove v and decrease k by 1.

We can perform these two rules in polynomial time using the block tree of $G[V(C) \cup \{v\}]$. Then we may assume that $G - (S_v \cup \{v\})$ contains at most $2k$ connected components such that the connected component C satisfies $G[V(C) \cup \{v\}] \notin \phi_{\mathcal{P} \cap \mathcal{B}_{2,d}}$. Thus, if v has degree at least ℓ, there are at least $\frac{\ell - 2(2k+1)(d^2-d+1)}{d-1} - 2k$ connected components of $G - (S_v \cup \{v\})$ such that the connected component C satisfies $G[V(C) \cup \{v\}] \in \phi_{\mathcal{P} \cap \mathcal{B}_{2,d}}$. As G is reduced under Reduction Rule 2, there is an edge between any such connected component C and S_v. We introduce a final reduction rule, which uses the α-expansion lemma [17].

Lemma 4.5 (α-expansion lemma). *Let α be a positive integer, and let F be a bipartite graph with vertex bipartition (X, Y) such that $|Y| \geq \alpha|X|$ and every vertex of Y has at least one neighbor in X. Then there exist non-empty subsets $X' \subseteq X$ and $Y' \subseteq Y$ and a function $\phi : X' \to \binom{Y'}{\alpha}$ such that*

- *$N_F(Y') \cap X = X'$,*
- *$\phi(x) \subseteq N_F(x)$ for each $x \in X'$, and*
- *the sets in $\{\phi(x) : x \in X'\}$ are pairwise disjoint.*

In addition, such a pair X', Y' can be computed in time polynomial in $\alpha|V(F)|$.

Reduction Rule 7 (Large degree rule). Let v be a vertex of G. If there is a set $\mathcal{C}$ of connected components of $G - (S_v \cup \{v\})$ such that $|\mathcal{C}| \geqslant 2d(2k+1)(d^2 - d + 1)$ and, for each $C \in \mathcal{C}$, we have $G[V(C) \cup \{v\}] \in \phi_{\mathcal{P} \cap \mathcal{B}_{2,d}}$, then do the following: (1) Construct an auxiliary bipartite graph H with bipartition $(S_v, \mathcal{C})$ where $w \in S_v$ and $C \in \mathcal{C}$ are adjacent in H if and only if w has a neighbor in C. (2) Compute sets $\mathcal{C}' \subseteq \mathcal{C}$ and $S_v' \subseteq S_v$ obtained by applying Lemma 4.5 to H with $\alpha = d$. (3) Remove all edges in G between v and each connected component C of $\mathcal{C}'$. (4) Add $d - 1$ internally vertex-disjoint paths of length 2 between v and each vertex $x \in S_v'$. (5) Remove all vertices of degree 1 in the resulting graph.

Lemma 4.6. *Reduction Rule 7 is safe.*

Lemma 4.7. *Reduction Rules 1 to 7 can be applied exhaustively in polynomial time.*

Proof (Proof of Theorem 4.1). We apply Reduction Rules 1 to 7 exhaustively. Note that this takes polynomial time, by Lemma 4.7. Suppose that (G, d, k) is the reduced instance, and $|V(G)| \geqslant 4dk(\ell - 1)(2d + 3)(d + 3)$ where $\ell = 2d^2(2k + 1)(d^2 - d + 3)$. Then, by Lemma 4.3, there exists a vertex v of degree at least ℓ.

By Proposition 4.4, v has at most $d - 1$ neighbors in each connected component of $G - (S_v \cup \{v\})$. Since $\ell = 2d^2(2k + 1)(d^2 - d + 3)$, the subgraph $G - (S_v \cup \{v\})$ contains at least $\frac{\ell - 2(2k+1)(d^2 - d + 1)}{d - 1} \geqslant 2d(2k + 1)(d^2 - d + 3)$ connected components. By Reduction Rules 5 and 6, $G - (S_v \cup \{v\})$ contains at least $2d(2k + 1)(d^2 - d + 1)$ connected components such that, for each connected component C, $G[V(C) \cup \{v\}] \in \phi_{\mathcal{P} \cap \mathcal{B}_{2,d}}$. Then we can apply Reduction Rule 7, contradicting our assumption. We conclude that $|V(G)| = \mathcal{O}(k^2 d^7)$. $\square$

One might ask whether the kernel with $\mathcal{O}(k^2 d^7)$ vertices can be improved upon. Regarding the k^2 factor, reducing it to linear in k would imply a linear kernel for FEEDBACK VERTEX SET. On the other hand, it is possible to reduce the d^7 factor depending on the block-hereditary class $\mathcal{P}$.

Theorem 4.8.

– BOUNDED BLOCK VD admits a kernel with $\mathcal{O}(k^2 d^6)$ vertices.
– BOUNDED COMPLETE BLOCK VD admits a kernel with $\mathcal{O}(k^2 d^3)$ vertices.
– BOUNDED CACTUS GRAPH VD admits a kernel with $\mathcal{O}(k^2 d^4)$ vertices.

References

1. Agrawal, A., Kolay, S., Lokshtanov, D., Saurabh, S.: A faster FPT algorithm and a smaller kernel for block graph vertex deletion. In: Kranakis, E., Navarro, G., Chávez, E. (eds.) LATIN 2016. LNCS, vol. 9644, pp. 1–13. Springer, Heidelberg (2016)

2. Cai, L.: Fixed-parameter tractability of graph modification problems for hereditary properties. Inf. Proc. Lett. **58**(4), 171–176 (1996)
3. Drange, P.G., Dregi, M.S., Hof, P.: On the computational complexity of vertex integrity and component order connectivity. In: Ahn, H.-K., Shin, C.-S. (eds.) ISAAC 2014. LNCS, vol. 8889, pp. 285–297. Springer, Heidelberg (2014). doi:10.1007/978-3-319-13075-0_23
4. El-Mallah, E.S., Colbourn, C.J.: The complexity of some edge deletion problems. IEEE Trans. Circ. Syst. **35**(3), 354–362 (1988)
5. Even, G., Naor, J., Zosin, L.: An 8-approximation algorithm for the subset feedback vertex set problem. SIAM J. Comput. **30**(4), 1231–1252 (2000)
6. Fomin, F., Lokshtanov, D., Misra, N., Saurabh, S.: Planar $\mathcal{F}$-deletion: approximation and optimal FPT algorithms. In: Foundations of Computer Science (FOCS), pp. 470–479 (2012)
7. Hopcroft, J., Tarjan, R.: Algorithm 447: efficient algorithms for graph manipulation. Commun. ACM **16**(6), 372–378 (1973)
8. Impagliazzo, R., Paturi, R., Zane, F.: Which problems have strongly exponential complexity? J. Comput. Syst. Sci. **63**(4), 512–530 (2001)
9. Joret, G., Paul, C., Sau, I., Saurabh, S., Thomassé, S.: Hitting and harvesting pumpkins. SIAM J. Discrete Math. **28**(3), 1363–1390 (2014)
10. Kim, E.J., Kwon, O.: A polynomial kernel for block graph deletion. In: Husfeldt, T., Kanj, I. (eds.) IPEC 2015. Leibniz International Proceedings in Informatics (LIPIcs), vol. 43, pp. 270–281. Schloss dagstuhl-leibniz-zentrum fuer informatik, Dagstuhl, Germany (2015)
11. Kim, E.J., Langer, A., Paul, C., Reidl, F., Rossmanith, P., Sau, I., Sikdar, S.: Linear kernels and single-exponential algorithms via protrusion decompositions. In: Fomin, F.V., Freivalds, R., Kwiatkowska, M., Peleg, D. (eds.) ICALP 2013. LNCS, vol. 7965, pp. 613–624. Springer, Heidelberg (2013). doi:10.1007/978-3-642-39206-1_52
12. Kociumaka, T., Pilipczuk, M.: Faster deterministic feedback vertex set. Inf. Proc. Lett. **114**(10), 556–560 (2014)
13. Kolay, S., Lokshtanov, D., Panolan, F., Saurabh, S.: Quick but odd growth of cacti. In: Husfeldt, T., Kanj, I. (eds.) 10th International Symposium on Parameterized and Exact Computation (IPEC 2015), pp. 258–269, no. 43. Schloss Dagstuhl-Leibniz-Zentrum fuer Informatik, Dagstuhl (2015)
14. Lewis, J.M., Yannakakis, M.: The node-deletion problem for hereditary properties is NP-complete. J. Comput. Syst. Sci. **20**(2), 219–230 (1980)
15. Misra, P., Raman, V., Ramanujan, M.S., Saurabh, S.: Parameterized algorithms for EVEN CYCLE TRANSVERSAL. In: Golumbic, M.C., Stern, M., Levy, A., Morgenstern, G. (eds.) WG 2012. LNCS, vol. 7551, pp. 172–183. Springer, Heidelberg (2012)
16. Reed, B.A., Smith, K., Vetta, A.: Finding odd cycle transversals. Oper. Res. Lett. **32**(4), 299–301 (2004)
17. Thomassé, S.: A quadratic kernel for feedback vertex set. In: Proceedings of the Twentieth Annual ACM-SIAM Symposium on Discrete Algorithms, SODA 2009, New York, NY, USA, 4–6 January 2009, pp. 115–119 (2009)
18. Thomassé, S.: A $4k^2$ kernel for feedback vertex set. ACM Trans. Algorithms **6**(2), 32 (2010)
19. Wahlström, M.: Half-integrality, LP-branching and FPT algorithms. In: Proceedings of the Twenty-Fifth Annual ACM-SIAM Symposium on Discrete Algorithms, pp. 1762–1781. SIAM (2014)

Harmonious Coloring: Parameterized Algorithms and Upper Bounds

Sudeshna Kolay[1], Ragukumar Pandurangan[1], Fahad Panolan[2],
Venkatesh Raman[1], and Prafullkumar Tale[1(✉)]

[1] The Institute of Mathematical Sciences, Chennai, India
{skolay,ragukumar,vraman,pptale}@imsc.res.in
[2] University of Bergen, Bergen, Norway
Fahad.Panolan@ii.uib.no

Abstract. A harmonious coloring of a graph is a partitioning of its vertex set into parts such that, there are no edges inside each part, and there is at most one edge between any pair of parts. It is known that finding a minimum harmonious coloring number is NP-hard even in special classes of graphs like trees and split graphs.

We initiate a study of parameterized and exact exponential time complexity of harmonious coloring. We consider various parameterizations like by solution size, by above or below known guaranteed bounds and by the vertex cover number of the graph. While the problem has a simple quadratic kernel when parameterized by the solution size, our main result is that the problem is fixed-parameter tractable when parameterized by the size of a vertex cover of the graph. This is shown by reducing the problem to multiple instances of fixed variable integer linear programming.

We also observe that it is $W[1]$-hard to determine whether at most $n - k$ or $\Delta + 1 + k$ colors are sufficient in a harmonious coloring of an n-vertex graph G, where Δ is the maximum degree of G and k is the parameter. Concerning exact exponential time algorithms, we develop a $2^n n^{\mathcal{O}(1)}$ algorithm for finding a minimum harmonious coloring in split graphs improving on the naive $2^{\mathcal{O}(n \log n)}$ algorithm.

1 Introduction and Motivation

Graph Coloring is the problem of partitioning the vertex set of a graph to satisfy some constraints. Coloring problems have been extensively studied in discrete mathematics and theoretical computer science. Given a coloring χ of a graph G, the set of vertices that receive the same color is said to be a *color class*. One of the most well-known coloring problems is the chromatic number problem that seeks the minimum number of colors required so that each color class induces an independent set (i.e. no pair of vertices in a set is adjacent), and it is one of Karp's 21 NP-complete problems from 1972 [18]. Lawler gave an algorithm for the problem running in time $2.4423^n n^{\mathcal{O}(1)}$ on an n-vertex graph [19]. Later, using the principle of inclusion-exclusion Björklund et al. [5] gave an algorithm

© Springer-Verlag GmbH Germany 2016
P. Heggernes (Ed.): WG 2016, LNCS 9941, pp. 245–256, 2016.
DOI: 10.1007/978-3-662-53536-3_21

running in time $2^n n^{\mathcal{O}(1)}$ on an n-vertex graph and this is the fastest known exact algorithm for the problem.

Different variants of the graph coloring problem have been studied in the literature. The ACHROMATIC NUMBER seeks the *maximum* number of colors required so that each color class induces an independent set, and there is at least one edge between every pair of color classes. A characterization for this problem was given in [14] using which one can obtain an FPT algorithm for ACHROMATIC NUMBER parameterized by the solution size (see Sect. 2.1 for definitions on parameterized complexity). The PSEUDO-ACHROMATIC NUMBER problem is a generalization of ACHROMATIC NUMBER, and does not demand that each color class induces an independent set. This problem is also FPT parameterized by the solution size [7]. Another related problem is the b-CHROMATIC NUMBER. Here the objective is to color the vertices with the same properties as that in ACHROMATIC NUMBER, but insist that in each color class there is a vertex that has a neighbor in every other color class. This problem was introduced in [2]. The problem is W[1]-hard when parameterized by the solution size [22].

In 1989, Hopcroft and Krishnamoorthy [15] introduced the notion of HARMONIOUS COLORING. A harmonious coloring of a graph is a partition of the vertex set into sets such that every set induces an independent set and additionally between any pair of sets, there is *at most* one edge. The minimum number of sets in such a partition is called the harmonious coloring number of the graph. Determining whether a graph has harmonious coloring using at most k colors is known to be NP-complete [15], even in trees [13], split graphs [3], interval graphs [3,6] and several other classes of graphs [3,4,6,12,13,16]. Polynomial time algorithms are known for some special classes of graphs [21], the most important being for trees of bounded degree [11].

In this paper, we initiate the parameterized complexity of the problem under natural parameterizations. With solution size k (the harmonious coloring number) as a parameter, there is a trivial kernel on $O(k^2)$ vertices and edges, and this is discussed in Sect. 4.1. In this section, we also discuss parameterized complexity of parameterizing above or below some known bounds for harmonious coloring number. As the problem is NP-complete on trees, the problem parameterized by the treewidth or feedback vertex set is trivially para NP-hard. Our main result is that the problem is fixed-parameter tractable when parameterized by the size of the minimum vertex cover of the graph. This is shown by solving several bounded variable integer linear programming (ILP) problems. The number of ILPs is upper bounded by a function of minimum vertex cover. This is developed in Sect. 4.2. In Sect. 5, we discuss exact exponential algorithms for harmonious coloring, and give an $2^{\mathcal{O}(n)}$ algorithm in split graphs, improving on the naive $2^{\mathcal{O}(n \log n)}$ algorithm. In Sect. 3, we develop improved upper bounds on the harmonious coloring number in terms of the vertex cover number and the maximum degree of the graph. Results marked with a $(\star)$ have their proofs in the full version of this paper.

2   Preliminaries

We use $\mathbb{N}$ and $\mathbb{Z}$ to denote the set of natural numbers and set of integers, respectively. For $n \in \mathbb{N}$ we use $[n]$ to denote $\{1, \ldots, n\}$. We use standard notations from graph theory [9]. By "graph" we mean simple undirected graph. The vertex set and edge set of a graph G are denoted as $V(G)$ and $E(G)$ respectively. The complement of a graph G, denoted by $\overline{G}$, has $V(G)$ as its vertex set and $\binom{V(G)}{2} \setminus E(G)$ as its edge set. Here, $\binom{V(G)}{2}$ denotes the family of two sized subsets of $V(G)$. The *neighborhood* of a vertex v is represented as $N_G(v)$, or, when the context of the graph is clear, simply as $N(v)$. The *closed neighborhood* of a vertex v, denoted by $N[v]$, is the subset $N(v) \cup \{v\}$. For set U, we define $N(U)$ as union of $N(v)$ all vertices v in U. If $U = \emptyset$ then $N(U) = \emptyset$. For two disjoint subsets $V_1, V_2 \subseteq V(G)$, $E(V_1, V_2)$ is set of edges where one end point is in V_1 and another is in V_2. An edge in the set $E(V_1, V_2)$ is said to be *going across*. A *trivial* component of graph is a component which does not contain any edge. A *non-trivial component* of a graph is a connected component of G that has at least two vertices. The function $d_G : V(G) \times V(G) \to \mathbb{N}$ corresponds to the minimum distance between a pair of vertices in the graph G. A d-degenerate graph is a graph G where $V(G)$ has an ordering in which any vertex has at most d neighbors with indices lower than that of the vertex. For a graph G, a set $S \subseteq V(G)$ is called a *vertex cover* of G if $G - S$ is an independent set. A graph G is called a *split graph* if $V(G)$ has a bipartition (V_1, V_2) such that $G[V_1]$ is an induced clique and $G[V_2]$ is an induced independent set. In this case, $(G[V_1], G[V_2])$ is called a *split partition* of G. No split graph contains a 4-cycle (C_4), a 5-cycle (C_5) or the complement of a 4-cycle ($2K_2$) as an induced subgraph. The finite set of graphs $\{C_4, C_5, 2K_2\}$ is said to be a *finite forbidden set* of the class of split graphs. Each graph in the finite forbidden set is referred to as a *forbidden structure*.

A function $h : V(G) \to [k]$, where k is a positive integer, is called a *coloring function*. For a coloring function h and for any $i \in [k]$, the vertex subset $h^{-1}(i)$ is called the ith *color class* of h. If no edge has both its end points in the same color class then coloring function is said to be *proper*. *Harmonious coloring* is a proper coloring with additional property that there is at most one edge across any two color classes. The minimum number of colors required for a harmonious coloring of a graph G is denoted by $hc(G)$. The restriction of a coloring function h to a subset $V' \subseteq V(G)$, denoted by $h|_{V'}$, is a coloring function such that $h|_{V'} : V' \to [k]$, and $h|_{V'}(u) = h(u)$ for each vertex $u \in V'$. In this case, h is said to be an *extension* of $h|_{V'}$. For a subset $V' \subseteq V(G)$, $h(V') = \{i | h^{-1}(i) \cap V' \neq \emptyset\}$.

The technical tool we use to prove that HARMONIOUS COLORING is fixed-parameter tractable (defined in next section) by size of vertex cover is the fact that INTEGER LINEAR PROGRAMMING is fixed-parameter tractable parameterized by the number of variables. An instance of INTEGER LINEAR PROGRAMMING consists of a matrix $A \in \mathbb{Z}^{m \times p}$, a vector $b \in \mathbb{Z}^m$ and a vector $c \in \mathbb{Z}^p$. The goal is to find a vector $x \in \mathbb{Z}^p$ which satisfies $Ax \leq b$ and minimizes the value of $c \cdot x$ (scalar product of c and x). We assume that an input is given in binary and thus the size of the input is the number of bits in its binary representation.

Proposition 1 ([17], [20]). *An* INTEGER LINEAR PROGRAMMING *instance of size L with p variables can be solved using $\mathcal{O}(p^{2.5p+o(p)} \cdot (L + \log M_x) \cdot \log(M_x \cdot M_c))$ arithmetic operations and space polynomial in $L + \log M_x$, where M_x is an upper bound on the absolute value a variable can take in a solution, and M_c is the largest absolute value of a coefficient in the vector c.*

2.1 Parameterized Complexity

The goal of parameterized complexity is to find ways of solving NP-hard problems more efficiently than brute force by associating a *small* parameter to each instance. Formally, a *parameterization* of a problem is assigning a positive integer parameter k to each input instance and we say that a parameterized problem is *fixed-parameter tractable (FPT)* if there is an algorithm that solves the problem in time $f(k) \cdot |I|^{\mathcal{O}(1)}$, where $|I|$ is the size of the input and f is an arbitrary computable function depending only on the parameter k. Such an algorithm is called an FPT algorithm and such a running time is called FPT running time. There is also an accompanying theory of hardness using which one can identify parameterized problems that are unlikely to admit FPT algorithms. The hard classes are $W[i], i \in \mathbb{N}$. For the purpose of this paper, it is enough to know that the INDEPENDENT SET problem is W[1]-hard [10].

A parameterized problem is said to be in the class para-NP if it has a nondeterministic algorithm with FPT running time. To show that a problem is para-NP-hard, we need to show that the problem is NP-hard when the parameter takes a value from a finite set of positive integers.

Another direction of research is in providing a refinement of the FPT class, through the concept of *kernelization*. A parameterized problem is said to admit a $h(k)$-*kernel* if there is a polynomial time algorithm (the degree of the polynomial is independent of k), called a *kernelization* algorithm, that reduces the input instance to an instance with size upper bounded by $h(k)$, while preserving the answer. If the function $h(k)$ is polynomial in k, then we say that the problem admits a polynomial kernel. For more on parameterized complexity, see the recent book [8].

3   Upper and Lower Bounds and Structural Results

In this section, we give some general upper bounds of harmonious coloring number based on other natural graph parameters and show some structural results which are used later in our algorithms.

Observation 1. *For a given graph G and two vertices u, v, if u and v belong to the same harmonious color class then $d_G(u, v) > 2$.*

Definition 1 (Identify). *For a graph G, identifying a vertex set U of $V(G)$ is the operation of deleting U, adding a new vertex w and the edge set $\{wx | x \notin U, \exists u \in U \text{ and } xu \in E(G)\}$.*

Observation 2 ($\star$). *For a graph G, let ϕ be an optimal harmonious coloring. Suppose the graph G' is formed by identifying a color class of ϕ. Then $\mathsf{hc}(G) = \mathsf{hc}(G')$.*

Lemma 1 ($\star$). *Let G be a graph without isolated vertices, X be a vertex cover of G, and let H be the auxiliary graph defined such that $V(H) = V(G - X)$ and for $u, v \in V(H)$, $uv \in E(H)$ if $d_G(u,v) = 2$. A coloring function h of G, where (1) $h(X) \cap h(V(G-X)) = \emptyset$, (2) $h(i) \neq h(j)$ for all $i \neq j \in X$, is a harmonious coloring of G if and only if $h|_{V(G-X)}$ is a proper coloring of H.*

Let $\Delta(G)$ denote the maximum degree of the graph, and $vc(G)$ denote the vertex cover number of G. We use Δ if the graph G is clear from the context. We show the following bound for general graphs.

Theorem 1. *For any graph G with $\Delta \geq 2, \Delta + 1 \leq \mathsf{hc}(G) \leq vc(G) + \Delta(\Delta - 1)$.*

Proof. By Observation 1, any two vertices in the same harmonious color class should be at a distance three or more from each other. This implies that for any vertex u, every vertex in its closed neighbourhood gets a separate color. Since this is true for a vertex with the highest degree, lower bound on harmonious coloring follows.

We first construct a harmonious coloring with $vc(G) + \Delta(\Delta - 1) + 1$ many colors and then apply a trick to save one color. Let X be a vertex cover of graph G. Construct a coloring $\phi : V(G) \to [vc(G) + \Delta(\Delta - 1) + 1]$ in the following fashion: Color each vertex in vertex cover X with separate color which will not be used for remaining vertices. Construct an auxiliary graph H as mentioned in Lemma 1. Notice that $\Delta(H) = \Delta(G)(\Delta(G) - 1)$. Graph H can be properly colored using $\Delta(H) + 1$ many colors ([9] p.115). Coloring $\phi|_{V(G-X)}$ is proper coloring of H and satisfies the premises of Lemma 1 hence it is harmonious coloring of G.

We now show how to save one color from this coloring using a similar idea from [1]. Let X be the vertex cover. If our greedy coloring above used only $\Delta(\Delta - 1)$ colors to color vertices of $V(G) \backslash X$, then we are already done. Otherwise, pick any vertex u in X. We recolor u using a color used by vertices in $V(G) \backslash X$. Let u be adjacent to $i \leq \Delta - 1$ vertices in X (If all neighbors of u are in X, then u can be moved out of X, without loss of generality). Hence there are at most $i(\Delta - 1)$ vertices in $V(G) \backslash X$ which are at distance two from vertex u. There are at most $\Delta - i$ vertices adjacent to u in $V(G) \backslash X$. Colors used by all these vertices can not be used to recolor vertex u because of Observation 1 but u can be colored with any other color. Thus the number of forbidden colors is $i(\Delta - 1) + \Delta - i = i(\Delta - 2) + \Delta$. But $i(\Delta - 2) + \Delta \leq (\Delta - 1)(\Delta - 2) + \Delta = \Delta(\Delta - 1) - \Delta + 2 \leq \Delta(\Delta - 1)$ when $\Delta \geq 2$ and hence we can always find a color to recolor vertex u reducing the upper bound by 1. $\square$

The upper bound is tight for C_4, a cycle on 4 vertices.

Theorem 2 ($\star$). *If G is a d-degenerate graph, then $\Delta + 1 \leq \mathsf{hc}(G) \leq vc(G) + d(\Delta - 1) + \Delta(d - 1) + 1$.*

The following corollary follows from Theorem 2 as a forest is 1-degenerate.

Corollary 1. *If G is a forest with at least one edge, then $\Delta + 1 \leq \mathsf{hc}(G) \leq vc(G) + \Delta$.*

The upper bounds in Theorem 1 and Corollary 1 improve respectively the bounds of Theorems 6 and 4 of [1].

4   Parameterized Complexity of Harmonious Coloring

4.1   'Standard' and 'Above/Below Guarantee' Parameterizations

In this subsection, we capture some easy observations on the parameterized complexity of harmonious coloring under some standard parameterizations. We start with the following theorem whose proof (given in the full version of this paper) follows from the observation that if the number of edges is 'large', then the harmonious coloring number has to be large.

Lemma 2 ($\star$). *Let G be a graph on n vertices and m edges.* HARMONIOUS COLORING, *parameterized by the number of colors used, is* FPT *with a quadratic kernel.*

The proof of the above theorem suggests that the harmonious coloring number of most graphs is large with respect to the number of vertices. The number of vertices n is a trivial upper bound and Theorem 1 gives a lower bound of $\Delta + 1$ for the harmonious coloring number of a graph. So the natural question is: is it FPT to determine whether one can harmoniously color using at most $n - k$ or $\Delta + k + 1$ colors where the parameter is k. We prove the following theorem.

Theorem 3 ($\star$). *(i) It is W[1]-hard to determine whether a given n-vertex graph has harmonious coloring number at most $n - k$ where k is the parameter. (ii) It is para-NP-hard to determine whether a given graph has a harmonious coloring number at most $\Delta + 1 + k$ where Δ is the maximum degree of the graph, and k is the parameter.*

4.2   Parameterization by Size of Vertex Cover

As the HARMONIOUS COLORING is NP-complete on trees, it is trivially para NP-hard when parameterized by the treewidth of the graph or the feedback vertex set size of the graph. In this section, we consider the structural parameterization by the well-studied vertex cover number of the graph. We describe an FPT algorithm for HARMONIOUS COLORING when parameterized by the size of a vertex cover of the input graph. We show that the problem reduces to several instances of INTEGER LINEAR PROGRAMMING. We assume that the input graph G has no isolated vertices. Otherwise, for any harmonious coloring of the input graph G, we can include the set of isolated vertices into any one of the color classes.

In case of structural parameters, sometimes it is necessary to demand a witness of the required structure as part of the input. However, when the size of a vertex cover is the parameter, this is not a serious demand. Suppose the input parameter is ℓ. We find a 2-approximation of the minimum vertex cover of the input graph G(pp 11,[23]). If the size of the approximate vertex cover is strictly more than 2ℓ, then we have verified that the input parameter does not correspond to a valid vertex cover number of G. Otherwise, the approximate vertex cover is of size 2ℓ and we can use this vertex cover as a witness. Thus, we may assume that we are solving the following problem.

VC-HARMONIOUS COLORING   **Parameter:** $|X|$
Input: A graph G, a vertex cover X of G, a non-negative integer k
Question: Is there a harmonious coloring of G with k colors?

The idea is to enumerate over all the possible harmonious coloring of $G[X]$ and for each harmonious coloring, verify whether it can be extended to G using a total of k colors. As we will see, the problem of extending harmonious coloring of $G[X]$ to the entire graph is equivalent to that of finding harmonious coloring of the graph such that each color class contains at most one vertex from the vertex cover. We first observe some properties of such a harmonious coloring.

In the remaining section, unless stated otherwise, G is the input graph with vertex cover X of size ℓ and $I = V(G) \setminus X$ is an independent set.

Observation 3 ($\star$). *For any harmonious coloring of G the size of a color class is at most ℓ.*

For each vertex u in I we associate a brand.

Definition 2. *The brand of a vertex v in I with respect to X is the set $N(v)$.*

The number of different brands is upper bounded by the number of nonempty subsets of X which is $2^\ell - 1$. For vertices u, v in I if brand$(u) \cap$ brand$(v) \neq \emptyset$ then $d_G(u, v) = 2$ and by Observation 1 these two vertices can not belong to the same harmonious color class. For $S \subseteq X$, we define set $I(S) = \{v \in I | \text{brand}(u) = S\}$.

Consider a harmonious coloring $h : V(G) \to [k]$ and two vertices u, v in I, such that brand$(u) = $ brand(v). Let $h(u) = i$ and $h(v) = j$. Define a coloring $\widetilde{h}$ on $V(G)$ as $\widetilde{h}(w) = h(w)$ for all w in $V(G) \setminus \{u, v\}$, and $\widetilde{h}(u) = j$ and $\widetilde{h}(v) = i$.

Observation 4 ($\star$). *For a given harmonious coloring h of G, let u, v be two vertices in I such that brand$(u) = $ brand(v). If coloring $\widetilde{h}$ is as defined above then $\widetilde{h}$ is also a harmonious coloring of G.*

Thus we can characterize a harmonious color class based on the brand of the vertices which are part of it. Once the brands which make up the color class are fixed, it does not matter which vertex having that brand is chosen for the color class. This leads us to the definition of a type of a potential color class.

Definition 3 (type). *A type Z with respect to X is a $\ell + 1$ sized tuple where the first entry is subset of X of cardinality at most 1, and each of the remaining ℓ entries is either $\emptyset$ or a distinct brand of a vertex in I.*

A type Z can be represented as $(Y; S_1, S_2, \ldots, S_\ell)$ where Y is either an empty set or a singleton set from X. All the entries in this tuple are subsets of X but we distinguish the first entry from the remaining entries. The number of different types is at most $\ell \cdot \binom{2^\ell}{\ell}$, which is at most $\ell \cdot 2^{\ell^2}$. Any color class C which contains at most one vertex from the vertex cover and at most ℓ vertices from the independent set can be labeled with some type.

Definition 4 (Color Class of type Z). *Let h be a harmonious coloring of G such that each color class contains at most one vertex from X, and let $Z = (Y; S_1, S_2, \ldots, S_\ell)$ be a type defined with respect to X. Color class C of h is of type Z if $C \cap X = Y$ and for every $u \in C \cap I$ there exists S_i in type Z such that* brand$(u) = S_i$.

Not all the types can be used to label a harmonious color class. We define the notion of valid types to filter out such types.

Definition 5 (Valid type). *A type $Z = (Y; S_1, S_2, \ldots, S_\ell)$ is said to be valid if all the sets in the family $\{N[Y], S_1, S_2, \ldots S_\ell\}$ are pairwise disjoint.*

The validity constraints imply that if a vertex set is labeled with a valid type Z, then for any u, v in that set, the minimum distance between u and v is strictly greater than 2. Only the valid types can be used to label harmonious color classes.

Definition 6 (Compatible types). *Two valid types $Z = (Y; S_1, S_2, \ldots, S_\ell)$ and $Z' = (Y'; S'_1, S'_2, \ldots, S'_\ell)$ are said to be compatible with each other if $|Y \cap (S'_1 \cup S'_2 \cup \cdots \cup S'_\ell)| + |Y' \cap (S_1 \cup S_2 \cup \cdots \cup S_\ell)| \leq 1$.*

The compatibility condition of types encodes that the number of edges running across two harmonious color classes is at most 1. Two harmonious color classes C and C' can be of type Z and Z' respectively only if these two types are compatible with each other.

Lemma 3. *Let C and C' are two disjoint sets of $V(G)$ of valid types $Z = (Y; S_1, S_2, \ldots, S_\ell)$ and $Z' = (Y'; S'_1, S'_2, \ldots, S'_\ell)$ respectively. $|E(C, C')| \leq 1$ if and only if Z and Z' are campatible with each other.*

Proof. ($\Rightarrow$) If $|E(C, C')| = 0$ then there is no edge across C' and C and hence $|Y \cap (S'_1 \cup S'_2 \cup \cdots \cup S'_\ell)| = |Y' \cap (S_1 \cup S_2 \cup \cdots \cup S_\ell)| = 0$ making types Z and Z' compatible. Consider the case when $|E(C, C')| = 1$. With out loss of generality, let $x \in C \cap X$ and $z' \in C'$ and xz' is the edge across C and C'. For any u in $C \setminus X$, $E(\{u\}, Y') = \emptyset$ implying $N(u) \cap Y' = \emptyset$ which is equivalent to $|Y' \cap (S_1 \cup S_2 \cup \cdots \cup S_\ell)| = 0$. Since xz' is the only edge across C and C', $|Y \cap (S'_1 \cup S'_2 \cup \cdots \cup S'_\ell)|$ is 0 or 1 depending on whether z' is in X or not. In either case, types Z and Z' are compatible.

($\Leftarrow$) If $|Y \cap (S'_1 \cup S'_2 \cup \cdots \cup S'_\ell)| = |Y' \cap (S_1 \cup S_2 \cup \cdots \cup S_\ell)| = 0$ then there is no edge across C and C' whose one end point is outside vertex cover X. Since Y and Y' has cardinality of at most 1, $|E(C, C')| \leq 1$. So now we are in a case where $|Y \cap (S'_1 \cup S'_2 \cup \cdots \cup S'_\ell)| + |Y' \cap (S_1 \cup S_2 \cup \cdots \cup S_\ell)| = 1$. Without loss of generality, assume that $|Y \cap (S'_1 \cup S'_2 \cup \cdots \cup S'_\ell)| = 1$. This imply that there is an edge whose one end point is in Y and another end point is in $C' \setminus Y'$. Also, $|Y' \cap (S_1 \cup S_2 \cup \cdots \cup S_\ell)| = 0$ implies that there is no edge with one end point incident on Y' and another end point in $C \setminus Y$. The only thing that remains to argue that in this situation $E(Y, Y') = \emptyset$. If this is not the case then $Y \cap N(Y') \neq \emptyset$. But there exists S'_i such that $Y \cap S'_i \neq \emptyset$. Since Y is singleton set, this implies $N(Y') \cap S'_i \neq \emptyset$ which contradicts the fact that type Z' is valid. Hence $E(Y, Y') = \emptyset$ which concludes the proof of $|E(C, C')| \leq 1$. $\square$

For a given graph G and a vertex cover X of G, we construct a set $\mathcal{Z}$ consisting of all types with respect to X which are valid. For every subset $\mathcal{Z}'$ of $\mathcal{Z}$ such that any two types in $\mathcal{Z}'$ are compatible with each other, we construct an instance $\mathcal{J}_{\mathcal{Z}'}$ of INTEGER LINEAR PROGRAMMING as follows.

We define a variable z_i as the number of color class of type Z_i used in the coloring. In the following objective function, we encode the aim of minimizing number of color classes used.

$$\text{minimize} \sum_{i=1}^{|\mathcal{Z}'|} z_i$$

For every $S \subseteq X$ and $j \in [|\mathcal{Z}'|]$ define

$b_j^S = 1$ if there is brand S in type Z_j; otherwise 0

There are exactly $|I(S)|$ many vertices of brand S.

$$\sum_{j=1}^{|\mathcal{Z}'|} z_j \cdot b_j^S = |I(S)| \qquad \forall S \subseteq X \tag{1}$$

For every $x \in X$ and $j \in [|\mathcal{Z}'|]$ define

$c_j^x = 1$ if $\{x\}$ is the first entry in type Z_j; otherwise 0

There can be at most one color class which contains vertex x in X.

$$\sum_{j=1}^{|\mathcal{Z}'|} z_j \cdot c_j^x = 1 \qquad \forall x \in X \tag{2}$$

Corollary 2. *An instance $\mathcal{J}_{\mathcal{Z}'}$ can be solved in time $2^{\mathcal{O}(2^{\ell^2} \cdot \ell^3)} n^{\mathcal{O}(1)}$.*

Proof. The number of variables in instance $\mathcal{J}_{\mathcal{Z}'}$ is $|\mathcal{Z}'|$ which is upper bounded by $\ell \cdot 2^{\ell^2}$. The maximum value, any variable z_i can take, is n and the largest value any coefficient in the objective function can take is 1. The coefficients in the constraints are upper bounded by n. The number of constraints is at most $2^\ell + \ell$. By Proposition 1, instance $\mathcal{J}_{\mathcal{Z}'}$ can be solved in time $2^{\mathcal{O}(2^{\ell^2} \cdot \ell^3)} n^{\mathcal{O}(1)}$. $\square$

Recall that for a given graph G and its vertex cover X, $\mathcal{Z}$ is the set of all valid types with respect to X and $\mathcal{Z}'$ is a subset of $\mathcal{Z}$ such that any two types in $\mathcal{Z}'$ are compatible with each other.

Lemma 4 ($\star$). *Given a graph G with a vertex cover X, an integer k, there exists a harmonious coloring of G with at most k colors and each color class contains at most one vertex from X if and only if there exists $\mathcal{Z}' \subseteq \mathcal{Z}$ such that the minimum value for an instance $\mathcal{I}_{\mathcal{Z}'}$ is at most k.*

This leads us to the main theorem of this section.

Theorem 4 ($\star$). HARMONIOUS COLORING, *parameterized by the size of a vertex cover of the input graph, is fixed-parameter tractable.*

While it is an interesting open problem to improve the bound of the FPT algorithm, we show that when the input graph is a forest, the bound can be substantially improved to show the following.

Theorem 5 ($\star$). *Given a forest G, a vertex cover X of size ℓ, we can find the minimum harmonious number, and the corresponding coloring of G in $2^{\mathcal{O}(\ell^2)}n^{\mathcal{O}(1)}$ time.*

The main reason for the improved bound is that the number of brands for vertices in $V(I)$ comes down to at most $2\ell - 1$ (from $2^\ell - 1$). Also, except for ℓ brands, all others have at most one vertex having that brand. Furthermore, we can run through some careful choices and avoid solving the integer linear programming. The details are in the full version of this paper.

5   Exact Algorithm on Split Graphs

As the number of vertices is a trivial upper bound for the harmonious coloring number, a naive algorithm to find the minimum harmonious number runs through all the n^n possible colorings to find the minimum number. It is know that HARMONIOUS COLORING on Split graphs is NP-Complete. In this section, we give an exact algorithm for HARMONIOUS COLORING on the class of split graphs improving on this $2^{n \log n}$ bound to $2^n n^{\mathcal{O}(1)}$. We make use of a relation between a harmonious coloring of a split graph and a proper coloring of an auxiliary graph to obtain our improved algorithm. We can relate the number of colors required for a harmonious coloring of the graph G with that for a harmonious coloring of its non-trivial component.

Observation 5 ($\star$). *Let G be an input split graph with $E(G) \neq \emptyset$ and let C be a non-trivial component of G. Then $\mathsf{hc}(G) = \mathsf{hc}(C)$.*

Observation 6 ([21]). *For any harmonious coloring h of G and a split-partition (K, I), each vertex in K must be given a distinct color.*

As a corollary to Lemma 1, we obtain the following relation in split graphs.

Corollary 3 (⋆). *Let G be a connected split graph with a split-partition (K, I), and let H be the auxiliary graph defined from G as in the statement of Lemma 1. A coloring function h is a harmonious coloring of G if and only if $(i) h(K) \cap h(I) = \emptyset$, (ii) each vertex of K gets distinct color, and $(iii) h|_I$ is a proper coloring of H.*

Theorem 6. *Given a split graph G, there is an algorithm, running in $2^n n^{\mathcal{O}(1)}$ time, that computes the minimum harmonious coloring of graph G.*

Proof. By Observation 5, we can assume that G is a connected graph. Let (K, I) be a split partition of G. By Observation 6, in any harmonious coloring of G, each vertex of K must get a distinct color. Also, by connectivity, each vertex in $V(I)$ must be adjacent to a vertex in $V(K)$. Hence, in any harmonious coloring of G, the vertices of $V(I)$ must be colored distinctly from the vertices of $V(K)$. From Corollary 3, the minimum proper coloring of the auxiliary graph H gives the minimum harmonious coloring of G extending the coloring of K. Thus, it is enough to find the minimum proper coloring of H, which can be done in time $2^n n^{\mathcal{O}(1)}$ using the algorithm of Björklund et al. [5]. $\qquad\square$

We obtain an improved FPT algorithm for split graphs as a corollary.

Corollary 4 (⋆). *Given a split graph G and a non-negative integer k, we can determine whether G has a harmoniously coloring with at most k colors in $2^{\mathcal{O}(k^2)} n^{\mathcal{O}(1)}$ time.*

6 Conclusions

We have shown that the harmonious coloring problem is fixed-parameter tractable when parameterized by the harmonious coloring number or the vertex cover number. While improving the bounds for our FPT algorithms is a natural open problem, we end with the following specific open problems.

- When parameterizing by k, the harmonious coloring number, can the kernel size of $O(k^2)$ be improved?
- When parameterizing by the vertex cover number ℓ, is there a $c^\ell n^{\mathcal{O}(1)}$ algorithm, for some constant c, at least on trees?

References

1. Aflaki, A., Akbari, S., Edwards, K., Eskandani, D., Jamaali, M., Ravanbod, H.: On harmonious colouring of trees. Electron. J. Comb. **19**(1), P3 (2012)
2. Appel, K., Haken, W.: Every planar map is four colorable. part i: discharging. Ill. J. Math. **21**(3), 429–490 (1977)
3. Asdre, K., Ioannidou, K., Nikolopoulos, S.D.: The harmonious coloring problem is NP-complete for interval and permutation graphs. Discrete Appl. Math. **155**(17), 2377–2382 (2007)

4. Asdre, K., Nikolopoulos, S.D.: NP-completeness results for some problems on sub-classes of bipartite and chordal graphs. Theor. Comput. Sci. **381**(1), 248–259 (2007)
5. Björklund, A., Husfeldt, T., Koivisto, M.: Set partitioning via inclusion-exclusion. SIAM J. Comput. **39**(2), 546–563 (2009)
6. Bodlaender, H.L.: Achromatic number is NP-complete for cographs and interval graphs. Inf. Proc. Lett. **31**(3), 135–138 (1989)
7. Chen, J., Kanj, I.A., Meng, J., Xia, G., Zhang, F.: On the pseudo-achromatic number problem. Theor. Comput. Sci. **410**(810), 818–829 (2009)
8. Cygan, M., Fomin, F.V., Kowalik, Ł., Lokshtanov, D., Marx, D., Pilipczuk, M., Pilipczuk, M., Saurabh, S.: Parameterized Algorithms, vol. 4. Springer, Switzerland (2015)
9. Diestel, R.: Graph Theory. Graduate Texts in Mathematics, vol. 101 (2005)
10. Downey, R.G., Fellows, M.R.: Parameterized Complexity. Springer, New York (1999)
11. Edwards, K.: The harmonious chromatic number of bounded degree trees. Comb. Probab. Comput. **5**(01), 15–28 (1996)
12. Edwards, K.: The harmonious chromatic number and the achromatic number. Surv. Comb. **16**, 13 (1997)
13. Edwards, K., McDiarmid, C.: The complexity of harmonious colouring for trees. Discrete Appl. Math. **57**(2), 133–144 (1995)
14. Hell, P., Miller, D.J.: Graph with given achromatic number. Discrete Math. **16**(3), 195–207 (1976)
15. Hopcroft, J.E., Krishnamoorthy, M.S.: On the harmonious coloring of graphs. SIAM J. Algebraic Discrete Meth. **4**(3), 306–311 (1983)
16. Ioannidou, K., Nikolopoulos, S.D.: Harmonious coloring on subclasses of colinear graphs. In: Rahman, M.S., Fujita, S. (eds.) WALCOM 2010. LNCS, vol. 5942, pp. 136–148. Springer, Heidelberg (2010)
17. Kannan, R.: Minkowski's convex body theorem and integer programming. Math. Oper. Res. **12**(3), 415–440 (1987)
18. Karp, R.M.: Reducibility among combinatorial problems. In: Miller, R.E., Thatcher, J.W., Bohlinger, J.D. (eds.) Complexity of Computer Computations, pp. 85–103. Springer, New York (1972)
19. Lawler, E.L.: A note on the complexity of the chromatic number problem. Inf. Proc. Lett. **5**(3), 66–67 (1976)
20. Lenstra Jr., H.W.: Integer programming with a fixed number of variables. Math. Oper. Res. **8**(4), 538–548 (1983)
21. Miller, Z., Pritikin, D.: The harmonious coloring number of a graph. Discrete Math. **93**(2–3), 211–228 (1991)
22. Panolan, F., Philip, G., Saurabh, S.: B-chromatic number: Beyond np-hardness. In: 10th International Symposium on Parameterized and Exact Computation, IPEC 2015, 16–18 September 2015, Patras, Greece, pp. 389–401 (2015)
23. Papadimitriou, C.H., Steiglitz, K.: Combinatorial Optimization: Algorithms and Complexity. Courier Corporation, New York (1982)

On Directed Steiner Trees with Multiple Roots

Ondřej Suchý[✉]

Department of Theoretical Computer Science, Faculty of Information Technology,
Czech Technical University in Prague, Prague, Czech Republic
ondrej.suchy@fit.cvut.cz

Abstract. We introduce a new Steiner-type problem for directed graphs named q-ROOT STEINER TREE. Here one is given a directed graph $G = (V, A)$ and two subsets of its vertices, R of size q and T, and the task is to find a minimum size subgraph of G that contains a path from each vertex of R to each vertex of T. The special case of this problem with $q = 1$ is the well known DIRECTED STEINER TREE problem, while the special case with $T = R$ is the STRONGLY CONNECTED STEINER SUBGRAPH problem.

We first show that the problem is W[1]-hard with respect to $|T|$ for any $q \geq 2$. Then we restrict ourselves to instances with $R \subseteq T$ (PEDESTAL version). Generalizing the methods of Feldman and Ruhl [SIAM J. Comput. 2006], we present an algorithm for this restriction with running time $O(2^{2q+4|T|} \cdot n^{2q+O(1)})$, i.e., this restriction is FPT with respect to $|T|$ for any constant q. We further show that we can, without significantly affecting the achievable running time, loosen the restriction to only requiring that in the solution there is a vertex v and a path from each vertex of R to v and from v to each vertex of T (TRUNK version).

Finally, we use the methods of Chitnis et al. [SODA 2014] to show that the PEDESTAL version can be solved in planar graphs in $O(2^{O(q \log q + |T| \log q)} \cdot n^{O(\sqrt{q})})$ time.

1 Introduction

Steiner type problems are one of the most fundamental problems in the network design. In general words the task is to connect a given set of points at the minimum cost. The study of these problems in graphs was initiated independently by Hakimi [16] and Levin [21]. In the classic STEINER TREE one is given a (weighted) undirected graph $G = (V, E)$ and a set T of its vertices (*terminals*) and the task is to find a minimum cost connected subgraph containing all the terminals.

In directed graphs, the notion of connectivity is more complicated. The notion which turns out to be the closest to the undirected STEINER TREE is that of DIRECTED STEINER TREE (DST), where one is given a (weighted) directed graph $G = (V, A)$, a set T of terminals, and additionally a root vertex r and the task is to find a minimum weight subgraph that provides a path from r to

---

O. Suchý—The research was supported by the grant 14-13017P of the Czech Science Foundation.

P. Heggernes (Ed.): WG 2016, LNCS 9941, pp. 257–268, 2016.
DOI: 10.1007/978-3-662-53536-3_22

each vertex of T. Another natural option is, given a digraph $G = (V, A)$ and a set T of terminals, to search for a minimum weight subgraph that provides a path between each pair of terminals in both directions. This is problem is called STRONGLY CONNECTED STEINER SUBGRAPH (SCSS). The most general problem allows to prescribe the demanded connection between the terminals. Namely, in DIRECTED STEINER NETWORK (DSN) one is given a digraph $G = (V, A)$ and a set of q pairs of vertices $\{(s_1, t_1), \ldots (s_q, t_q)\}$ and is asked to find a minimum weight subgraph H of G that contains a directed path from s_i to t_i for every i.

Obviously, DSN is a generalization of both DST and SCSS. In this paper we consider a special case of DSN, which is still a very natural generalization of both DST and SCSS, namely the following problem:

---

q-ROOT STEINER TREE (q-RST)
Input: A directed graph $G = (V, A)$, two subsets of its vertices $R, T \subseteq V$ with $|R| = q$, and a positive integer k.
Question: Is there a set $S \subseteq V$ of size at most k such that in $G[R \cup S \cup T]$ there is a directed path from r to t for every $r \in R$ and every $t \in T$?

If $q = 1$, then q-RST problem is equal to (unweighted) DST. On the other hand, if we let $T = R$, then the problem is equivalent to (unweighted) SCSS on the terminal set T. We study the problem from a multivariate perspective, examining the influence of various parameters on the complexity of the problem. We focus on the following parameters: number of roots $q = |R|$, number of terminals $|T|$, and to a limited extent also to the budget k. Thorough the paper we denote $n = |V|$ and $m = |A|$. Before we present our results, let us summarize what is known about the problems.

Known Results: STEINER TREE is NP-hard [14] and remains so even in very restricted planar cases [13]. As the NP-hardness can be easily transferred also to DST and SCSS, the problems were studied from approximation perspective. However, in general terms, the problems are also hard to approximate. The best known approximation factor for DST and SCSS is $O(|T|^\epsilon)$ for any fixed $\epsilon > 0$ [3]. On the other hand, the problems cannot be approximated to within a factor of $O(\log^{2-\epsilon} n)$ for any $\epsilon > 0$, unless NP has quasi-polynomial time Las Vegas algorithms [17]. For the most general DSN problem the best known ratio is $n^{2/3+\epsilon}$ for any $\epsilon > 0$ and the problem cannot be approximated to within $O(2^{\log^{1-\epsilon} n})$ for any $\epsilon > 0$, unless NP has quasi-polynomial time algorithms [1]. We refer to surveys, e.g., [20], for more information on the numerous polynomial-time approximation results for Steiner-type problems.

From the perspective of parameterized algorithms [6,7] the problems are mostly studied with respect to the number of terminals. It follows from the classical result of Dreyfus and Wagner [8] (independently found by Levin [21]), that STEINER TREE and also DST can be solved in $O(3^{|T|} \cdot n^{O(1)})$ time. The algorithm was subsequently improved [2,9,12] with the latest algorithm of Nederlof [23] achieving $O(2^{|T|} \cdot n^{O(1)})$ time and polynomial space complexity.

For the SCSS and DSN with q terminals and q terminal pairs, Feldman and Ruhl [10] showed that the problems can be solved roughly in $O(n^{2q-1})$ and $O(n^{4q})$ time, respectively. We cannot expect fixed parameter tractability for these problems, since the problems are W[1]-hard with respect to this parameter [15] (and even with respect to the total size of the sought graph). In fact, unless the Exponential Time Hypothesis (ETH)[1] [18] fails, SCSS cannot be solved in $f(q)n^{o(q/\log q)}$ time on general graphs and DSN cannot be solved in $f(q)n^{o(q)}$ time even on planar DAGs [4]. Chitnis et al. [4] also showed that on planar graphs SCSS can be solved within $2^{O(q \log q)}n^{O(\sqrt{q})}$ time, but it is still W[1]-hard and cannot be solved within $f(q)n^{o(\sqrt{q})}$ time, unless ETH fails. These results hold for any computable function f.

With respect to the less studied parameter "number of nonterminals in the solution", representing one possible measure of the solution size, all the problems are on general graphs W[2]-hard by an easy reduction from SET COVER (see, e.g., Guo et al. [15]). Chitnis et al. [5] considered FPT-approximations for the problems. Notable SCSS admits a factor-2 FPT-approximation. On planar graphs, only DST was studied with respect to this parameter, achieving fixed parameter tractability [19].

An independent paper on the topics similar to the topics covered in this paper appeared in the proceeding of ICALP 2016 [11].

Our Contribution: In this paper our aim is to generalize the positive results for DST and SCSS also to q-RST. Unfortunately, as our first result, we show that q-RST is still too general to achieve this goal. Namely, we show that for any constant $q \geq 2$ the q-RST is W[1]-hard with respect to $|T|$ even on directed acyclic graphs and cannot be solved within $f(|T|)n^{o(|T|/\log|T|)}$ time, unless ETH fails. In fact the same results hold even if we replace $|T|$ by $(k+|T|)$, the total number of vertices in the resulting subgraph (minus q).

Then, we restrict the problem further to its special case by requiring $R \subseteq T$. In fact, for better readability we require the solution to provide a path from each $r \in R$ to each vertex $t \in R \cup T$ and assume $T \cap R = \emptyset$. We call the resulting problem q-ROOT STEINER TREE WITH PEDESTAL (q-RST-P). Observe that it still generalizes DST as well as SCSS.

We show that we can generalize the algorithm of Feldman and Ruhl [10] for SCSS to q-RST-P, using an algorithm for DST as a subroutine. The running time of our algorithm is $O(2^{2q+4|T|} \cdot n^{2q+O(1)})$, i.e., the problem is FPT with respect to $|T|$ for any constant q and the exponent of the polynomial depends linearly on q. The lower bounds for SCSS indicate that this dependency on q is almost optimal. In fact if $T = \emptyset$, then our algorithm is exactly the algorithm of Feldman and Ruhl, while if $q = 1$, the algorithm boils down to a single call to the DST subroutine.

The algorithm of Feldman and Ruhl is based on a token game, where the tokens trace the path required in the solution. The solution of the instance is

---

[1] ETH states that there is a positive constant c such that no algorithm can solve n-variable 3-SAT in $O(2^{cn})$ time.

then represented by a sequence of moves of the tokens between two specified configurations. We first enrich the game by introducing new tokens that trace the path to vertices of T while using the original tokens to trace paths between the vertices in R. We call this game cautious.

We then show that the solutions can be represented by move sequences with further interesting properties. These allow us to group the moves and reduce the number of intermediate configurations. The resulting game, which we call accelerated, has moves very similar to the original game of Feldman and Ruhl, but each move is now equipped by a subset of vertices of T that is also reached in this move. We use this similarity for further results in our paper.

The crucial property of the problem that allows us to come up with the algorithm is that there is a vertex such that every path required by the solution can be dragged through this specific vertex (allowing the vertices to repeat on the path). To illustrate this, we introduce another variant of the problem named q-ROOT STEINER TREE WITH TRUNK (q-RST-T), which is the same as q-RST, but the solution is further required to contain a vertex which has a path from each vertex in R and to each vertex in T. We show that this problem can be solved in similar running time as q-RST-P, namely $O(2^{2q+4|T|} \cdot n^{3q+O(1)})$. Qualitatively similar running time can be also achieved if the special vertex provides all but a constant number of the paths required by the problem.

We further generalize the result of Chitnis et al. [4] giving the improved algorithm for SCSS in planar graphs to obtain an algorithm for q-RST-P in planar graphs with running time $O(2^{O(q \log q + |T| \log q)} \cdot n^{O(\sqrt{q})})$.

While the hardness result applies to the decision variant, the algorithms directly apply to the (cardinality) optimization case. Moreover, it is straightforward to generalize them to the case of vertex weights (we might want to use different, more suitable, DST algorithm as a subroutine, based on the actual range of the weights). In order to use arc weights, one just has to subdivide each arc and give the weight of the arc to the newly created vertex. Thus our algorithms also apply to vertex weighted and arc weighted variants of the problems. Nevertheless, for ease of presentation, we formulate all our results only for the cardinality case.

Organization of the paper: In Sect. 2 we present the hardness result for the unrestricted version of q-RST. Section 3 describes the games and the algorithm for q-RST-P. This is generalized to q-RST-T in Sect. 4. The improved algorithm for q-RST-P in planar graphs is contained in Sect. 5. We conclude the paper with outlook in Sect. 6.

Due to space constraints many proofs and explanations had to be deferred to the full version of the paper, preprint of which is available on arXiv [25].

2   Unrestricted Case

This section is devoted to the proof of the following theorem.

Theorem 1 ($\star$).[2] *q-RST is W[1]-hard with respect to $|T|$ even on directed acyclic graphs for every $q \geq 2$. Moreover, there is no algorithm for q-RST on directed acyclic graphs running in $f(|T|)n^{o(\frac{|T|}{\log|T|})}$ time for any constant $q \geq 2$, unless ETH fails.*

Our starting point are the known results for the following problem.

PARTITIONED SUBGRAPH ISOMORPHISM (PSI)
Input: Undirected graphs $H = (V_H, E_H)$ and $G = (V_G, E_G)$ and a coloring function $col : V_H \to V_G$.
Question: Is there an injection $\phi : V_G \to V_H$ such that for every $i \in V_G$, $col(\phi(i)) = i$ and for every $\{i, j\} \in E_G$, $\{\phi(i), \phi(j)\} \in E_H$?

PSI is known to be W[1]-hard [24]. We also need the following lemma.

Lemma 1 ([22, Corollary 6.3]). PARTITIONED SUBGRAPH ISOMORPHISM *cannot be solved in time $f(k)|V_H|^{o(\frac{k}{\log k})}$ where f is an arbitrary function and $k = |E_G|$ is the number of edges in the smaller graph G unless ETH fails.*

We provide a parameterized reduction from PSI parameterized by $|E_G|$ to q-RST parameterized by $|T|$. Let us start with the case $q = 2$.

Let $(H = (V_H, E_H), G = (V_G, E_G), col)$ be an instance of PSI and let there be some strict linear orders $<$ on the vertices in V_H and in V_G such that if $u < v$, then $col(u) < col(v)$. We also assume that for every edge $\{u, v\} \in E_h$ we have $\{col(u), col(v)\} \in E_G$ and that G is connected and has at least one edge.

We start by constructing the directed graph $G' = (V', A')$ We let $V' = R \cup V_H \cup E' \cup F \cup T$, where $R = \{r_V, r_E\}$, $E' = \{a_{u,v} \mid \{u, v\} \in E_H, u < v\}$, $F = \{b_{u,v}, b_{v,u} \mid \{u, v\} \in E\}$, and $T = \{t_{i,j}, t_{j,i} \mid \{i, j\} \in E_G, i < j\}$.

The set of arcs is constructed as follows. We add arcs from r_V to all vertices in V_H and from r_E to all vertices in E'. For every edge $\{u, v\}$ where $u < v$, we add the following set of arcs: an arc from $a_{u,v}$ to $b_{u,v}$ and an arc from $a_{u,v}$ to $b_{v,u}$; an arc from u to $b_{u,v}$ and an arc from v to $b_{v,u}$; and an arc from $b_{u,v}$ to $t_{col(u),col(v)}$ and an arc from $b_{v,u}$ to $t_{col(v),col(u)}$.

To finish the construction we let $k' = 3|E_G| + \ell$. Note that we have $|T| = 2|E_G|$, i.e., the new parameter depends linearly on the original one and the constructed graph is a directed acyclic graph.

The proof that the instance (G', R, T, k') of 2-RST is equivalent to the original one is deferred to the full paper. For the case $q > 2$ it is enough to add $q - 2$ vertices to R, each having an arc only to r_V.

3   Restriction to Solutions with Pedestal

Having shown in the previous section that we cannot show q-RST FPT like we got for DST (which is 1-RST), in this section we restrict ourselves further. To this end, we modify the definition of our problem in the sense that we do not

---

[2] Proofs of statements marked with ($\star$) were deferred to the full version of the paper.

require to obtain a path from each vertex of R only to each vertex of T, but also to each other vertex of R.

---

q-ROOT STEINER TREE WITH PEDESTAL (q-RST-P)
Input: A directed graph $G = (V, A)$, two subsets of its vertices $R, T \subseteq V$ with $|R| = q$.
Task: Find a minimum size of a set $S \subseteq V$ such that in $G[R \cup S \cup T]$ there is a directed path from r to t for every $r \in R$ and every $t \in R \cup T$.

Note that this variant of q-RST could be also modeled by requiring $R \subseteq T$. However, to simplify the description, we assume $R \cap T = \emptyset$.

Theorem 2. *For every $q \geq 1$ the problem q-RST-P is fixed-parameter tractable with respect to $|T|$. Namely there is an algorithm solving it in $O(2^{2q+4|T|} \cdot n^{2q+O(1)})$ time, where the constants hidden in the $O()$ notations are independent of $|T|$ and q.*

The rest of this section is devoted to the proof of this theorem.

The q-RST-P problem with the set T empty is exactly the SCSS problem (with q terminals). This problem was shown to be polynomial time solvable for every constant q by Feldman and Ruhl [10] using a modeling by a token game. The cost of an optimal strategy for that game equals cost of the smallest solution to the SCSS instance.

We first slightly modify this game to model the problem for arbitrary T in Subsect. 3.1. We show there that optimal strategies for this game have some interesting properties which we can further use. Then, in Subsect. 3.2, we introduce a new game with more powerful moves which allows us to make many moves of the original game at once. Finally, in Subsect. 3.3, we show that the optimal strategies for the new game can be computed in the claimed running time.

3.1 Cautious Token Game

In this subsection we show how to modify the original token of Feldman and Ruhl in order to model the q-RST-P problem. We fix a vertex $r_0 \in R$ and let $R' = R \setminus \{r_0\}$. For a solution S the graph $G[R \cup S \cup T]$ will contain a path from r_0 to t for each vertex t in $R' \cup T$. These paths together form an out-tree rooted at r_0 which is called the *backward tree*. Also there is a path from each of the vertices in R' to r_0, and these together form an in-tree rooted at r_0, called the *forward tree*.

The game traces the two trees by having three types of tokens, where two of them behave similarly. First, we have an F-token at each of the vertices of R' and this token moves forward along the arcs of graph G. Second, we have a B-token at each vertex of R', moving backward against the direction of the arcs of G. The third type of tokens we use (different from Feldman and Ruhl) are D-tokens which are originally placed one on each of the vertices of T and move similarly as B-tokens.

The purpose of the tokens is to trace the forward and backward tree. Hence, whenever two tokens of the same type arrive at the same vertex we can merge them to one token. This is also the case for B-tokens and D-tokens, and in case a B-token merges with a D-token we let the merged token be a B-token. The purpose of introducing the D-tokens is to show that these are somewhat less important for the game than B-tokens and, hence, they can be treated in a different way in the new game we will introduce in the next subsection.

The state of the game can be described by tree subset of vertices (F, B, D) representing the set of vertices occupied by F-tokens, B-tokens, and D-tokens, respectively. Note that $|F| \leq q, |B| \leq q, |D| \leq |T|$ during the whole game. Hence we take $F, B \in \binom{V}{\leq q}$ and $D \in \binom{V}{\leq |T|}$ (here and on $\binom{V}{\leq q}$ is the set of subsets of V of size at most q).

The allowed moves are the following:

(1) *Single moves for respective tokens:* For every arc $(u, v) \in A$ and all sets $F, B \in \binom{V}{\leq q}$ and $D \in \binom{V}{\leq |T|}$ we introduce the following moves:

   (a) If $u \in F$, then we have a move $(F, B, D) \overset{c}{\to} ((F \setminus \{u\}) \cup \{v\}, B, D)$, where the cost c of the move is 1 if $v \notin F \cup B \cup D$ and 0 otherwise.

   (b) If $v \in B$, then we have a move $(F, B, D) \overset{c}{\to} (F(B \setminus \{v\}) \cup \{u\}, D \setminus \{u\})$, where the cost c of the move is 1 if $u \notin F \cup B \cup D$ and 0 otherwise.

   (c) If $v \in D$, then we have a move $(F, B, D) \overset{c}{\to} (F, B(D \setminus \{v\}) \cup (\{u\} \setminus B))$, where the cost c of the move is 1 if $u \notin F \cup B \cup D$ and 0 otherwise.

(2) *Flipping:* For all sets $F, B \in \binom{V}{\leq q}$ and $D \in \binom{V}{\leq |T|}$ we introduce the following moves:

   (a) if $F' \subseteq F$, $B' \subseteq B$, $D' \subseteq D$, $f \in F'$, and $b \in B'$, then we have a move $(F, B, D) \overset{c}{\to} ((F \setminus F') \cup \{b\}(B \setminus B') \cup \{f\}, D \setminus (D' \cup \{f\}))$, where c is the number of vertices on a shortest walk from f to b going through all vertices in $F' \cup B' \cup D'$. Here each vertex is counted each time it is visited, but vertices in $F' \cup B' \cup D'$ are counted once less.

   (b) if $F' \subseteq F$, $B' \subseteq B$, $B' \neq \emptyset$, $D' \subseteq D$, $f \in F'$, and $d \in D'$, then we have a move $(F, B, D) \overset{c}{\to} ((F \setminus F') \cup \{d\}(B \setminus B') \cup \{f\}, D \setminus (D' \cup \{f\}))$, where c is the number of vertices on a shortest walk from f to d going through all vertices in $F' \cup B' \cup D'$. Here, again, each vertex is counted each time it is visited, but vertices in $F' \cup B' \cup D'$ are counted once less.

   (c) if $F' \subseteq F$, $D' \subseteq D$, $f \in F'$, and $d \in D'$, then we have a move $(F, B, D) \overset{c}{\to} ((F \setminus F') \cup \{d\}, B(D \setminus D') \cup (\{f\} \setminus B))$, where c is the number of vertices on a shortest walk from f to d going through all vertices in $F' \cup D'$. As in the previous cases, each vertex is counted each time it is visited, but vertices in $F' \cup D'$ are counted once less.

The original game of Feldman and Ruhl has only three types of moves: Single moves for F-tokens (exactly as (1-a)), single moves for B-tokens (similar as (1-b) and (1-c)) and flipping (all of (2)). If we did not distinguish the B-tokens and D-tokens (and consider all of them as B-tokens), then we would get exactly this three types of moves. We make use of this fact in the proof of the equivalence of costs of optimal strategies for this game and sizes of solutions for the q-RST-P instance.

We distinguish the B- and D-tokens since we aim to show, e.g., that there is an optimal strategy for the game not using any moves of type (2-c). During the whole game the moves ensure that the invariant $D \cap B = \emptyset$ is maintained. This could be easily achieved by taking $D = D \setminus B$ after each move, however, we prefer to be more specific in taking out only the vertices which could actually newly appear in the intersection.

Now we would like to claim, that the game represents the instance (G, R, T) of q-RST-P. Namely, that the minimum size of a solution to (G, R, T) is exactly one less than the minimum cost of moves to get from (R', R', T) to $(\{r_0\}, \{r_0\}, \emptyset)$ in the cautious token game. The easier direction is summarized by the following lemma (see also Lemma 3.1 of [10]):

Lemma 2. *If there is a move sequence from (R', R', T) to $(\{r_0\}, \{r_0\}, \emptyset)$ of total cost c, then there is a set $S \subseteq V$ of size at most $c - 1$ such that in $G[R \cup S \cup T]$ there is a directed path from r to t for every $r \in R$ and every $t \in R \cup T$. Moreover, given the sequence, the corresponding set S is easy to find.*

The proof of this lemma follows from the definition of the moves of the game. If we let S be the set of newly encountered vertices in the moves of the sequence excluding r_0 we get $|S| + 1 \leq c$, as the cost of each move is an upper bound on the number of newly encountered vertices including r_0.

The next lemma provides the counterpart. The aim is to construct a move sequence, where all intermediate position of tokens are in $H = G[R \cup S \cup T]$. Let us call a move of type (2) *path-driven*, if the minimum size walk in the definition of the cost of the move can be taken as a simple path. We subsequently show that the move sequence can be selected such that: all moves of type (2) are path-driven; there are no moves of type (2-c); and if a D-token meets with an F-token, then it stays on place until it is merged with some B-token. More details can be found in the full version of the paper.

Lemma 3 ($\star$). *If there is a solution $S \subseteq V$ of size at most $c - 1$ for (G, R, T), then there is a move sequence from (R', R', T) to $(\{r_0\}, \{r_0\}, \emptyset)$ of total cost at most c in which all type (2) moves are path-driven and, moreover, there are no moves of type (2-c). Furthermore, in this sequence of moves, whenever after some move an F-token and a D-token sit together on a vertex $v \in V$, then the next move touching the D-token is either of type (2), or single-move (type (1-b)) of some B-token merging with the D-token.*

3.2   Accelerated Token Game

In the accelerated game the D-tokens stay at their places until the move in which they should be merged with a B-token. The moves then also include the costs of moving the D-tokens from their original places to the vertex where they get merged with the B-token. Therefore, we now represent the positions of the D-tokens only as subsets of T.

To define the costs of the moves we use the following notion. Let $ST(r, X)$ be the minimum number of vertices in a set S such that in $G[\{r\} \cup X \cup S]$ there is a path from r to every $x \in X$. I.e., this is a variation of DST with root r and terminals X.

We have the following moves (we number the moves from (3), as not to confuse them with the moves of the cautious game).

(3) *Single moves:* For every arc $(u, v) \in A$ and all sets $F, B \in \binom{V}{\leq q}$ and $D \subseteq T$ we introduce the following moves:

    (a) If $u \in F$, then we have a move $(F, B, D) \xrightarrow{c} ((F \setminus \{u\}) \cup \{v\}, B, D)$, where the cost c of the move is 1 if $v \notin F \cup B \cup D$ and 0 otherwise.

    (b) If $v \in B$ and $D' \subseteq D$ then we have a move $(F, B, D) \xrightarrow{c} (F(B \setminus \{v\}) \cup \{u\}, D \setminus (D' \cup \{u\}))$, where the cost c is $ST(u, D' \cup \{v\}) + 1$ if $u \notin F \cup B \cup D$ and $ST(u, D' \cup \{v\})$ otherwise.

(4) *Flipping:* For all sets $F, B \in \binom{V}{\leq q}$ and $D \subseteq T$ we introduce the following moves:

    (a) if $F' \subseteq F$, $B' \subseteq B$, $D' \subseteq D$, $f \in F'$, and $b \in B'$, then we have a move $(F, B, D) \xrightarrow{c} ((F \setminus F') \cup \{b\}(B \setminus B') \cup \{f\}, D \setminus (D' \cup \{f\}))$, where c is as described below.

    (b) if $F' \subseteq F$, $B' \subseteq B$, $B' \neq \emptyset$, $D' \subseteq D$, $D' \neq \emptyset$, $f \in F'$, and v is an arbitrary vertex in $V \setminus B$, then we have a move $(F, B, D) \xrightarrow{c} ((F \setminus F') \cup \{v\}(B \setminus B') \cup \{f\}, D \setminus (D' \cup \{f\}))$, where c is as described below.

(5) *Finishing:* For all sets $D \subseteq T$ we have a move $(\{r_0\}, \{r_0\}, D) \xrightarrow{c} (\{r_0\}, \{r_0\}, \emptyset)$, where $c = ST(r_0, D)$.

Let us now explain the intuition behind the moves. It is clear for (3-a). We consider the move (3-b) to move the B-token from v as well as the D-tokens from D' to u. The move (4-a) moves all B- and D-tokens from $B' \cup D'$ to f at the same time moving the F-tokens from F' to b. The move (4-b) does the same thing, except that the F-tokens are taken to a vertex v. The move (5) moves the D-tokens from D to r_0.

We would like to define the cost of moves of type (4) as the minimum number of vertices in a subgraph that provides a walk from f to b (or from f to v) through all vertices in $F' \cup B'$ and at the same time a path from f to each vertex of D', where the vertices in $F' \cup B' \cup D'$ again do not count. In fact the solution will again use this type of moves only when there is a simple path from f to b (or from f to v) through all vertices in $F' \cup B'$.

As this condition is complicated to test and the desired cost is complicated to compute, we will define a cost of a move which provides an upper bound on the desired cost, and coincides with the desired cost whenever the optimal walk is actually a simple path.

We define the cost of the type (4-a) moves c to be the minimum over all bijections $\phi : \{2, \ldots, |F' \cup B'| - 1\} \to (F' \cup B') \setminus \{f, b\}$ (representing the order of the vertices of $(F' \cup B') \setminus \{f, b\}$ along the walk) and all mappings $\psi : D' \to (F' \cup B') \setminus \{b\}$ (representing the part of the path at which the particular D-token joins it) of the sum $\sum_{i=1}^{|F' \cup B'| - 1} ST(\phi(i), \psi^{-1}(\phi(i)) \cup \{\phi(i+1)\})$, where $\phi(1) = f$ and $\phi(|F' \cup B'|) = b$.

Similarly, the cost of a (4-b) move is the minimum over all bijections ϕ : $\{2,\ldots,|F'\cup B'|\} \to (F'\cup B')\setminus\{f\}$ and all mappings $\psi : D' \to (F'\cup B')$ of the sum $\sum_{i=1}^{|F'\cup B'|} ST(\phi(i),\psi^{-1}(\phi(i)) \cup \{\phi(i+1)\})$, where $\phi(1) = f$ and $\phi(|F'\cup B'|) = v$. Here, the cost is increased by one if $v \notin (D \cup F)$.

The following two lemmata show that solving the instance of q-RST-P again corresponds to finding a cheapest possible move sequence from (R', R', T) to $(\{r_0\}, \{r_0\}, \emptyset)$ in the accelerated token game.

Lemma 4. *If there is a move sequence of the accelerated game from (R', R', T) to $(\{r_0\}, \{r_0\}, \emptyset)$ of total cost c, then there is a solution $S \subseteq V$ for (G, R, T) of size at most $c - 1$.*

Lemma 5 ($\star$). *If there is a solution $S \subseteq V$ for (G, R, T) of size at most $c - 1$, then there is a move sequence of the accelerated game from (R', R', T) to $(\{r_0\}, \{r_0\}, \emptyset)$ of total cost at most c.*

3.3 The Algorithm

The algorithm first computes the so-called "game graph" for the accelerated game. The vertex set V' of this directed graph is formed by all possible configurations of the tokens in the game, i.e., by all triples (F, B, D), where $F, B \in \binom{V}{\leq q}$ and $D \subseteq T$. The arcs A' of this graph correspond to legal moves of the accelerated game, i.e., moves of type (3), (4), and (5). The length of each arc is equal to the cost of the corresponding move of the game.

Once the game graph is constructed, we simply find the shortest path from the vertex (R', R', T) to the vertex $(\{r_0\}, \{r_0\}, \emptyset)$. Since each arc of the graph corresponds to a move of the game, the shortest path corresponds to an optimal sequence of moves to get from the configuration (R', R', T) to the configuration $(\{r_0\}, \{r_0\}, \emptyset)$. By Lemmatas 4 and 5 such a sequence of cost c exists if and only if there is a solution of size $c - 1$ for the instance (G, R, T) of q-RST-P.

We defer the running time analysis of the algorithm to the full version of the paper.

4   Restriction to Solutions with Trunk

In this section we relax the conditions on the solutions of the problem, namely, we consider the following problem:

---

q-ROOT STEINER TREE WITH TRUNK (q-RST-T)
Input: A directed graph $G = (V, A)$, two subsets of its vertices $R, T \subseteq V$ with $|R| = q$.
Task: Find a minimum size of a set $S \subseteq V$ such that in $G[R \cup S \cup T]$ there is a vertex v, a directed path from r to v for every $r \in R$, and a directed path from v to t for every $t \in T$.

We show that this problem can be solved in asymptotically similar time as q-RST-P.

Theorem 3 ($\star$). *For every $q \geq 1$ the problem q-RST-T is fixed-parameter tractable with respect to T. Namely, there is an algorithm solving it in $O(2^{2q+4|T|} \cdot n^{3q+O(1)})$ time.*

Suppose now that we restrict to solutions such that there is a vertex v that is in all but a constant number h of the required paths. More formally, for all but h (given) pairs $(r, t) \in R \times T$ the solution contains a path from r to v and at the same time a path from v to t. We can solve also this variant of the problem in $O(2^{O(q+|T|)} \cdot n^{O(q+h)})$ time.

5   Planar Graphs

In this section we show how to modify the method of Chitnis et al. [4] for SCSS in planar graphs (and graphs excluding a fixed minor) to show the following result.

Theorem 4 ($\star$). *q-ROOT STEINER TREE WITH PEDESTAL in planar graphs and graphs excluding a fixed minor can be solved in $O(2^{O(q \log q + |T| \log q)} \cdot n^{O(\sqrt{q})})$ time.*

6   Conclusion and Future Directions

We have shown that there is a nice special case of DSN that allows for as effective algorithms as were known for DST and SCSS, even with respect to planar graphs. We characterized that the crucial property of the solution to allow this is the existence of a vertex over which almost all paths required by the problem definition "factorize". An interesting open question is what is the complexity of q-RST (the unrestricted variant) in planar graphs.

Another interesting question is tied to the other parameterization of the problems. We are not aware of any result determining the complexity of SCSS in planar graphs with respect to the parameterization the number of nonterminals in the solution.

References

1. Berman, P., Bhattacharyya, A., Makarychev, K., Raskhodnikova, S., Yaroslavtsev, G.: Approximation algorithms for spanner problems and directed Steiner forest. Inf. Comput. **222**, 93–107 (2013)
2. Björklund, A., Husfeldt, T., Kaski, P., Koivisto, M.: Fourier meets Möbius: fast subset convolution. In: STOC 2007, pp. 67–74. ACM (2007)
3. Charikar, M., Chekuri, C., Cheung, T.Y., Dai, Z., Goel, A., Guha, S., Li, M.: Approximation algorithms for directed Steiner problems. J. Algorithms **33**(1), 73–91 (1999)
4. Chitnis, R., Hajiaghayi, M., Marx, D.: Tight bounds for planar strongly connected Steiner subgraph with fixed number of terminals (and extensions). In: SODA 2014, pp. 1782–1801. SIAM (2014)

5. Chitnis, R., Hajiaghayi, M.T., Kortsarz, G.: Fixed-parameter and approximation algorithms: a new look. In: Gutin, G., Szeider, S. (eds.) IPEC 2013. LNCS, vol. 8246, pp. 110–122. Springer, Heidelberg (2013)
6. Cygan, M., Fomin, F.V., Kowalik, L., Lokshtanov, D., Marx, D., Pilipczuk, M., Pilipczuk, M., Saurabh, S.: Parameterized Algorithms. Springer, Switzerland (2015)
7. Downey, R.G., Fellows, M.R.: Fundamentals of Parameterized Complexity. Texts in Computer Science. Springer, London (2013)
8. Dreyfus, S.E., Wagner, R.A.: The Steiner problem in graphs. Networks 1, 195–207 (1972)
9. Erickson, R.E., Monma, C.L., Veinott Jr., A.F.: Send-and-split method for minimum-concave-cost network flows. Math. Oper. Res. 12(4), 634–664 (1987)
10. Feldman, J., Ruhl, M.: The directed Steiner network problem is tractable for a constant number of terminals. SIAM J. Comput. 36(2), 543–561 (2006)
11. Feldmann, A.E., Marx, D.: The complexity landscape of fixed-parameter directed Steiner network problems. In: ICALP 2016. LIPIcs, vol. 55, pp. 27:1–27:4. Dagstuhl (2016). http://dx.doi.org/10.4230/LIPIcs.ICALP.2016.27
12. Fuchs, B., Kern, W., Mölle, D., Richter, S., Rossmanith, P., Wang, X.: Dynamic programming for minimum Steiner trees. Theor. Comput. Syst. 41(3), 493–500 (2007)
13. Garey, M.R., Johnson, D.S.: The rectilinear Steiner tree problem is NP-complete. SIAM J. Appl. Math. 32(4), 826–834 (1977)
14. Garey, M.R., Johnson, D.S.: Computers and Intractability: A Guide to the Theory of NP-Completeness. Freeman, New York (1979)
15. Guo, J., Niedermeier, R., Suchý, O.: Parameterized complexity of arc-weighted directed Steiner problems. SIAM J. Discrete Math. 25(2), 583–599 (2011)
16. Hakimi, S.L.: Steiner's problem in graphs and its implications. Networks 1, 113–133 (1971)
17. Halperin, E., Kortsarz, G., Krauthgamer, R., Srinivasan, A., Wang, N.: Integrality ratio for group Steiner trees and directed Steiner trees. SIAM J. Comput. 36(5), 1494–1511 (2007)
18. Impagliazzo, R., Paturi, R.: On the complexity of k-SAT. J. Comput. Syst. Sci. 62(2), 367–375 (2001)
19. Jones, M., Lokshtanov, D., Ramanujan, M.S., Saurabh, S., Suchý, O.: Parameterized complexity of directed Steiner tree on sparse graphs. In: Bodlaender, H.L., Italiano, G.F. (eds.) ESA 2013. LNCS, vol. 8125, pp. 671–682. Springer, Heidelberg (2013)
20. Kortsarz, G., Nutov, Z.: Approximating minimum cost connectivity problems. In: Handbook of Approximation Algorithms and Metaheuristics, chapt. 58. CRC (2007)
21. Levin, A.Y.: Algorithm for the shortest connection of a group of graph vertices. Sov. Math. Dokl. 12, 1477–1481 (1971)
22. Marx, D.: Can you beat treewidth? Theor. Comput. 6(1), 85–112 (2010)
23. Nederlof, J.: Fast polynomial-space algorithms using inclusion-exclusion. Algorithmica 65(4), 868–884 (2013)
24. Pietrzak, K.: On the parameterized complexity of the fixed alphabet shortest common supersequence and longest common subsequence problems. J. Comput. Syst. Sci. 67(4), 757–771 (2003)
25. Suchý, O.: On directed Steiner trees with multiple roots. CoRR abs/1604.05103(2016). http://arxiv.org/abs/1604.05103

A Faster Parameterized Algorithm for GROUP FEEDBACK EDGE SET

M.S. Ramanujan[✉]

Vienna University of Technology, Vienna, Austria
ramanujan@ac.tuwien.ac.at

Abstract. In the GROUP FEEDBACK EDGE SET (ℓ) (GROUP FES(ℓ)) problem, the input is a group-labeled graph G over a group Γ of order ℓ and an integer k and the objective is to test whether there exists a set of at most k edges intersecting every non-null cycle in G. The study of the parameterized complexity of GROUP FES(ℓ) was motivated by the fact that it generalizes the classical EDGE BIPARTIZATION problem when $\ell = 2$. Guillemot [IWPEC 2008, Discrete Optimization 2011] initiated the study of the parameterized complexity of this problem and proved that it is fixed-parameter tractable (FPT) parameterized by k. Subsequently, Wahlström [SODA 2014] and Iwata et al. [2014] presented algorithms running in time $\mathcal{O}(4^k n^{\mathcal{O}(1)})$ (even in the oracle access model) and $\mathcal{O}(\ell^{2k} m)$ respectively. In this paper, we give an algorithm for GROUP FES(ℓ) running in time $\mathcal{O}(4^k k^3 \ell(m + n))$. Our algorithm matches that of Iwata et al. when $\ell = 2$ (upto a multiplicative factor of k^3) and gives an improvement for $\ell > 2$.

1 Introduction

In a covering problem we are given a universe of elements U, a family $\mathcal{F}$ ($\mathcal{F}$ could be given implicitly) and an integer k and the objective is to check whether there exists a subset of U of size at most k which intersects all the elements of $\mathcal{F}$. Several natural problems on graphs can be framed as a covering problem. One of the most well-studied covering problems are the *feedback set* problems. In these problems, the family $\mathcal{F}$ is a succinctly defined subset of the set of cycles in the given graph. For instance, in the FEEDBACK VERTEX SET problem, the objective is to decide whether there exists a vertex subset S (also called a transversal) of size at most k which intersects *all* cycles in the graph. That is, $\mathcal{F}$ is the set of all cycles in the input graph. In the classical ODD CYCLE TRANSVERSAL (EDGE BIPARTIZATION) problem, the objective is to decide whether there exists a vertex subset (respectively edge subset) S of size at most k which intersects all *odd* cycles.

Yet another kind of feedback set problem deals with *gain-graphs* or *group-labeled graphs* and is called GROUP FEEDBACK VERTEX SET. Group-labeled graphs are generalizations of the well-studied class of *signed-graphs* introduced by Harary [11]. These are directed graphs where the arcs are labeled by elements of a group Γ and whenever multiplying the arc labels in order around a cycle

© Springer-Verlag GmbH Germany 2016
P. Heggernes (Ed.): WG 2016, LNCS 9941, pp. 269–281, 2016.
DOI: 10.1007/978-3-662-53536-3_23

results in an element other than 1_Γ – the identity element of the group, the cycle under consideration is called a *non-null* cycle.

In the GROUP FEEDBACK EDGE SET (GROUP FES) problem, the objective is to check whether there is a set of at most k arcs that intersect all non-null cycles in a given Γ-labeled graph for some finite group Γ. In the GROUP FES(ℓ) problem, we require Γ to have order ℓ. Similarly, in the GROUP FVS problem, the objective is to hit all non-null cycles with at most k vertices and in GROUP FVS(ℓ), the group is required to have order ℓ. Although group-labeled graphs and the GROUP FVS problem has been studied from a graph-theoretic point of view (see for example [9,14,21]) the study of the parameterized complexity of both versions of this problem was first initiated by Guillemot [10]. Formally, a *parameterization* of a problem is the assignment of an integer k to each input instance and we say that a parameterized problem is FPT if there is an algorithm that solves the problem in time $f(k) \cdot |I|^{\mathcal{O}(1)}$, where $|I|$ is the size of the input instance and f is an arbitrary computable function depending only on the parameter k. For more background, the reader is referred to the books [3,7,8,18]. Guillemot [10] showed that GROUP FVS(ℓ) is FPT parameterized by k and ℓ and GROUP FES(ℓ) is FPT parameterized by k by giving algorithms running in time $\mathcal{O}^*((4\ell+1)^k)$ and $\mathcal{O}^*((8k+1)^k)$ respectively (the $\mathcal{O}^*()$ notation subsumes polynomial factors).

The first single-exponential FPT algorithm for GROUP FES was given by Wahlström [20] who extended the branching algorithm of Guillemot to a more sophisticated LP-guided branching algorithm based on newly developed tools from the theory of valued constraint satisfaction. We remark that this algorithm, which runs in time $\mathcal{O}^*(4^k)$ was designed for the more general vertex version of the problem and improved upon the work of Cygan et al. [4] who obtained the first FPT algorithm for GROUP FVS parameterized only by k. In fact, the algorithm of Wahlström as well as that of Cygan et al. works in the oracle access model where the group is not given via its multiplication table but in the form of a polynomial time oracle. However, this algorithm relies on solving linear programs and hence has a dependence on the input-size that is far from linear even when the group-size ℓ is constant. A recently studied generalization of GROUP FES(ℓ) is the UNIQUE LABEL COVER(ℓ) problem where the input is a graph labeled by permutations of $[\ell]$, that is, elements of the symmetric group S_ℓ and the objective is to delete at most k edges such that the resulting graph can be labeled by elements of $[\ell]$ in a way that 'respects' the permutations of $[\ell]$ on the arcs. The fact that UNIQUE LABEL COVER(ℓ) generalizes GROUP FES(ℓ) follows from Cayley's Theorem which states that every finite group of order ℓ is isomorphic to a subgroup of S_ℓ.

Chitnis et al. [2] were the first to prove that UNIQUE LABEL COVER(ℓ) is FPT parameterized by ℓ and k. In fact they showed that under standard complexity hypotheses, to obtain fixed-parameter tractability, parameterizing by both ℓ and k is unavoidable. Subsequently, Wahlström [20] improved upon this result by giving an algorithm that runs in time $\mathcal{O}^*(\ell^{2k})$. Following this work, Iwata et al. [13] gave an algorithm for UNIQUE LABEL COVER(ℓ) that runs in time $\mathcal{O}(\ell^{2k}m)$

where m is the number of edges in the input graph. This was the first linear-time FPT algorithm for this problem and implies an algorithm for GROUP FES(ℓ) running in time $\mathcal{O}(\ell^{2k}m)$. Hence, prior to this work, the $\mathcal{O}^*(4^k)$ algorithm in [20] and the $\mathcal{O}(\ell^{2k}m)$ algorithm in [13] were the best known FPT algorithms for GROUP FES(ℓ) with respect to dependence on the parameter and the input and group sizes respectively. In this paper, we give an algorithm for GROUP FES(ℓ) that comes close to matching the best of both algorithms. We obtain an algorithm that has a dependence of $4^{k+\mathcal{O}(\log k)}$ on k and a dependence of $\mathcal{O}(\ell(m+n))$ on the input and group-sizes. In fact this algorithm outperforms that of Iwata et al. [13] for all $\ell > 2$. We now give a formal description of the problem under consideration and state our result.

GROUP FES(ℓ)   **Parameter:** k
Input: A Γ-labeled graph (G, Λ) where $|\Gamma| = \ell \geq 2$, integer k.
Question: Is there a set $X \subseteq A(G)$ of size at most k such that $G - X$ has no non-null cycles?

Theorem 1. GROUP FES(ℓ) can be solved in time $\mathcal{O}(4^k k^3 \ell(m+n))$ where m and n denote the number of arcs and vertices in the input graph respectively.

Methodology. We closely follow the template developed for solving graph separation problems via important separators in [1,16], those via LP-guided branching in [5,10,12,15], the Valued CSP-based algorithms in [13,20] and the skew-symmetric branching algorithm for 2-SAT DELETION in [19]. The common thread connecting these algorithms is that they all begin by proving a 'persistence lemma' or Nemhauser-Trotter-type theorem. In these lemmas, one proves that the solution to an appropriate linear program or a maximum-flow question on an appropriate network can be used to 'fix a configuration' for vertices which satisfy certain properties. For instance, the classical Nemhauser-Trotter Theorem [17] for VERTEX COVER states that if an optimal solution to the standard relaxation of the Vertex Cover Integer Linear Program (ILP) assigns 0 or 1 to a vertex then there is also an optimal solution to the ILP which does the same with respect to this vertex.

However, in this work the persistence lemma we prove will be based directly on the solution to a max-flow question in a network as opposed to using the solution to a linear program. The reason behind this is that the structural properties of GROUP FES(ℓ) closely resemble those of the classical EDGE MULTIWAY CUT problem while the vertex version, GROUP FVS(ℓ) is closely related to the NODE MULTIWAY CUT problem (see [4,10]). As a result, we are able to design our persistence lemma using an appropriate analogue of the classical notion of 'isolating cuts' from [6]. Once we prove this lemma, we design a natural reduction rule based on it and describe a subroutine that runs in polynomial time (with a linear dependence on ℓ and $m+n$) and either finds a valid application of the reduction rule or an arc on which a naive branching step will decrease a predetermined measure for the input. Finally, we remark that while the structure of our algorithm strongly resembles that of the algorithms in the works cited above, the fact that we are dealing with groups of order greater than 2 while trying to simultaneously optimize the dependence of the running time in 'three-dimensions' – parameter, input-size and group-size, poses non-trivial obstacles when it comes to actually implementing each step.

2   Preliminaries

Let Γ be a group with identity element 1_Γ. A Γ-labeled graph is a pair (G, Λ) where G is a digraph with at most one arc between every pair of vertices and $\Lambda : A(G) \to \Gamma$. If $(u, v) \in A(G)$, then we denote by $\Lambda(v, u)$ the group element $\Lambda(u, v)^{-1}$. For a digraph G, we denote by $\tilde{G}$ the underlying undirected graph. Let $P = v_1, \ldots, v_\ell$ be a path in $\tilde{G}$. We denote by $\Lambda(P)$ the group element $\Lambda(v_1, v_2) \cdot \Lambda(v_2, v_3) \cdots \Lambda(v_{\ell-1}, v_\ell)$. Let $C = v_1, \ldots, v_\ell, v_1$ be a cycle in $\tilde{G}$. We denote by $\Lambda(C)$ the group element $\Lambda(v_1, v_2) \cdot \Lambda(v_2, v_3) \cdot \Lambda(v_{\ell-1}, v_\ell) \cdots \Lambda(v_\ell, v_1)$. We call C *non-null* if $\Lambda(C) \neq 1_\Gamma$. Note that even though different choices of the vertex v_1 in the same cycle may lead to different values for $\Lambda(C)$, it is easy to see that if for one choice of v_1 the value of $\Lambda(C)$ is not 1_Γ then for no choice of v_1 is it 1_Γ.

For an undirected graph H and vertex set $Z \subseteq V(H)$, we denote by $\delta(Z)$ the set of edges which have exactly one endpoint in Z. We denote by $E(Z)$ the edges of H which have both endpoints in Z. This notation also extends to directed graphs as $A(Z)$. For a set X of edges in an undirected graph or arcs in a directed graph, we denote by $V(X)$ the set of endpoints of the edges or arcs in X. For a vertex subset X, $N(X)$ denotes the set of neighbors of X and $N[X]$ denotes the set $X \cup N(X)$. For an undirected graph G and disjoint vertex sets X and Y, a path is called an X-Y path if it has one endpoint in X and the other in Y and a set $S \subseteq E(G)$ is said to be an X-Y separator if there is no X-Y path in the graph $G - S$. We denote the vertices in the components of $G - S$ which intersect X by $R(X, S)$. We denote by $\lambda_G(X, Y)$ the size of the smallest X-Y separator in G. Due to space constraints, proofs of Lemmas marked [⋆] have been omitted from the extended abstract and can be found in the full version of the paper.

3   Consistent Labelings and the Auxiliary Graph

In this section, we begin by recalling known results on group-labeled graphs that exclude a non-null cycle. Following that, we will associate an auxiliary graph with every instance of GROUP FES(ℓ).

Definition 1. *Let (G, Λ) be a Γ-labeled graph and let $\Psi : V(G) \to \Gamma$. We say that Ψ is a **consistent labeling** for this graph if for all $(u, v) = a \in A(G)$, $\Psi(u) \cdot \Lambda(a) = \Psi(v)$.*

Lemma 1 [10]. *Let (G, Λ) be a Γ-labeled graph. There is no non-null cycle in G if and only if G has a consistent labeling.*

Observation 2. *Let (G, Λ) be a Γ-labeled graph. If G has a consistent labeling, then for every $v \in V(G)$ and $g \in \Gamma$, there is a consistent labeling ψ_g^v for G such that $\psi_g^v(v) = g$.*

Definition 2. *Let (G, Λ) be a Γ-labeled graph and let $\Psi : V(G) \rightarrow \Gamma$ be a consistent labeling for G. For a set $Z \subseteq V(G)$ and a function $\tau : Z \rightarrow \Gamma$, we say that Ψ **agrees with** τ on Z if for every $v \in Z$, $\Psi(v) = \tau(v)$.*

We consider a slightly more general formulation of the GROUP FES(ℓ) problem, where the input contains (G, Λ, k), a set $\hat{Z}$ such that $G[\hat{Z}]$ is connected and a function $\tau : \hat{Z} \rightarrow \Gamma$ such that τ *is a consistent labeling for* $G[\hat{Z}]$ and the objective is to find a solution *given that* if there is a solution then there is one whose deletion leaves a graph which has a consistent labeling that *agrees* with τ on $\hat{Z}$. That is, we may assume that we are looking for a solution whose deletion allows a consistent labeling that 'extends' τ. Clearly this formulation is more general since we can simply set $\hat{Z} = \emptyset$ to begin with and leave τ undefined. For a given $\hat{Z}$ and $\tau : \hat{Z} \rightarrow \Gamma$, we denote by $\hat{Z}_\tau$ the set $\{z_\alpha | z \in \hat{Z}, \alpha = \tau(z)\}$. We now define the auxiliary graph associated with the instance. We will be performing almost all of our computations in this graph.

The Auxiliary Graph and Some Properties. For an instance $I = (G, \Lambda, k, \hat{Z}, \tau)$ of GROUP FES(ℓ), we define an associated auxiliary graph H_I as follows. The vertex set of H_I is $\{v_g | v \in V(G), g \in \Gamma\}$. The vertex v_g represents the existence of an (eventual) consistent labeling of G where v is assigned the group element g. The edge set of H_I is defined as follows. For every arc $a = (u, v) \in A(G)$ and for every $g \in \Gamma$, there is an edge $(u_g, v_{g \cdot \Lambda(a)})$. Observe that corresponding to a, there are exactly ℓ edges in H_I and furthermore, these form a matching. Therefore, H_I has ℓn vertices and ℓm edges, where n and m are the number of vertices and edges in G respectively. Note that the graph H_I in fact only depends on (G, Λ). However, we choose to denote the graph as H_I in order to facilitate an easier presentation in the description of the algorithm. Moving forward, we will characterize the dependencies between vertices when subjected to certain constraints. Before we do so, we need the following definitions and observation.

Definition 3. *Let $I = (G, \Lambda, k, \tau)$ be an instance of GROUP FES(ℓ). For $v \in V(G)$, we use $[v]$ to denote the set $\{v_g | g \in \Gamma\}$. For a subset $S \subseteq V(G)$, we use $[S]$ to denote the set $\bigcup_{v \in S}[v]$. Similarly, for an arc $a = (u, v) \in A(G)$, we use $[a]$ to denote the set $\{(u_i, v_j)\}_{i \in \Gamma, j = \Lambda(a) \cdot i}$ of edges in H_I and for a subset $X \subseteq A(G)$, we use $[X]$ to denote the set $\bigcup_{a \in X}[a]$. For the sake of convenience, we also reuse the same notation in the following way. For every $v \in V(G)$ and $\alpha \in \Gamma$, we denote by $[v_\alpha]$ the set $[v]$. Similarly, for every $a = (u, v) \in A(G)$ and $\alpha, \beta \in \Gamma$ such that $e = (u_\alpha, v_\beta) \in E(H_I)$, we denote by $[e]$ the set $[a]$. This definition extends in a natural way to sets of vertices and edges of the auxiliary graph H_I. For a set $S \subseteq V(H_I) \cup E(H_I)$, we denote by S^{-1} the set $\{s | s \in V(G) \cup A(G) : [s] \cap S \neq \emptyset\}$. For an arc $a \in A(G)$ and edge $e \in [a]$, we also use e^{-1} to denote the arc a.*

Observation 3 *Let $I = (G, \Lambda, k, \hat{Z}, \tau)$ be an instance of* GROUP FES(ℓ). *Then the following statements hold. (a) For every $v \in V(G)$, for every distinct $g_1, g_2 \in \Gamma$, v_{g_1} and v_{g_2} have no common neighbors in H_I. (b) For a set $S \subseteq A(G)$, $H_I - [S] = H_{I'}$ where $I' = (G - S, \Lambda, 0, \hat{Z}, \tau)$. (c) If Ψ is a consistent labeling for G, then for any $u, v \in V(G)$, if u_g is in the same connected component as $v_{g'}$ in H_I where $g = \psi(v)$ then $\Psi(u) = g'$.*

Definition 4. *Let $I = (G, \Lambda, k, \hat{Z}, \tau)$ be an instance of* GROUP FES(ℓ). *We say that a set $Z \subseteq V(H_I) \cup E(H_I)$ is* **regular** *if $|Z \cap [v]| \leq 1$ for any $v \in V(G)$ and $|Z \cap [a]| \leq 1$ for any $a \in A(G)$. We say that Z is* **irregular** *otherwise. That is, regular sets contain at most 1 copy of any vertex and arc of G.*

Now that we have defined the notion of regularity of sets, we prove the following lemma which shows that the auxiliary graph displays a certain symmetry with respect to regular paths. This will allow us to transfer arguments which involve a regular path between vertices v_{g_1} and u_{g_2} to one between vertices v_{g_3} and u_{g_4} where $g_1 \neq g_3$ and $g_2 \neq g_4$.

Lemma 2 [⋆]. *Let $I = (G, \Lambda, k, \hat{Z}, \tau)$ be an instance of* GROUP FES(ℓ). *Let P be a regular path in H_I from v_g to $u_{g'}$ for some $u, v \in V(G)$ and $g, g' \in \Gamma$. Let $V(P)$ denote the set of vertices of G in P and let U denote the set $[V(P)]$. Then, there is a set $\mathcal{P} = \{P_r\}_{r \in \Gamma}$ of ℓ vertex disjoint regular paths in H_I and a partition $\mathcal{U}$ of U into sets $\{U_r\}_{r \in \Gamma}$ such that for each $\gamma \in \Gamma$, $V(P_\gamma) = U_\gamma$ and P_γ is a path from v_γ to $u_{\gamma \cdot \Lambda(P^{-1})}$.*

Observation 4 *Let $I = (G, \Lambda, k, \hat{Z}, \tau)$ be an instance of* GROUP FES(ℓ). *Then, the following statements hold. (a) The set $\hat{Z}_\tau$ is regular and furthemore, $H_I[\hat{Z}_\tau]$ is connected. (b) If $S \subseteq A(G)$ is such that $[S]$ intersects all $\hat{Z}_\tau - [\hat{Z}] \setminus \hat{Z}_\tau$ paths in H_I then $R(\hat{Z}_\tau, [S])$ is regular. (c) If $S \subseteq A(G)$ is a minimal set such that $G - S$ has a consistent labeling that agrees with τ on $\hat{Z}$, then $[S]$ is disjoint from $A(\hat{Z})$.*

Using the observations and structural lemmas proved so far, we will now give a forbidden-structure characterization of 'solved' YES instances of GROUP FES(ℓ), that is instances where $k = 0$.

Lemma 3 [⋆]. *Let $I = (G, \Lambda, 0, \hat{Z}, \tau)$ be a YES instance of* GROUP FES(ℓ) *where G is connected. Let $v \in V(G)$ and $g \in \Gamma$. Then, there is a consistent labeling Ψ such that $\Psi(v) = g$ if and only if there is no $g' \in \Gamma$ such that v_g and $v_{g'}$ are in the same connected component of H_I.*

In the next lemma, we extend the statement of the previous lemma to include a description of general YES instances of the GROUP FES(ℓ) problem.

Lemma 4. *Let $I = (G, \Lambda, k, \hat{Z}, \tau)$ be an instance of* GROUP FES(ℓ). *Then, I is a YES instance if and only if there is a set $S \subseteq A(G)$ of size at most k such that for every vertex $v \in V(G)$, there is no path in $H_I - [S]$ from v_g to $v_{g'}$ for any $g' \neq g$. Furthermore, if $G - S$ has a consistent labeling that agrees with τ on $\hat{Z}$ then, $[S]$ intersects all $\hat{Z}_\tau - ([\hat{Z}] \setminus \hat{Z}_\tau)$ paths in H_I.*

Proof. We first argue both directions in the first part of the statement. In the forward direction, suppose that I is a YES instance and let $S \subseteq A(G)$ be a solution. That is, $G - S$ has a consistent labeling. Now, suppose that for some distinct $g, g' \in \Gamma$, there is a path in H_I from v_g to $v_{g'}$ disjoint from $[S]$. But by Observation 3(b), this path exists in $H_{I'}$ where $I' = (G - S, \Lambda, 0, \hat{Z}, \tau)$. However, since $G - S$ has a consistent labeling, this contradicts Lemma 3, completing the argument in the forward direction.

In the converse direction, suppose that $S \subseteq A(G)$ such that $[S]$ intersects all paths from v_g to $v_{g'}$ for every $v \in V(G)$ and $g \neq g'$ in the graph H_I. Let $I' = (G - S, \Lambda, 0, \hat{Z}, \tau)$. By Observation 3(b), we know that there are no v_g-$v_{g'}$ paths in $H_{I'}$ for any $v \in V(G)$ and distinct $g, g' \in \Gamma$. But applying Lemma 3 on each connected component of $G - S$ (since the premise of this lemma requires connectivity of the graph), we conclude that $G - S$ has a consistent labeling. This completes the argument in the converse direction.

For the second statement, suppose that Ψ is a consistent labeling of G that agrees with τ on $\hat{Z}$. Suppose that for some $u, v \in \hat{Z}$ and $g, g' \in \Gamma$ where $u_g \in \hat{Z}_\tau$ and $v_{g'} \notin \hat{Z}_\tau$, v_g is in the same component of $H_I - [S]$ as $u_{g'}$. Then, Observation 3(c) implies that $\Psi(u) = g'$, a contradiction to our assumption that Ψ agrees with τ on $\hat{Z}$. This completes the proof of the lemma.   □

Using the above lemma, we will interpret the GROUP FES(ℓ) problem as a parameterized cut-problem. Furthermore, observe that due to this lemma, the size of a minimum $\hat{Z}_\tau - [Z] \setminus \hat{Z}_\tau$ separator in H_I is a natural lower bound on the number of edges of H_I which correspond to the arcs in a solution for the given instance. Note that although for a solution $S \subseteq A(G)$, a naive upper bound on the number of edges in $[S]$ that are required to hit all $\hat{Z}_\tau - [Z] \setminus \hat{Z}_\tau$ paths in H_I is ℓk, we will prove shortly that the actual bound is much tighter and is in fact, independent of the group-size. This is a crucial difference between the vertex and edge variants of this problem as a similar property does not exist in the vertex version. We conclude this subsection by stating the following consequence of Lemma 4 and Observation 4(b).

Lemma 5 [⋆]. *Let $I = (G, \Lambda, k, \hat{Z}, \tau)$ be an instance of GROUP FES(ℓ) and let $S \subseteq A(G)$ be a solution for this instance such that $G - S$ has a consistent labeling that agrees with τ on $\hat{Z}$. Then, $R(\hat{Z}_\tau, [S])$ is regular.*

Closed and Open Edges. Here, we will examine the set of edges crossing a regular set and divide them into 'closed' and 'open' edges. These notions are crucial for the description of our reduction rule. Intuitively, an edge e crossing a regular set Z in H_I is considered open if its image in G, e^{-1} also crosses Z^{-1} in G. Otherwise, it is considered closed. We now define these notions in a way that is most convenient for us to invoke in our proofs.

Definition 5. *Let $Z \subseteq V(H_I)$ be a regular set in H_I and let $u, v \in V(G)$ and $\alpha, \beta \in \Gamma$ be such that $e = (u_\alpha, v_\beta) \in \delta(Z)$. We call e a **closed edge** in H_I with respect to Z if $Z \cap [u]$ and $Z \cap [v]$ are both non-empty. Otherwise, we say that e is an **open edge** in H_I with respect to Z.*

Observation 5. *Let $Z \subseteq V(H_I)$ be a regular set in H_I and let $u, v \in V(G)$ and $\alpha, \beta \in \Gamma$ be such that $e = (u_\alpha, v_\beta) \in \delta(Z)$. Then, the following statements hold.*

(a) If e is closed with respect to Z then e^{-1} has both endpoints in Z^{-1}.

(b) If e is closed with respect to Z then H_I has no edge between $Z \cap [u]$ and $Z \cap [v]$.

(c) If e is open with respect to Z then, $e^{-1} \in \delta_{\hat{G}}(Z^{-1})$.

We now state and prove a crucial fact regarding the *number* of open and closed edges crossing any regular set in H_I.

Lemma 6 [$\star$]. *Let $I = (G, \Lambda, \hat{Z}, k, \tau)$ be an instance of GROUP FES(ℓ) and let $Y \subseteq V(H_I)$ be a regular set. Then, for every edge $e \in \delta(Y)$, $|\delta(Y) \cap [e^{-1}]| = 1$ if e is open with respect to Y and $|\delta(Y) \cap [e^{-1}]| = 2$ otherwise.*

Figure 1 illustrates the structure of the edges crossing a regular set, guaranteed by Lemma 6. As a consequence of Lemma 6 and Observation 4(b), we have the following.

Lemma 7. *Let $I = (G, \Lambda, k, \hat{Z}, \tau)$ be an instance of GROUP FES(ℓ) and let $S \subseteq A(G)$ be a set of size at most k such that $[S]$ intersects all $\hat{Z}_\tau - [\hat{Z}] \setminus \hat{Z}_\tau$ paths in H_I. Then, $|\delta(R(\hat{Z}_\tau, [S]))| \leq 2k$.*

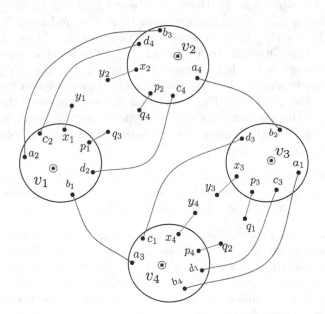

Fig. 1. An illustration of all the edges crossing the regular set Z. The four sets in the figure denote the four regular 'copies' of Z^{-1} in the graph H_I. Here, the edges $(a_2, b_3), (a_3, b_1), (c_2, d_4), (c_4, d_2)$ are closed and the edges $(x_1, y_1), (p_1, q_3)$ are open with respect to Z which is the set containing v_1

Lemma 7 implies that even though $[S]$ can contain up to ℓk edges, at most $2k$ of these are required to separate $\hat{Z}_\tau$ from $[\hat{Z}] \setminus \hat{Z}_\tau$. Motivated by this fact, we define the following measure on instances of GROUP FES(ℓ) which captures the gap between the budget k which is the upper bound on the size of the solution S and the size of the smallest $\hat{Z}_\tau - [\hat{Z}] \setminus \hat{Z}_\tau$ separator in H_I which is a lower bound on $2|S|$. Lemma 7 implies that for a YES instance I, the gap $\mu(I)$ is always non-negative. Furthermore, it will be easy to see that for any given instance I, we can check whether $\mu(I)$ is non-negative in time $\mathcal{O}(k\ell(m + n))$.

Definition 6. Let $I = (G, \Lambda, k, \hat{Z}, \tau)$ be an instance of GROUP FES(ℓ). The gap $\mu(I)$ is defined as $\mu(I) = k - \frac{1}{2}(\lambda_{H_I}(\hat{Z}_\tau, [Z] \setminus \hat{Z}_\tau))$.

Proof of Persistence and Description of the Reduction Rule. We are now ready to prove the *Persistence* Lemma which plays a major role in the design of the algorithm. In essence this lemma says that if we find a minimum $\hat{Z}_\tau - [\hat{Z}] \setminus \hat{Z}_\tau$ separator S such that $R(\hat{Z}_\tau, S)$ is regular, then we can *correctly fix* the labels of all vertices which have exactly one copy in $R(\hat{Z}_\tau, S)$ and look for a solution whose deletion allows a consistent labeling that is an extension of this one. It will be shown later that once we fix the labels of these vertices, the subsequent exhaustive branching steps will decrease the gap $\mu(I)$ by at least $\frac{1}{2}$ and since $\mu(I)$ is always required to be non-negative, the *depth* of the search tree will be bounded by $2\mu(I)$.

Lemma 8 [⋆](Persistence Lemma). Let $I = (G, \Lambda, k, \hat{Z}, \tau)$ be a YES instance of GROUP FES(ℓ). Let $X \subseteq A(G)$ be a minimal set of size at most k such that $G - X$ has a consistent labeling Ψ agreeing with τ on $\hat{Z}$. Let T denote the set $[\hat{Z}] \setminus \hat{Z}_\tau$. Let S be a minimum $\hat{Z}_\tau - T$ separator in H_I and let $Z = R(\hat{Z}_\tau, [S])$. Then, there is a solution for the given instance disjoint from $A(Z^{-1})$.

Having proved the Persistence Lemma, we proceed to describe the reduction rule based on it. Before we do so, we need to prove the following lemmas.

Lemma 9 [⋆]. Let $I = (G, \Lambda, k, \hat{Z}, \tau)$ be a YES instance of GROUP FES(ℓ) and let S be a minimum $\hat{Z}_\tau - [\hat{Z}] \setminus \hat{Z}_\tau$ separator in H_I and let $Y = R(\hat{Z}_\tau, [S])$. Then, $\delta(Y)$ is also a minimum $\hat{Z}_\tau - [\hat{Z}] \setminus \hat{Z}_\tau$ separator in H_I. Furthermore, if S is not regular, then there is an edge $e \in \delta(Y)$ which is closed with respect to Y.

Lemma 10 [⋆]. Let $I = (G, \Lambda, k, \hat{Z}, \tau)$ be a YES instance of GROUP FES(ℓ) and let S be a minimum $\hat{Z}_\tau - [\hat{Z}] \setminus \hat{Z}_\tau$ separator in H_I. If S is not regular, then there is an edge $e \in [S]$ which is closed with respect to $R(\hat{Z}_\tau, S)$ and a solution for the given instance containing e^{-1}. Moreover, given the instance and the set S, the arc e^{-1} can be computed in time $\mathcal{O}(\ell(m + n))$.

This leads us to the following reduction rule.

Reduction Rule 1. Given an instance $I = (G, \Lambda, k, \hat{Z}, \tau)$ of GROUP FES(ℓ) and a set S which is both irregular and a minimum $\hat{Z}_\tau - [\hat{Z}] \setminus \hat{Z}_\tau$ separator in H_I, execute the algorithm of Lemma 10 to compute the arc a and return the instance $I' = (G - \{a\}, \Lambda, k - 1, \hat{Z}, \tau)$.

The correctness as well as the fact that it can be applied in time $\mathcal{O}(\ell(m+n))$ follows from Lemma 10. However, observe that in order to apply this rule, we *need* the set S. As a result, we cannot apply this rule exhaustively and we will only apply it selectively during our algorithm. For this, we will also have a slightly modified version of the above reduction rule. We remark that this rule is introduced only in order to maintain a linear dependence on ℓ at certain points in the algorithm.

Reduction Rule 2. *Given an instance* $I = (G, \Lambda, k, \hat{Z}, \tau)$ *of* GROUP FES*(ℓ) and an arc* $a \in A(G)$ *such that there is an irregular minimum* $\hat{Z}_\tau - [\hat{Z}] \setminus \hat{Z}_\tau$ *separator in* H_I *that contains 2 edges of* $[a]$, *return the instance* $I' = (G - \{a\}, \Lambda, k-1, \hat{Z}, \tau)$.

Observe that in the above reduction rule, we have skipped the intermediate step of computing a from the irregular minimum separator. This is because sometimes, we can detect the arc a *faster* than we can compute the irregular minimum separator. We conclude this subsection by arguing that applying these reduction rules to a given instance of GROUP FES(ℓ) does not *increase* the gap. That is, if I' is the instance resulting from I by an application of these rules, then it should not be the case that $\mu(I') > \mu(I)$. In order to prove this (Lemma 12), we need the next lemma.

Lemma 11 [⋆]. *Let* $I = (G, \Lambda, k, \hat{Z}, \tau)$ *be an instance of* GROUP FES*(ℓ). Let* $v \in V(G)$, $g \in \Gamma$ *and* Y *be a regular set containing* $\hat{Z}_\tau$ *such that* $\delta(Y)$ *is a minimum* $\hat{Z}_\tau - [\hat{Z}] \setminus \hat{Z}_\tau$ *separator and let* $r = |\delta(Y)|$. *Let* $e \in \delta(Y)$ *be an edge closed with respect to* Y *and* $a = e^{-1}$. *Then, the size of a minimum* $\hat{Z}_\tau - [\hat{Z}] \setminus \hat{Z}_\tau$ *separator in* $H_I - [a]$ *is exactly* $r - 2$.

Lemma 12 [⋆]. *Let* $I = (G, \Lambda, k, \hat{Z}, \tau)$ *be an instance of* GROUP FES*(ℓ) and let* $I' = (G - \{a\}, \Lambda, k-1, \hat{Z}, \tau)$ *be the instance obtained from* I *by an application of Reduction Rule 1 or Reduction Rule 2 on the arc* a. *Then,* $\mu(I) = \mu(I')$.

Proof. Recall that $\mu(I) = k - \frac{1}{2}(\lambda_{H_I}(\hat{Z}_\tau, [Z] \setminus \hat{Z}_\tau))$ and $\mu(I') = (k-1) - \frac{1}{2}(\lambda_{H_{I'}}(\hat{Z}_\tau, [Z] \setminus \hat{Z}_\tau))$. Let $r = \lambda_{H_I}(\hat{Z}_\tau, [Z] \setminus \hat{Z}_\tau)$ and $r' = \lambda_{H_{I'}}(\hat{Z}_\tau, [Z] \setminus \hat{Z}_\tau)$. Lemma 11 implies that $r' = r - 2$. Hence, $\mu(I') = (k-1) - \frac{1}{2}(r-2) = k - \frac{1}{2}r = \mu(I)$. This completes the proof of the lemma. □

Having stated the reduction rules, we are almost ready to describe the algorithm. Before we do so, we need two subroutines. The first subroutine simply checks whether we can apply Reduction Rule 2 for a given arc. The second one is the main subroutine which either allows us to say No or apply one of the two reduction rules or compute a pair of 'branchable' arcs or computes an arc which is part of *some* 'farthest' minimum isolating cut separating $\hat{Z}_\tau$ from the rest of $[\hat{Z}]$.

Lemma 13 [⋆]. *There is an algorithm that, given an instance* $I = (G, \Lambda, k, \hat{Z}, \tau)$ *of* GROUP FES*(ℓ) and an arc* $a \in A(G)$, *runs in time* $\mathcal{O}(k\ell(m+n))$ *and correctly* *concludes either that there exists a minimum* $\hat{Z}_\tau - [\hat{Z}] \setminus \hat{Z}_\tau$ *separator containing at least 2 edges of* $[a]$ *or that no such separator exists. Furthermore, in the latter case,* $\lambda_{H_I - [a]}(\hat{Z}_\tau, [\hat{Z}] \setminus \hat{Z}_\tau) \geq \lambda_{H_I}(\hat{Z}_\tau, [\hat{Z}] \setminus \hat{Z}_\tau)\text{-}1.$

Lemma 14 [⋆](**Main Subroutine**). *There is an algorithm that, given an instance* $I = (G, \Lambda, k, \hat{Z}, \tau)$ *of* GROUP FES(ℓ)*, runs in time* $\mathcal{O}(k^2\ell(m + n))$ *and*

(a) *correctly concludes that the given instance is a* NO *instance or*
(b) *returns an irregular minimum* $\hat{Z}_\tau - [\hat{Z}] \setminus \hat{Z}_\tau$ *separator of size at most* $2k$ *or an arc* $a \in A(G)$ *such that two edges of* $[a]$ *appear in such a separator or*
(c) *returns a pair of edges* $e_1, e_2 \in E(H_I)$ *such that there is always a solution intersecting the set* $\{e_1^{-1}, e_2^{-1}\}$ *and there is no irregular minimum* $\hat{Z}_\tau - [\hat{Z}] \setminus \hat{Z}_\tau$ *separator containing at least 2 edges from* $[e_1]$ *or* $[e_2]$ *or*
(d) *returns a regular set* $Z \supseteq \hat{Z}_\tau$ *such that* $\delta(Z)$ *is a minimum* $\hat{Z}_\tau - [\hat{Z}] \setminus \hat{Z}_\tau$ *separator and an edge* $e^\star = (u_\alpha, v_\beta) \in \delta(Z)$ *where* $u_\alpha \in Z$*, such that there is no* $(Z \cup \{v_\beta\}) - [Z \cup \{v_\beta\})] \setminus (Z \cup \{v_\beta\})$ *separator of size at most* $|\delta(Z)|$ *and there is no minimum* $\hat{Z}_\tau - [\hat{Z}] \setminus \hat{Z}_\tau$ *separator containing at least 2 edges from* $[e^\star]$*.*

4 Description of the **FPT** Algorithm for GROUP FES(ℓ)

Let $I = (G, \Lambda, k, \hat{Z}, \tau)$ be the given instance of GROUP FES(ℓ). The algorithm we describe is a recursive algorithm with $k = 0$ as the base case. If $k = 0$ then we return YES if and only if G already has a consistent labelling. Furthermore, if $\lambda_{H_I}(\hat{Z}_\tau, [\hat{Z}] \setminus \hat{Z}_\tau) = 0$ (there is no path in H_I from $\hat{Z}_\tau$ to $[\hat{Z}] \setminus \hat{Z}_\tau$), then we recurse on the instance $(G, \Lambda, k, \emptyset, \tau')$, where τ' is undefined. The correctness of this operation follows from applying Lemma 3 to the connected component of G containing $\hat{Z}$. Finally, if $\hat{Z} = \emptyset$, then we pick an arbitrary vertex v in a component of G which *does not* already have a consistent labeling, pick an arbitrary $g \in \Gamma$ and recurse on the instance $(G, \Lambda, k, \{v\}, \tau')$ where $\tau'(v) = g$. The correctness of this step follows from Observation 2. We now describe the steps executed by the algorithm when $k > 0$ and none of the aforementioned conditions hold. We begin by executing the main subroutine (Lemma 14) on the given instance and describe subsequent steps of the algorithm based on the output of this subroutine.

Case (a): In this case, we simply return NO.

Case (b): In this case, we apply Reduction Rule 1 or Reduction Rule 2 as appropriate and recurse on the resulting instance.

Case (c): In this case, we branch on the arcs a_1, a_2 where $a_1 = [e_1^{-1}]$ and $a_2 = [e_2^{-1}]$. That is, we recursively call the algorithm on the tuples $I_1 = (G - a_1, \Lambda, k - 1, \hat{Z}, \tau)$ and $I_2 = (G - a_2, \Lambda, k - 1, \hat{Z}, \tau)$.

Case (d): Let $Z \supseteq \hat{Z}_\tau$ be the returned regular set and let $e^\star = (u_\alpha, v_\beta) \in \delta(Z)$ be the edge such that $u_\alpha \in Z$, there is no $(Z \cup \{v_\beta\}) - [Z \cup \{v_\beta\})] \setminus (Z \cup \{v_\beta\})$ separator of size at most $|\delta(Z)|$ and there is no minimum $\hat{Z}_\tau - [\hat{Z}] \setminus \hat{Z}_\tau$ separator that contains more than one edge from $[e^\star]$. Let $a \in A(G)$ such that $e^\star \in [a]$.

We now branch by either adding a to the solution or by choosing not to pick a in the solution. Formally speaking, we recursively call the algorithm on the tuples $I_1 = (G - a, \Lambda, k - 1, \hat{Z}, \tau)$ and $I_2 = (G, \Lambda, k, \hat{Z} \cup \{v\}, \tau')$ where τ' is the same as τ on $\hat{Z}$ and $\tau'(v) = \beta$.

This completes the description of the algorithm. The correctness of the algorithm follows from the correctness of Lemmas 8 and 14, and the fact that the branching is exhaustive. It remains to analyze the running time. Observe that the time taken at each step of the recursion is dominated by the time required to execute the subroutine of Lemma 14 which runs in time $\mathcal{O}(k^2 \ell(m + n))$. Furthermore, along any root to leaf path, Reduction Rule 1 applies at most k times. Hence, the running time is bounded by the product of $\mathcal{O}(k^3 \ell(m + n))$ and the number of root to leaf paths in the search tree resulting from a run of the algorithm on the input instance $I = (G, \Lambda, k, \hat{Z}, \tau)$.

To complete the proof of Theorem 1, we prove by induction on $\mu(I)$ that the number of leaves in the search tree resulting from a run on input I is bounded by $4^{\mu(I)}$.

5   Concluding Remarks

We have presented an FPT algorithm for GROUP FES(ℓ) that for finite groups of fixed size has linear dependence on the input-size and matches the best known parameter dependence upto polynomial factors. For this, we had to assume that the multiplication table of the group is explicitly known. A natural question that remains is whether it is possible to obtain a linear time FPT algorithm in the oracle model assuming constant query time. Finally, we leave open the question of improving upon the 4^k dependence on the parameter for GROUP FES(ℓ) even at the cost of superlinear (but still polynomial) dependence on the input-size and group-size.

Acknowledgments. The author acknowledges support from the Austrian Science Fund (FWF), project P26696 X-TRACT.

References

1. Chen, J., Liu, Y., Lu, S.: An improved parameterized algorithm for the minimum node multiway cut problem. Algorithmica **55**(1), 1–13 (2009)
2. Chitnis, R.H., Cygan, M., Hajiaghayi, M., Pilipczuk, M., Pilipczuk, M.: Designing FPT algorithms for cut problems using randomized contractions. In: 53rd Annual IEEE Symposium on Foundations of Computer Science, FOCS 2012, New Brunswick, NJ, USA, 20–23 October 2012, pp. 460–469 (2012)
3. Cygan, M., Fomin, F.V., Kowalik, L., Lokshtanov, D., Marx, D., Pilipczuk, M., Pilipczuk, M., Saurabh, S.: Parameterized Algorithms. Springer, Switzerland (2015)
4. Cygan, M., Pilipczuk, M., Pilipczuk, M.: On group feedback vertex set parameterized by the size of the cutset. Algorithmica **74**(2), 630–642 (2016)

5. Cygan, M., Pilipczuk, M., Pilipczuk, M., Wojtaszczyk, J.O.: On multiway cut parameterized above lower bounds. TOCT **5**(1), 3 (2013)
6. Dahlhaus, E., Johnson, D.S., Papadimitriou, C.H., Seymour, P.D., Yannakakis, M.: The complexity of multiterminal cuts. SIAM J. Comput. **23**(4), 864–894 (1994)
7. Downey, R.G., Fellows, M.R.: Parameterized Complexity. Springer, New York (1999)
8. Flum, J., Grohe, M.: Parameterized Complexity Theory. Texts in Theoretical Computer Science. An EATCS Series. Springer, Berlin (2006)
9. Geelen, J., Gerards, B.: Excluding a group-labelled graph. J. Comb. Theory Ser. B **99**(1), 247–253 (2009)
10. Guillemot, S.: FPT algorithms for path-transversal and cycle-transversal problems. Discrete Optim. **8**(1), 61–71 (2011)
11. Harary, F.: On the notion of balance of a signed graph. Michigan Math. J. **2**(2), 143–146 (1953)
12. Iwata, Y., Oka, K., Yoshida, Y.: Linear-time FPT algorithms via network flow. In: SODA, pp. 1749–1761 (2014)
13. Iwata, Y., Wahlström, M., Yoshida, Y.: Half-integrality, lp-branching and FPT algorithms (2013). CoRR, abs/1310.2841
14. Kawarabayashi, K., Wollan, P.: Non-zero disjoint cycles in highly connected group labelled graphs. J. Comb. Theory, Ser. B **96**(2), 296–301 (2006)
15. Lokshtanov, D., Narayanaswamy, N.S., Raman, V., Ramanujan, M.S., Saurabh, S.: Faster parameterized algorithms using linear programming. ACM Trans. Algorithms **11**(2), 15:1–15:31 (2014)
16. Marx, D.: Parameterized graph separation problems. Theor. Comput. Sci. **351**(3), 394–406 (2006)
17. Nemhauser, G.L., Trotter Jr., L.E.: Properties of vertex packing and independence system polyhedra. Math. Program. **6**, 48–61 (1974)
18. Niedermeier, R.: Invitation to Fixed-Parameter Algorithms. Oxford Lecture Series in Mathematics and its Applications, vol. 31. Oxford University Press, Oxford (2006)
19. Ramanujan, M.S., Saurabh, S.: Linear time parameterized algorithms via skew-symmetric multicuts. In: SODA, pp. 1739–1748 (2014)
20. Wahlström, M.: Half-integrality, lp-branching and FPT algorithms. In: Proceedings of the Twenty-Fifth Annual ACM-SIAM Symposium on Discrete Algorithms, SODA 2014, Portland, Oregon, USA, 5–7 January 2014, pp. 1762–1781 (2014)
21. Wollan, P.: Packing cycles with modularity constraints. Combinatorica **31**(1), 95–126 (2011)

Sequence Hypergraphs

Kateřina Böhmová[1]([✉]), Jérémie Chalopin[2], Matúš Mihalák[1,3],
Guido Proietti[4,5], and Peter Widmayer[1]

[1] Institute of Theoretical Computer Science, ETH Zürich, Zürich, Switzerland
{katerina.boehmova,widmayer}@inf.ethz.ch
[2] LIF, CNRS & Aix-Marseille University, Marseille, France
jeremie.chalopin@lif.univ-mrs.fr
[3] Department of Knowledge Engineering, Maastricht University,
Maastricht, The Netherlands
matus.mihalak@maastrichtuniversity.nl
[4] DISIM, Università degli Studi dell'Aquila, L'Aquila, Italy
[5] IASI, CNR, Roma, Italy
guido.proietti@univaq.it

Abstract. We introduce *sequence hypergraphs* by extending the concept of a directed edge (from simple directed graphs) to hypergraphs. Specifically, every hyperedge of a sequence hypergraph is defined as a sequence of vertices (imagine it as a directed path). Note that this differs substantially from the standard definition of directed hypergraphs. Sequence hypergraphs are motivated by problems in public transportation networks, as they conveniently represent transportation lines. We study the complexity of some classic algorithmic problems, arising (not only) in transportation, in the setting of sequence hypergraphs. In particular, we consider the problem of finding a *shortest st-hyperpath*: a minimum set of hyperedges that "connects" (allows to travel to) t from s; finding a *minimum st-hypercut*: a minimum set of hyperedges whose removal "disconnects" t from s; or finding a *maximum st-hyperflow*: a maximum number of hyperedge-disjoint st-hyperpaths.

We show that many of these problems are APX-hard, even in acyclic sequence hypergraphs or with hyperedges of constant length. However, if all the hyperedges are of length at most 2, we show, these problems become polynomially solvable. We also study the special setting in which for every hyperedge there also is a hyperedge with the same sequence, but in the reverse order. Finally, we briefly discuss other algorithmic problems (e.g., finding a minimum spanning tree, or connected components).

Keywords: Colored graphs · Labeled problems · Oriented hypergraphs · Algorithms · Complexity

1 Introduction

Consider a public transportation network, e.g. a bus network, where each bus line is specified as a fixed sequence of stops. Clearly, one can travel in the network

© Springer-Verlag GmbH Germany 2016
P. Heggernes (Ed.): WG 2016, LNCS 9941, pp. 282–294, 2016.
DOI: 10.1007/978-3-662-53536-3_24

by taking a bus and following the stops in the order fixed by the corresponding line. Note that we think of a line as a sequence of stops in one direction only, since there might be one-way streets or other obstacles that cause that the bus can travel the stops in a single direction only. Then, interesting questions arise: How can one travel from s to t using the minimum number of lines? How many lines must break down, so that t is not reachable from s? Are there two ways to travel from s to t that both use different lines?

These kind of questions are traditionally modeled by algorithmic graph theory, but we lacked a model that would capture all the necessary aspects of the problems formulated as above. We propose the following non-standard, but a very natural way to extend the concept of directed graphs to hypergraphs.

A hypergraph $\mathcal{H} = (\mathcal{V}, \mathcal{E})$ with an ordering of the vertices of every hyperedge is called a *sequence hypergraph*. Formally, the sequence hypergraph $\mathcal{H}$ consists of the set of vertices $\mathcal{V} = \{v_1, v_2, \ldots, v_n\}$, and the set of *(sequence) hyperedges* $\mathcal{E} = \{E_1, E_2, \ldots, E_k\}$, where each hyperedge $E = (v_{i_1}, v_{i_2}, \ldots, v_{i_l})$ is defined as a sequence of vertices without repetition. We remark that this definition substantially differs from the commonly used definition of directed hypergraphs [1,2,13], where each directed hyperedge is a pair (From, To) of disjoint subsets of $\mathcal{V}$.[1] We note that the order of vertices in a sequence hyperedge does not imply any order of the vertices of other hyperedges. Furthermore, the sequence hypergraphs do not impose any global order on $\mathcal{V}$.

There is another way to look at sequence hypergraphs coming from our motivation in transportation. For a sequence hypergraph $\mathcal{H} = (\mathcal{V}, \mathcal{E})$, we construct a *directed colored multigraph* $G = (V, E, c)$ as follows. The set of vertices V is identical to $\mathcal{V}$, and for a hyperedge $E_i = (v_1, v_2, \ldots, v_l)$ from $\mathcal{E}$, the multigraph G contains $l - 1$ edges (v_j, v_{j+1}) for $j = 1, \ldots, l - 1$, all colored with color $c(E_i)$. Therefore, each edge of G is colored by one of the $k = |\mathcal{E}|$ colors $\mathcal{C} = \{c(E_1), c(E_2), \ldots, c(E_k) \mid E_i \in \mathcal{E}\}$. Clearly, the edges of each color form a directed path in G. We refer to G as the *underlying colored graph* of $\mathcal{H}$.

In this paper, we study some standard graph algorithmic problems in the setting of sequence hypergraphs. In particular, we consider the problem of finding a *shortest st-hyperpath*: an st-path that uses the minimum number of sequence hyperedges; finding a *minimum st-hypercut*: an st-cut that uses the minimum number of sequence hyperedges; or finding a *maximum st-hyperflow*: a maximum number of hyperedge-disjoint st-hyperpaths. We note that the shortest st-hyperpath problem was already considered by Böhmová et al. [5] in the setting of finding good routes in public transportation networks (studied under a quite different terminology), who mainly focused on the problem of listing shortest paths in public transportation networks, but also showed that minimizing the number of lines in an st-path is hard to approximate.

In the present paper we show that the shortest st-hyperpath can be found in polynomial time if the given sequence hypergraph is acyclic. On the other hand, we show that both maximum st-hyperflow and minimum st-hypercut are APX-

---

[1] To avoid confusion with directed hypergraphs, we prefer the term *sequence hypergraphs* to refer to the hypergraphs with hyperedges formed as sequences of vertices.

hard to find even in acyclic sequence hypergraphs. We then consider sequence hypergraphs with sequence hyperedges of constant length (defined as the number of vertices minus one). We note that the shortest st-hyperpath problem remains hard to approximate even with hyperedges of length at most 5, and we show that the maximum st-hyperflow problem remains APX-hard even with hyperedges of length at most 3. On the other hand, we show that if all the hyperedges are of length at most 2, all 3 problems become polynomially solvable. We also study the complexity in a special setting in which for each hyperedge there also is a hyperedge with the same sequence, but in the opposite direction. We show that the shortest st-hyperpath problem becomes polynomially solvable, but both maximum st-hyperflow and minimum st-hypercut are NP-hard to find also in this setting, and we give a 2-approximation algorithm for the minimum st-hypercut problem. Finally, we briefly study the complexity of other algorithmic problems (finding minimum spanning tree, or connected components) in sequence hypergraphs. For a summary of the results see Table 1. The table also shows known results for related labeled graphs (discussed below).

Table 1. Summary of the complexity of some classic problems in the setting of colored (labeled) graphs and sequence hypergraphs. The last row indicates whether the sizes of the maximum st-flow and the minimum st-cut equal in the considered setting. The cells in gray indicate our contribution

| | Colored/labeled graphs | | Sequence hypergraphs | | | |
	General	Span 1	General	Acyclic	Backward	Length≤ 2
Shortest st-path	APX-hard [8,17]	P [8]	APX-hard [5]	P	P	P
Minimum st-cut	APX-hard [8,23]	P [8]	APX-hard	APX-hard	NP-hard	P
Maximum st-flow	APX-hard [18]	P [8]	APX-hard	APX-hard	NP-hard	P
MaxFlow-MinCut Duality	× [8]	√ [8]	×	×	×	√

Related Work. Recently, there has been a lot of research concerning optimization problems in (multi)graphs with colored edges, where the cost of a solution is measured by the number of colors used, e.g., one may ask for an st-path using the minimum number of colors. The motivation comes from applications in optical or other communication networks, where a group of links (i.e., edges) can fail simultaneously and a goal is to find resilient solutions. Similar situation may occur in economics, when certain commodities are sold (and priced) in bundles.

Formally, *colored graphs* or *labeled graphs*, are (mostly undirected) graphs where each edge has one color, and in general there is no restriction on a set of edges of the same color. Some of the studies consider a slightly different definition of a colored graphs, where to each edge corresponds a set of colors instead of a single color. Since the computational complexity problems may differ in the two models, the transformations between the two models have been investigated [9].

The *minimum label path* problem, which asks for an st-path of a minimum number of colors, is NP-hard and hard to approximate [6–8,15,17,22]. The *2 label disjoint paths* problem, which asks for a pair of st-paths such that the sets of colors appearing on the two paths are disjoint, is NP-hard [18]. The *minimum label cut* problem, which asks for a set of edges of minimum number of colors that forms an *st*-cut, is NP-hard and hard to approximate [8,23]. The *minimum label spanning tree* problem, which asks for a spanning tree using edges of minimum number of colors, is NP-hard and hard to approximate [17,20].

Hassin et al. [17] give a $\log(n)$-approximation algorithm for the minimum label spanning tree problem and a $\sqrt{n}$-approximation algorithm for the minimum label path problem. Zhang et al. [23] give a $\sqrt{m}$-approximation algorithm for the minimum label cut problem. Fellows et al. study the parameterized complexity of minimum label problems [12]. Coudert et al. [8,9] consider special cases when the *span* is 1, i.e., each color forms a connected component; or when the graph has a *star property*, i.e., the edges of every color are adjacent to one vertex.

Note that, since most of these results consider undirected labeled graphs, they provide almost no implications on the complexity of similar problems in the setting of sequence hypergraphs. In our setting, not only we work with *directed* label graphs, but we also require edges of each color to form a directed path, which implies a very specific structure that, to the best of our knowledge, has not been considered in the setting of labeled graphs.

On the other hand, we are not the first to define hypergraphs with hyperedges specified as sequences of vertices. However, this type of hypergraphs are usually not explored from an algorithmic graph theory point of view. In fact, mostly, these hypergraphs are taken merely as a tool, convenient to capture certain relations, but they are not studied further. We shortly list a few articles where sequence hypergraphs appeared, but we do not give details, since there is very little relation to our area of study. Berry et al. [4] introduce and describe the basic architecture of a software tool for (hyper)graph drawing. Wachman et al. [21] present a kernel for learning from *ordered hypergraphs*, a formalization that captures relational data as used in Inductive Logic Programming. Erdös et al. [11] study Sperner-families and as an application of a derived result they study the maximum number of edges of a so called *directed Sperner-hypergraph*.

Finally, a special case of sequence hypergraphs arose as a generalization to tournaments [3,16]: A k-hypertournament can be seen as a sequence hypergraph where for every subset of k vertices there is exactly one sequence hyperedge. Gutin et al. [16] studied the Hamiltonicity of k-hypertournaments.

2 On the Shortest *st*-Hyperpath

In this section, we briefly discuss the complexity of the shortest *st*-hyperpath problem in general sequence hypergraphs and in acyclic sequence hypergraphs.

Definition 1 (*st*-hyperpath). *Let s and t be two vertices of a sequence hypergraph $\mathcal{H} = (\mathcal{V}, \mathcal{E})$. A set of hyperedges $P \subseteq \mathcal{E}$ forms a hyperpath from s to t, if the underlying (multi)graph G' of the sequence subhypergraph $\mathcal{H}' = (\mathcal{V}, P)$ contains an st-path, and P is minimal with respect to inclusion. We call such an st-path an underlying path of P.*

The length *of an st-hyperpath P is defined as the number of hyperedges in P.*
The number of switches *of an st-hyperpath P is the minimum number of changes*
between the hyperedges of P, when following an underlying st-path of P.

We note that each hyperpath may have multiple underlying paths. Also note
that, even though the number of switches of an st-hyperpath P gives an upper
bound on the length of P, the actual length of P can be much smaller than the
number of switches of P (see Fig. 1a).

Fig. 1. In both figures, the grey-dotted curve, and the black curve depict two sequence
hyperedges. (a) The length of the st-hyperpath is 2, but the number of switches is 7.
(b) The st-hyperpath consists of two sequence hyperedges that also form a hypercycle.

Proposition 1. *Given a sequence hypergraph, and two vertices s and t, an st-*
hyperpath minimizing the number of switches can be found in polynomial time.

This can be done, e.g., by a modified Dijkstra algorithm (starting from s,
following the outgoing sequence hyperedges and for each vertex storing the minimum number of switches necessary to reach it).

On the other hand, by a reduction from the set cover problem, Böhmová
et al. [5] showed the following result (in a slightly different setting).

Theorem 1 ([5]). *Shortest st-hyperpath in sequence hypergraphs is NP-hard to*
approximate within a factor of $(1 - \epsilon) \ln n$, unless $P = NP$.

However, if the given sequence hypergraph is acyclic, we show that the shortest st-hyperpath can be found in polynomial time.

Definition 2 (acyclic sequence hypergraph). *A set of hyperedges $O \subseteq \mathcal{E}$*
forms a hypercycle, *if there are two vertices $a \neq b$ such that O forms both*
a hyperpath from a to b, and a hyperpath from b to a. A sequence hypergraph
without hypercycles is called acyclic.

Observe that an st-hyperpath may also be a hypercycle (see Fig. 1b).

Definition 3 (edges of a hyperedge). *Let $E = (v_1, v_2, \ldots, v_k)$ be a hyper-*
edge of a sequence hypergraph $\mathcal{H}$. We call the set of directed edges $\{e_i =$
(v_i, v_{i+1}) for $i = 1, \ldots, k - 1\}$ the edges *of E. The edges of E are exactly the*
edges of color $c(E)$ in the underlying colored graph of $\mathcal{H}$. The length of a hyper-
edge is defined as the number of its edges.

For a fixed order $V^O = (v_1, v_2, \ldots, v_n)$ of vertices $\mathcal{V}$, an edge e of a hyperedge
E is called a forward *edge with respect to V^O, if its orientation agrees with the*
order V^O. Similarly, e is a backward *edge, if its orientation disagrees with V^O.*

Theorem 2. *The problem of finding the shortest st-hyperpath in acyclic sequence hypergraphs can be solved in polynomial time.*

Proof (Sketch). Observe that for every st-hyperpath, there is an underlying path where all the edges of each hyperedge appear consecutively. Thus, finding the shortest st-hyperpath P in $\mathcal{H}$ is the same as finding a hyperpath minimizing the number of switches, which can be done in polynomial time by Proposition 1. $\square$

3   On the Maximum st-Hyperflow

We consider the problem of finding a number of hyperedge-disjoint st-hyperpaths. Capturing a similar relation as in graphs (between a set of k edge-disjoint st-paths and an st-flow of size k, when all the capacities are 1), for simplicity and brevity, we refer to a set of hyperedge-disjoint st-hyperpaths as an st-hyperflow.

Definition 4 (st-hyperflow). *Let s and t be two vertices of a sequence hypergraph $\mathcal{H} = (\mathcal{V}, \mathcal{E})$. Let $\mathcal{F} \subseteq 2^{\mathcal{E}}$ be a set of pairwise hyperedge-disjoint st-hyperpaths $\mathcal{F} = \{P_1, \ldots, P_k\}$. Then, $\mathcal{F}$ is an st-hyperflow of size $|\mathcal{F}| = k$.*

We show that deciding whether the given sequence hypergraph contains an st-hyperflow of size 2 is NP-hard, and thus finding a maximum st-hyperflow is inapproximable within a factor $2 - \epsilon$ unless P=NP. This remains true even for acyclic sequence hypergraphs with all the hyperedges of length at most 3.

Theorem 3. *Given an acyclic sequence hypergraph $\mathcal{H} = (\mathcal{V}, \mathcal{E})$ with all hyperedges of length at most 3, and two vertices s and t, it is NP-complete to decide whether there are two hyperedge-disjoint st-hyperpaths.*

Proof. We construct a reduction from the NP-complete 3-SAT problem [14]. Let I be an instance of the 3-SAT problem, given as a set of m clauses $C = \{c_1, \ldots, c_m\}$ over a set $X = \{x_1, \ldots, x_n\}$ of Boolean variables. The goal is to find an assignment to the variables of X that satisfies all clauses of C.

From I we construct a sequence hypergraph $\mathcal{H} = (\mathcal{V}, \mathcal{E})$ as follows (cf. Figure 2 along with the construction). The set of vertices $\mathcal{V}$ consists of $2 + (m + 1) + (n + 1) + \sum_{c_i \in C} |c_i|$ vertices: a source vertex s, and a target vertex t; a vertex c_i for each clause $c_i \in C$ and a dummy vertex c_{m+1}; a vertex x_j for each variable $x_j \in X$ and a dummy vertex x_{n+1}; and finally a vertex $x_j c_i$ for each pair (x_j, c_i) such that $x_j \in c_i$, and similarly, $\overline{x_j} c_i$ for each $\overline{x_j} \in c_i$. Let us fix an arbitrary order C^O of the clauses in C. The set of hyperedges $\mathcal{E}$ consists of $4 + 2n + |I|$ hyperedges: There are 2 *source hyperedges* (s, c_1) and (s, x_1), and 2 *target hyperedges* (c_{m+1}, t) and (x_{n+1}, t). There are $2n$ *auxiliary hyperedges* $(x_i, x_i c_k)$ and $(x_i, \overline{x_i} c_{k'})$ for $i = 1, \ldots, n$, where c_k, or $c_{k'}$ is always the first clause (with respect to C^O) containing x_i, or $\overline{x_i}$, respectively. In case there is no clause containing x_i (or $\overline{x_i}$), the corresponding auxiliary hyperedge is (x_i, x_{i+1}). Finally, there are $|I|$ *lit-in-clause hyperedges* as follows. For each

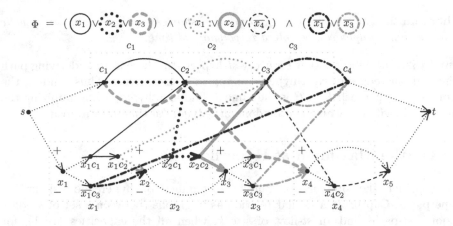

Fig. 2. Deciding st-hyperflow of size 2 is as hard as 3-SAT.

appearance of a variable x_j in a clause c_i as a positive literal there is one lit-in-clause hyperedge $(c_i, c_{i+1}, x_j c_i, x_j c_k)$, where c_k is the next clause (with respect to C^O) after c_i where x_j appears as a positive literal (in case, there is no such c_k, then the hyperedge ends in x_{j+1} instead). Similarly, if x_j is in c_i as a negative literal, there is one lit-in-clause hyperedge $(c_i, c_{i+1}, \overline{x_j} c_i, \overline{x_j} c_k)$, where c_k is the next clause containing the negative literal $\overline{x_j}$ (or it ends in x_{j+1}).

Clearly, each hyperedge is of length at most 3. We now observe that the constructed sequence hypergraph $\mathcal{H}$ is acyclic. All the hyperedges of $\mathcal{H}$ agree with the following order: the source vertex s; all the vertices $c_i \in C$ ordered according to C^O, and the dummy vertex c_{m+1}; the vertex x_1 followed by all the vertices $x_1 c_i$ ordered according to C^O, and then followed by the vertices $\overline{x_1} c_i$ again ordered according to C^O; the vertex x_2 followed by all $x_2 c_i$ and then all $\overline{x_2} c_i$; ...; the vertex x_n followed by all $x_n c_i$ and then all $\overline{x_n} c_i$; and finally the dummy vertex x_{n+1} and the target vertex t.

We show that the formula I is satisfiable if and only if the sequence hypergraph $\mathcal{H}$ contains two hyperedge-disjoint st-hyperpaths. There are 3 possible types of st-paths in the underlying graph of $\mathcal{H}$: first one leads through all the vertices $c_1, c_2, \ldots, c_{m+1}$ in this order; second one leads through all the vertices $x_1, x_2, \ldots, x_{m+1}$ in this order and between x_j, x_{j+1} it goes either through all the $x_j c_*$ vertices or through all the $\overline{x_j} c_*$ vertices; and the third possible st-path starts the same as the first option and ends as the second one. Based on this observation, notice that there can be at most 2 hyperedge-disjoint st-hyperpaths: necessarily, one of them has an underlying path of the first type, while the other one has an underlying path of the second type.

From a satisfying assignment A of I we can construct the two disjoint st-hyperpaths as follows. The underlying path of one hyperpath leads from s to t via the vertices $c_1, c_2, \ldots, c_{m+1}$, and to move from c_i to c_{i+1} it uses a lit-in-clause hyperedge that corresponds to a pair (l, c_i) such that l is one of the literals that satisfy the clause c_i in A. The second hyperpath has an underlying path of the second type, it leads via $x_1, x_2, \ldots, x_{n+1}$ and from x_j to x_{j+1} it uses the vertices containing only the literals that are not satisfied by the assignment A. Thus, the second hyperpath uses only those lit-in-clause hyperedges that corresponds to pairs containing literals that are not satisfied by A. This implies that the two constructed st-hyperpaths are hyperedge-disjoint.

Let P and Q be two hyperedge-disjoint st-hyperpaths of $\mathcal{H}$. Let P has an underlying path p of the first type and Q has an underlying path q of the second type. We can construct a satisfying assignment for I by setting to FALSE the literals that occur in the vertices on q. Then, the hyperpath P suggests how the clauses of I are satisfied by this assignment.   □

4   On the Minimum st-Hypercut

Quite naturally, we define an st-hypercut of a sequence hypergraph $\mathcal{H}$ as a set C of hyperedges, whose removal from $\mathcal{H}$ leaves s and t disconnected.

Definition 5 (st-hypercut). *Let s and t be two vertices of a sequence hypergraph $\mathcal{H} = (\mathcal{V}, \mathcal{E})$. A set of hyperedges $X \subseteq \mathcal{E}$ is an st-hypercut, if the subhypergraph $\mathcal{H}' = (\mathcal{V}, \mathcal{E} \setminus X)$ does not contain any hyperpath from s to t. The size of an st-hypercut X is $|X|$, that is the number of hyperedges in X.*

For directed (multi)graphs, the famous MaxFlow-MinCut Duality Theorem [10] states that the size of a maximum st-flow is equal to the size of a minimum st-cut. In sequence hypergraphs, this duality does not hold, even in acyclic sequence hypergraphs as Fig. 3 shows. But, of course, the size of an st-hyperflow is a lower bound on the size of an st-hypercut.

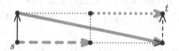

Fig. 3. Acyclic sequence hypergraph with minimum st-hypercut of size 2, and no two hyperedge-disjoint st-hyperpaths.

We showed maximum st-hyperflow to be APX-hard even in acyclic sequence hypergraphs. It turns out that also minimum st-hypercut problem in acyclic sequence hypergraphs is APX-hard.

Theorem 4. *Minimum st-hypercut in acyclic sequence hypergraphs is hard to approximate within a factor $2 - \epsilon$ under UGC, or within a factor $7/6 - \epsilon$ unless P=NP.*

Proof. We construct an approximation preserving reduction from the vertex cover problem, which has the claimed inapproximability [19].   □

5 Sequence Hypergraphs with Hyperedges of Length ≤ 2

We have seen that some of the classic, polynomially solvable problems in (directed) graphs become APX-hard in sequence hypergraphs. Note that this often remains true even if all the hyperedges are of constant length. In particular, the shortest st-hyperpath problem is APX-hard even if all the hyperedges are of length at most 5 (the proof given in [5] needs just a slight modification); Fig. 2 illustrates that the duality between minimum st-hypercut and maximum st-hyperflow breaks already with a single hyperedge of length 3; and Theorem 3 holds even if all hyperedges are of length at most 3.

It is an interesting question to investigate the complexity of the problems for hyperedge lengths smaller than 5 or 3. We show that, if all the hyperedges of the given sequence hypergraph are of length at most 2, the shortest st-hyperpath, the minimum st-hypercut, and the maximum st-hyperflow can all be found in polynomial time.

Theorem 5. *The shortest st-hyperpath problem in sequence hypergraphs with hyperedges of length at most 2 can be solved in polynomial time.*

The proof is based on similar ideas as in the proof of Theorem 2.

Theorem 6. *The maximum st-hyperflow problem and the minimum st-hypercut problem can be solved in polynomial time in sequence hypergraphs with hyperedges of length at most 2. The size of the maximum st-hyperflow then equals the size of the minimum st-hypercut.*

Proof. Let $\mathcal{H} = (\mathcal{V}, \mathcal{E})$ be a sequence hypergraph with hyperedges of length at most 2, and let s and t be two of its vertices. Then, using a standard graph algorithms we can find a maximum st-flow f in the underlying directed multigraph G of $\mathcal{H}$ with edge capacities 1. Thus, the flow f of size $|f|$ gives us a set of $|f|$ edge-disjoint st-paths $p_1, \ldots, p_{|f|}$ in G (note that any directed cycles in f can be easily removed).

We iteratively transform $p_1, \ldots, p_{|f|}$ into a set of st-paths such that all the edges of each hyperedge appear on only one of these paths. Let $E = (u, v, w)$ be a hyperedge that lies on two different paths, i.e., $(u, v) \in p_i$ and $(v, w) \in p_j$, for some $i, j \in [|f|]$. Then, p_i consists of an su-path, edge (u, v), and a vt-path. Similarly, p_j consists of an sv-path, edge (v, w), and a wt-path. Since all these paths and edges are pairwise edge-disjoint, by setting p_i to consist of the su-path, edge (u, v), edge (v, w), and the wt-path; and at the same time setting p_j to consist of the sv-path, and the vt-path, we again obtain two edge-disjoint st-paths p_i and p_j. However, now the hyperedge E is present only on p_i. At the same time, since each hyperedge is of length at most 2, all the edges of a hyperedge appear on any st-path consecutively, and any hyperedge that was present on only one of p_i, p_j, is not affected by the above rerouting and still is present on one of the two paths only.

Thus, the rerouting decreased the number of hyperedges present on more than one paths, and after at most $|\mathcal{E}|$ iterations of this transformation we obtain $|f|$

hyperedge-disjoint st-paths, which gives us an st-hyperflow F of size $|F| = |f|$. It is easy to observe that the size of the hyperflow is bounded from above by the size of the flow in the underlying multigraph. Thus, we obtained a maximum st-hyperflow in $\mathcal{H}$.

Since in directed multigraphs the size of the minimum cut equals the size of the maximum flow [10], it follows that we can find $|F|$ edges $e_1, \ldots, e_{|F|}$ of G that forms a minimum cut of G. Observe that each of these edges corresponds to exactly one hyperedge. Thus, we obtain a set C of at most $|F|$ hyperedges that forms an st-hypercut. Since any st-hypercut is bounded from below by the size of the hyperflow, C is a minimum st-hypercut of size $|C| = |F|$.   □

6 Sequence Hypergraphs with Backward Hyperedges

We consider a special class of sequence hypergraphs where for every hyperedge, there is the exact same hyperedge, but oriented in the opposite direction.

Definition 6 (backward hyperedges). *Let $E = (v_1, v_2, \ldots, v_k)$ be a hyperedge of a sequence hypergraph $\mathcal{H} = (\mathcal{V}, \mathcal{E})$. We say that E' is a backward hyperedge[2] of E, if $E' = (v_k, \ldots, v_2, v_1)$. If for every E of $\mathcal{E}$, there is exactly one backward hyperedge in $\mathcal{E}$, we refer to $\mathcal{H}$ as sequence hypergraph with backward hyperedges.*

Such a situation arise naturally in urban public transportation networks, for instance most of the tram lines have also a "backward" line (which has the exact same stops as the "forward" line, but goes in the opposite order). We study the complexity of shortest st-hyperpath, minimum st-hypercut, and maximum st-hyperflow under this setting. We show that, in this setting, we can find a shortest st-hyperpath in polynomial time. On the other hand, we show that minimum st-hypercut and maximum st-hyperflow remain NP-hard, and we give a 2-approximation algorithm for the minimum st-hypercut. The positive results are based on existing algorithms for standard hypergraphs, the negative results are obtained by a modification of the hardness proofs in Sects. 3 and 4.

Theorem 7. *The shortest st-hyperpath problem in sequence hypergraphs with backward hyperedges is in P.*

Proof. Let $\mathcal{H} = (\mathcal{V}, \mathcal{E})$ be a sequence hypergraph with backward hyperedges, and let s and t be two vertices of $\mathcal{H}$. We construct a (standard) hypergraph $\mathcal{H}^* = (\mathcal{V}^* = \mathcal{V}, \mathcal{E}^*)$ from $\mathcal{H}$ in such a way that for each sequence hyperedge E of $\mathcal{E}$, $\mathcal{E}^*$ contains a (non-oriented) hyperedge E^* that corresponds to the set of vertices of E. Note that E and its backward hyperedge E' consist of the same set of vertices, thus the corresponding E^* and E'^* are the same. A shortest st-hyperpath[3] P^* in (the standard) hypergraph $\mathcal{H}^*$ can be found in

---

[2] Note, if E' is a backward hyperedge of E, also E is a backward hyperedge of E'.

[3] An st-hyperpath P^* and its underlying path are defined as in sequence hypergraphs.

polynomial time. Observe that the size of P^* gives us a lower bound $|P^*|$ on the length of the shortest path in the sequence hypergraph $\mathcal{H}$.

In fact, we can construct from P^* an st-hyperpath in $\mathcal{H}$ of size $|P^*|$ as follows. Let us fix p^* to be an underlying path of P^*. Let $(s = v_1, v_2, \ldots, v_{|P^*|+1} = t)$ be a sequence of vertices, subsequence of p^*, such that for each $i = 1, \ldots, |P^*|$, there is a hyperedge E^* in P^* that contains both v_i and v_{i+1}, and v_i is the first vertex of E^* seen on p^*, and v_{i+1} is the last vertex of E^* seen on p^*. Since every hyperedge E^* of $\mathcal{E}^*$ corresponds to the set of vertices of some hyperedge E of $\mathcal{E}$, there is a sequence of sequence hyperedges $(E_1, E_2, \ldots, E_{|P^*|})$, $E_i \in \mathcal{E}$, such that v_i, v_{i+1} are vertices in E_i. Since $\mathcal{H}$ is sequence hypergraph with backward hyperedges, for every hyperedge E of $\mathcal{E}$ and a pair its of vertices v_i, v_{i+1} of E, there is an $v_i v_{i+1}$-hyperpath in $\mathcal{H}$ of size 1, which consists of E or its backward hyperedge E'. Therefore, there is an st-hyperpath of size $|P^*|$ in $\mathcal{H}$.   □

Theorem 8. *The maximum st-hyperflow problem in sequence hypergraphs with backward hyperedges is NP-hard.*

Theorem 9. *The minimum st-hypercut problem in sequence hypergraphs with backward hyperedges is NP-hard.*

Theorem 10. *The minimum st-hypercut problem in sequence hypergraphs with backward hyperedges can be 2-approximated.*

7   On Other Algorithmic Problems

We briefly consider some other standard graph algorithmic problems.

Definition 7 (rooted spanning hypergraph). *Let $\mathcal{H} = (\mathcal{V}, \mathcal{E})$ be a sequence hypergraph. We define s-rooted spanning hypergraph T as a subset of $\mathcal{E}$ such that for every $v \in \mathcal{V}$, T is an sv-hyperpath. The size of T is defined as $|T|$.*

Theorem 11. *Minimum s-rooted spanning hypergraph in acyclic sequence hypergraphs is NP-hard to approximate within a factor of $(1 - \epsilon) \ln n$, unless $P = NP$.*

Definition 8 (strongly connected component). *Let $\mathcal{H} = (\mathcal{V}, \mathcal{E})$ be a sequence hypergraph. We say that a set $C \subseteq \mathcal{E}$ forms a strongly connected component if for every two vertices $u, v \in \mathcal{V}'$, $\mathcal{V}'$ being all the vertices of $\mathcal{V}$ present in C, the set C is a uv-hyperpath. We say that the vertices in $\mathcal{V}'$ are covered by C.*

Clearly, we can decide in polynomial time whether the given set of hyperedges C forms a strongly connected component as follows. Consider the underlying graph G of $\mathcal{H}$ induced by the set of sequence hyperedges C and find a maximum strongly connected component there. If this component spans the whole G, then C is a strongly connected component in $\mathcal{H}$.

Theorem 12. *Given a sequence hypergraph* $\mathcal{H} = (\mathcal{V}, \mathcal{E})$, *it is NP-hard to find a* minimum *number of hyperedges that form a strongly connected component C so that a) C is any non-empty set, or b) all the vertices in $\mathcal{V}$ are covered by C.*

Theorem 13. *Given a sequence hypergraph* $\mathcal{H} = (\mathcal{V}, \mathcal{E})$, *finding a* maximum *number of hyperedges that form a strongly connected component C so that a) C is any non-empty set, or b) all the vertices in $\mathcal{V}$ are covered by C, is polynomial-time solvable.*

Acknowledgements. Kateřina Böhmová is supported by a Google Europe Fellowship in Optimization Algorithms. Jérémie Chalopin was partially supported by the ANR project MACARON (anr-13-js02-0002).

References

1. Ausiello, G., Franciosa, P.G., Frigioni, D.: Directed hypergraphs: problems, algorithmic results, and a novel decremental approach. In: Restivo, A., Ronchi Della Rocca, S., Roversi, L. (eds.) ICTCS 2001. LNCS, vol. 2202, p. 312. Springer, Heidelberg (2001)
2. Ausiello, G., Giaccio, R., Italiano, G.F., Nanni, U.: Optimal traversal of directed hypergraphs. Technical report (1992)
3. Barbut, E., Bialostocki, A.: A generalization of rotational tournaments. Discrete Math. **76**(2), 81–87 (1989)
4. Berry, J., Dean, N., Goldberg, M., Shannon, G., Skiena, S.: Graph drawing and manipulation with link. In: DiBattista, G. (ed.) GD 1997. LNCS, vol. 1353, pp. 425–437. Springer, Heidelberg (1997)
5. Böhmová, K., Mihalák, M., Pröger, T., Sacomoto, G., Sagot, M.-F.: Computing and listing st-paths in public transportation networks. In: Kulikov, A.S., Woeginger, G.J. (eds.) CSR 2016. LNCS, vol. 9691, pp. 102–116. Springer, Heidelberg (2016). doi:10.1007/978-3-319-34171-2_8
6. Broersma, H., Li, X., Woeginger, G., Zhang, S.: Paths and cycles in colored graphs. Australas. J. Comb. **31**, 299–311 (2005)
7. Carr, R.D., Doddi, S., Konjevod, G., Marathe, M.V.: On the red-blue set cover problem. In: SODA 2000. vol. 9, pp. 345–353. Citeseer (2000)
8. Coudert, D., Datta, P., Pérennes, S., Rivano, H., Voge, M.E.: Shared risk resource group complexity and approximability issues. Parallel Process. Lett. **17**(02), 169–184 (2007)
9. Coudert, D., Pérennes, S., Rivano, H., Voge, M.E.: Combinatorial optimization in networks with Shared Risk Link Groups. Research report, INRIA, October 2015
10. Elias, P., Feinstein, A., Shannon, C.E.: A note on the maximum flow through a network. IRE Trans. Inf. Theor. **2**(4), 117–119 (1956)
11. Erdős, P.L., Frankl, P., Katona, G.O.: Intersecting sperner families and their convex hulls. Combinatorica **4**(1), 21–34 (1984)
12. Fellows, M.R., Guo, J., Kanj, I.A.: The parameterized complexity of some minimum label problems. In: Paul, C., Habib, M. (eds.) WG 2009. LNCS, vol. 5911, pp. 88–99. Springer, Heidelberg (2010)
13. Gallo, G., Longo, G., Pallottino, S., Nguyen, S.: Directed hypergraphs and applications. Discrete Appl. Math. **42**(2), 177–201 (1993)

14. Garey, M.R., Johnson, D.S.: Computers and Intractability: A Guide to the Theory of NP-Completeness. W.H. Freeman & Co., New York (1979)
15. Goldberg, P.W., McCabe, A.: Shortest paths with bundles and non-additive weights is hard. In: Spirakis, P.G., Serna, M. (eds.) CIAC 2013. LNCS, vol. 7878, pp. 264–275. Springer, Heidelberg (2013)
16. Gutin, G., Yeo, A.: Hamiltonian paths and cycles in hypertournaments. J. Graph Theor. **25**(4), 277–286 (1997)
17. Hassin, R., Monnot, J., Segev, D.: Approximation algorithms and hardness results for labeled connectivity problems. J. Comb. Opt. **14**(4), 437–453 (2007)
18. Hu, J.Q.: Diverse routing in optical mesh networks. IEEE Trans. Commun. **51**(3), 489–494 (2003)
19. Khot, S., Regev, O.: Vertex cover might be hard to approximate to within 2-epsilon. J. Comput. Syst. Sci. **74**(3), 335–349 (2008)
20. Krumke, S.O., Wirth, H.C.: On the minimum label spanning tree problem. Inf. Process. Lett. **66**(2), 81–85 (1998)
21. Wachman, G., Khardon, R.: Learning from interpretations: a rooted kernel for ordered hypergraphs. In: Proceedings of the 24th International Conference on Machine Learning, pp. 943–950. ACM (2007)
22. Yuan, S., Varma, S., Jue, J.P.: Minimum-color path problems for reliability in mesh networks. In: INFOCOM 2005, vol. 4, pp. 2658–2669. IEEE (2005)
23. Zhang, P., Cai, J.Y., Tang, L.Q., Zhao, W.B.: Approximation and hardness results for label cut and related problems. J. Comb. Opt. **21**(2), 192–208 (2011)

On Subgraphs of Bounded Degeneracy in Hypergraphs

Kunal Dutta[1] and Arijit Ghosh[2(✉)]

[1] DataShape, Inria Sophia Antipolis – Méditerranée, Méditerranée, France
[2] ACM Unit, Indian Statistical Institute, Kolkata, India
arijitiitkgpster@gmail.com

Abstract. A k-uniform hypergraph has degeneracy bounded by d if every induced subgraph has a vertex of degree at most d. Given a k-uniform hypergraph $H = (V(H), E(H))$, we show there exists an induced subgraph of size at least

$$\sum_{v \in V(H)} \min \left\{ 1, c_k \left(\frac{d+1}{d_H(v)+1} \right)^{1/(k-1)} \right\},$$

where $c_k = 2^{-\left(1+\frac{1}{k-1}\right)} \left(1 - \frac{1}{k}\right)$ and $d_H(v)$ denotes the degree of vertex v in the hypergraph H. This extends and generalizes a result of Alon-Kahn-Seymour (Graphs and Combinatorics, 1987) for graphs, as well as a result of Dutta-Mubayi-Subramanian (SIAM Journal on Discrete Mathematics, 2012) for linear hypergraphs, to general k-uniform hypergraphs. We also generalize the results of Srinivasan and Shachnai (SIAM Journal on Discrete Mathematics, 2004) from independent sets (0-degenerate subgraphs) to d-degenerate subgraphs. We further give a simple non-probabilistic proof of the Dutta-Mubayi-Subramanian bound for linear k-uniform hypergraphs, which extends the Alon-Kahn-Seymour (Graphs and Combinatorics, 1987) proof technique to hypergraphs. Our proof combines the *random permutation* technique of Bopanna-Caro-Wei (see e.g. *The Probabilistic Method*, N. Alon and J. H. Spencer; Dutta-Mubayi-Subramanian) and also Beame-Luby (SODA, 1990) together with a new *local density* argument which may be of independent interest. We also provide some applications in discrete geometry, and address some natural algorithmic questions.

Keywords: Degenerate graphs · Independent sets · Hypergraphs · Random permutations

1 Introduction

For $k \geq 2$, a k-uniform hypergraph is a pair $(V(H), E(H))$ where $E(H) \subseteq \binom{V(H)}{k}$. We will call $V(H)$ and $E(H)$ the vertex set and edge set of H respectively. When there is no chance of confusion, we will use V and E to denote $V(H)$ and

© Springer-Verlag GmbH Germany 2016
P. Heggernes (Ed.): WG 2016, LNCS 9941, pp. 295–306, 2016.
DOI: 10.1007/978-3-662-53536-3_25

$E(H)$. For a vertex $v \in V(H)$, degree $d_H(v)$ of $V(H)$ will denote $|\{e : e \in E(H), v \in e\}|$. For readability, $k - 1$ will be denoted by t.

For a subset $I \subseteq V(H)$, the induced k-uniform hypergraph $H(I)$ of I denotes the hypergraph $(I, E(H) \cap \binom{I}{k})$. A hypergraph is *linear* if every pair of vertices are contained in at most a single hyperedge, i.e. any pair of hyperedges intersect in at most one vertex. A hypergraph $H = (V, E)$ is *d-degenerate* if the induced hypergraph of all subsets of V has a vertex of degree at most d, i.e., for all $I \subseteq V$, there exists $v \in I$ such that $d_{H(I)}(v) \leq d$. For a k-uniform hypergraph $H = (V, E)$, we will denote by $\alpha_{k,d}(H)$ the size of a maximum-sized subset of V whose induced hypergraph is d-degenerate, i.e.,

$$\alpha_{k,d}(H) = \max\{|I| : I \subseteq V, H(I) \text{ is } d\text{-degenerate}\}.$$

Observe that $\alpha(H) := \alpha_{k,0}(H)$ is the *independence number* of the hypergraph H.

1.1   Previous Results

Turán [Tur41] gave a lower bound on the independence number of graphs: $\alpha(G) \geq \frac{n}{d+1}$ where d is the average degree of vertices in G.

Caro [Car79] and Wei [Wei81] independently showed that for graphs

$$\alpha(G) \geq \sum_{v \in V(G)} \frac{1}{d_G(v) + 1},$$

see [AS08]. This degree-sequence based bound improves on the original average-degree based lower bound of Turán, and matches it in the case when all degrees are equal.

For hypergraphs, Spencer [Spe72] gave a bound on the independence number, based on the average degree d: $\alpha(H) \geq c_k \left(\frac{n}{d^{1/t}}\right)$, where c_k is independent of n and d. Caro and Tuza [CT91] generalized the Caro-Wei result to the case of hypergraphs:

Theorem 1. *For all k-uniform hypergraph H, we have*

$$\alpha(H) \geq \sum_{v \in V} \frac{1}{\binom{d_H(v)+1/t}{d_H(v)}}.$$

The above theorem directly implies the following corollary, which gives Spencer's bound:

Corollary 1. *For all $k \geq 2$, There exists $d_k > 0$ such that all k-uniform hypergraphs H satisfy*

$$\alpha(H) \geq d_k \sum_{v \in V} \frac{1}{(1 + d_H(v))^{1/t}}.$$

Further, Thiele [Thi99] obtained a lower bound on the independence number of arbitrary (non-uniform) hypergraphs, in terms of the *degree rank*, a generalization of the degree sequence.

On the algorithmic side, Srinivasan and Shachnai [SS04], used the random permutation method of Beame and Luby [BL90] and also Bopanna-Caro-Wei (see e.g. [AS08, DMS12]), together with the FKG correlation inequality, to obtain a randomized parallel algorithm for independent sets, which matched the asymptotic form of the Caro-Tuza bound. Dutta, Mubayi and Subramanian [DMS12] also used the Bopanna-Caro-Wei method; using elementary techniques they obtained degree-sequence based lower bounds on the independence numbers of K_r-free graphs and linear k-uniform hypergraphs, which generalized the earlier average-degree based bounds of Ajtai et al. [AKS80], Shearer [She83, She95] and Duke, Lefmann and Rödl [DLR95], in terms of degree sequences. [1]

Average Degree vs. Degree-Sequence. In general, a bound using the degree sequence would be intuitively expected to be better than a bound using just the average degree, since it has more information about the graph. For the above bounds on the independence numbers, this essentially follows from the convexity of the function $x^{-1/t}$. Dutta-Mubayi-Subramanian [DMS12] gave constructions of hypergraphs which show that the bounds based on the degree-sequence can be stronger than those based on the average degree by a polylogarithmic (in the number of vertices) factor.

Large d-degenerate Subgraphs. Compared to independent sets, d-degenerate subgraphs have been less well-investigated. However, it includes as special cases zero-degenerate subgraphs i.e. independent sets, as well as 1-degenerate subgraphs, i.e. maximum induced forests, whose complements are the well-known hitting set and feedback vertex set problems respectively. The best known result on this question is that of Alon, Kahn and Seymour [AKS87], who proved the following lower bound for $\alpha_{2,d}(G)$: [2]

Theorem 2 [AKS87]. *For all graphs $G = (V, E)$ we have*

$$\alpha_{2,d}(G) \geq \sum_{v \in V} \min\left\{1, \frac{d+1}{d_G(v)+1}\right\}.$$

This bound is sharp for every G which is a disjoint union of cliques. Moreover, they gave a polynomial time algorithm that finds in G an induced d-degenerate subgraph of at least this size.

---

[1] Their proof also yields an elementary proof of the main bound of Srinivasan and Shachnai [SS04] without using correlation inequalities, though they do not state this explicitly.

[2] Alon, Kahn and Seymour [AKS87] actually defined a d-degenerate graph as one where every subgraph has a vertex of degree *less than* d, whereas we use the more usual definition in which every subgraph has a vertex of degree *at most* d.

On the algorithmic side, Pilipczuk and Pilipczuk [PP12] addressed the question of finding a maximum d-degenerate subgraph of a graph, giving the first algorithm with running time $o(2^n)$. Zaker [Zak13] studied a more general version of degeneracy and gave upper and lower bounds for finding the largest subgraph of a 2-uniform graph having a given generalized degeneracy.

The proof of Dutta-Mubayi-Subramanian [DMS12] implies the following lower bound on $\alpha_{k,d}$ for linear hypergraphs (though not explicitly stated in their paper):

Theorem 3 [DMS12]. *Let $G = (V, E)$ be a linear k-uniform hypergraph, and for all $v \in V$, $d_G(v)$ denote the degree of v in G. Then*

$$\alpha_{k,d}(G) \geq w(G) := \sum_{v \in V} w_G(v), \tag{1}$$

where

$$w(v) = \begin{cases} 1 & \text{if } d_G(v) \leq d \\ \frac{1}{1+(t(d+1))^{-1}} \frac{\binom{d_G(v)}{d+1}}{\binom{d_G(v)+1/t}{d_G(v)-d-1}} & \text{if } d_G(v) > d. \end{cases} \tag{2}$$

Our Results

We first give a completely different and an extremely simple proof of Theorem 3 using a weight function. Our proof follows along the lines of the proof of Theorem 2 due to Alon, Kahn and Seymour [AKS87].

Next, we extend Theorem 3 to the case of general hypergraphs:

Theorem 4. *Let $G = (V, E)$ be a k-uniform hypergraph, and for all $v \in V$, $d_G(v)$ denote the degree of v in G. Then*

$$\alpha_{k,d}(G) \geq \sum_{v \in V} \min \left\{ 1, c_k \left(\frac{d+1}{d(v)+1} \right)^{1/t} \right\}, \tag{3}$$

where $t = k - 1$ and $c_k = 2^{-1-1/(k-1)}$. There exists a randomized algorithm that can extract a d-degenerate set of above size in expectation.

Our proof uses the *random permutation method* [AS08] of Bopanna-Caro-Wei, together with a new local density argument, avoiding advanced correlation inequalities. As a consequence, we obtain a simpler proof as well as a generalization of the result of Srinivasan-Shachnai [SS04].

As an application of Theorem 4 we will prove the following result in incidence geometry, which generalizes a result of Payne and Wood [PW13] on the maximum size of a subset, out of n points in the plane, such that no three points in the subset are collinear.

Lemma 1. *1. Let P be a set of n points in the plane such that for any line l in the plane $|l \cap P| \leq \ell$. For $d \leq O(n \log \ell + \ell^2)$ there exists a subset $S \subseteq P$ with at most $d|S|$ collinear triples in S and*

$$|S| = \Omega\left(\sqrt{\frac{d\,n^2}{n \log \ell + \ell^2}}\right).$$

And if $\ell \leq O(\sqrt{n})$, then

$$|S| = \Omega\left(\sqrt{\frac{d\,n}{\log \ell}}\right).$$

2. Let P be a set of n points in the plane such that for any line l in the plane $|l \cap P| \leq \ell$. Let $k \geq 4$ be a constant and $d \leq O(\ell^{k-3}n + \ell^{k-1})$. Then there exists a subset $S \subseteq P$ of size

$$\Omega\left(n\left(\frac{d}{\ell^{k-3}n + \ell^{k-1}}\right)^{1/(k-1)}\right)$$

such that S has at most $d|S|$ collinear k-tuples in S. And if $\ell \leq O(\sqrt{n})$, then

$$|S| = \Omega\left(\left(\frac{n^{k-2}d}{\ell^{k-3}}\right)^{1/(k-1)}\right).$$

The proof of Lemma 1 uses the following lemma by Payne and Wood [PW13], proved using Szemerédi-Trotter theorem [ST83] on incidence geometry.

Lemma 2 [PW13].

1. *Let P be a set of n points in the plane such that for any line l in the plane $|l \cap P| \leq \ell$. Then the number of collinear 3-tuples in P is at most $O(n^2 \log \ell + n\ell^2)$.*
2. *Let P be a set of n points in the plane such that for any line l in the plane $|l \cap P| \leq \ell$. Then, for $k \geq 4$, the number of collinear k-tuples in P is at most $O(\ell^{k-3}n^2 + \ell^{k-1}n)$.*

Proof (Lemma 1). Let H be a k-uniform hypergraph with $V(H) = P$, and $\{p_1, \ldots, p_k\} \in E(H)$ if there exists a line a line l in the plane with $\{p_1, \ldots, p_k\} \in l$. Lemma 2 bounds the size of $E(H)$. The result now follows directly from Theorem 4.

The rest of the paper is organised as follows: Sect. 2 has the simpler proof of Theorem 3, and Sect. 3 has the proof of Theorem 4. Finally in the Conclusions section there are some remarks and open questions.

2   Linear Hypergraphs

In this section we will give an alternative proof of the Theorem 3. The proof will follow exactly along the lines of the proof by Alon, Kahn and Seymour [AKS87] of Theorem 2.

First observe that

$$\frac{\binom{d_G(v)}{r}}{\binom{d_G(v)+1/t}{d_G(v)-r}} = \frac{1}{\left(1 + \frac{1}{t(r+1)}\right) \cdots \left(1 + \frac{1}{td_G(v)}\right)}$$

This implies that $w(v)$ is decreasing in $d_G(v)$ for all values of $d_G(v) \geq r$. Also, observe that

$$\frac{\binom{d_G(v)-1}{r}}{\binom{d_G(v)-1+1/t}{d_G(v)-1-r}} = \left(1 + \frac{1}{t\,d_G(v)}\right)\frac{\binom{d_G(v)}{r}}{\binom{d_G(v)+1/t}{d_G(v)-r}} \tag{4}$$

The alternative proof will be by induction on the number n of vertices of the k-uniform hypergraph G. The base case of $n = 1$ follows trivially. Assuming the result holds for $n - 1$, we will now show that the result also holds for n.

Case 1. If we have a vertex $v \in V(G)$ with $d_G(v) \leq d$, then consider the hypergraph $H = G(V')$ where $V' = V\backslash\{v\}$. Observe that $\alpha_{k,d}(G) = \alpha_{k,d}(H)+1$. Since $\forall u \in V'$, we have from Eq. (4), $w_G(u) \leq w_H(u)$. This implies

$$w(H) = \sum_{u \in V'} w_H(u) \geq \sum_{u \in V'} w_G(u) = w(G) - 1.$$

The last inequality follows from the fact that $w_G(v) = 1$ since $d_G(v) \leq d$. Using the induction hypothesis $\alpha_{k,d}(G) \geq w(G)$ and the fact that $\alpha_{k,d}(G) = \alpha_{k,d}(H) + 1$, we get

$$\alpha_{k,d}(G) = \alpha_{k,d}(H) + 1 \geq w(H) + 1 \geq w(G).$$

Case 2. Now we will consider the case where $d_G(v) > d$, $\forall v \in V(G)$. Let $\Delta = \max_{u \in V(G)} d_G(u)$, and let $v \in V(G)$ be a vertex with $d_G(v) = \Delta$. Let $u_1, \ldots, u_l$, where $l = t\Delta$, be the neighbors of v in G. Note that $l = t\Delta$ follows from the fact that G is a linear hypergraph. We will now show that $w(H) \geq w(G)$, where $H = G(V')$ and $V' = V \setminus \{v\}$. We will now show $w(H) \geq w(G)$.

$$w(H) = \sum_{u \in V'} w_H(u)$$

$$= w(G) - w_G(v) - \sum_{i=1}^{l} w_G(u_i) + \sum_{i=1}^{l} w_H(u_i)$$

$$= w(G) - w_G(v) + \sum_{i=1}^{l} \frac{w_G(u_i)}{t \, d_G(u_i)}$$

$$\geq w(G)$$

The second last inequality follows from the facts that $d_H(u_i) = d_G(u_i) - 1$ (as G is a linear hypergraph) and Eq. (4). The last inequality follows from the facts that $d_G(u) \leq \Delta$ for all $u \in V$ and $w_G(u_i) \geq w_G(v)$ (direct consequence of Eq. (4)). From induction hypothesis we have

$$\alpha_{k,d}(H) \geq w(H) \geq w(G).$$

This completes the proof of Theorem 3 since $\alpha_{k,d}(G) \geq \alpha_{k,d}(H)$.

3 General K-uniform Hypergraphs

In this section we shall prove a lower bound on $\alpha_{k,d}(H)$ for general k-uniform hypergraph H in terms of its degree sequence. We will give a very simple randomized algorithm to obtain an d-degenerate subgraph of a k-uniform hypergraph, whose analysis in expectation will yield the desired bound in Theorem 4.

3.1 Details of the Algorithm

Before we can give the details of the algorithm, we will need some definitions.

Definition 1. *Let σ be an ordering of the vertices of H.*

- *Fix a vertex $v \in V(H)$. Call a hyperedge $e \in E(H)$ with $v \in e$ a backward edge with respect to σ, if $\forall u \in e \setminus \{v\}$, $\sigma(u) < \sigma(v)$.*
- *We will denote by $b_\sigma(v)$ the number of backward edges of the vertex v with respect to the ordering σ.*

Algorithm 1. RandPermute

Input: $H := (V, E)$ and d;
// H is a k-uniform hypergraph
Random odering: Let σ be a random ordering of the vertex set V;
Initialization: $I \leftarrow \emptyset$;
// I will be the degenerate subset we output;
for $v \in V$ **do**
    Compute: $b_\sigma(v)$;
    if $b_\sigma(v) \leq d$ **then**
        $I \leftarrow I \cup \{v\}$;
    end if
end for
Output: I;

3.2   Analysis of the Algorithm

Theorem 4 directly follows from the following result.

Claim.

$$E\left[|I|\right] \geq \sum_{v \in V} \min\left\{1, \; c_k \left(\frac{d+1}{d_H(v)+1}\right)^{1/t}\right\},$$

where $c_k = c_k = 2^{-\left(1+\frac{1}{k-1}\right)}\left(1 - \frac{1}{k}\right) = 2^{-(1+o_k(1))}$.

Proof. For all vertices $v \in V$, we will denote by $N(v)$ the neighbors of v in H.

Given an arbitrary vertex $v \in V$, and a random ordering of the vertices σ, we need to bound $\Pr\left[v \in I\right]$ from below. Since the event of v being selected depends on the relative ordering of the vertices in $N(v)$, therefore, the probability v being selected in I in a random ordering is the number of orderings for which v is selected, divided by $(|N(v)|+1)!$. Let σ be an ordering of the vertices of V, such that v is selected in I in the ordering σ. Given a vertex $v \in V$, consider now $L_v := (V(L_v), E(L_v))$, the $(k-1)$-uniform *link hypergraph* on the neighbourhood of v, defined as follows:

$$V(L_v) := N(v), \quad \text{and}$$
$$E(L_v) := \{S \subset V(L_v) : S \cup \{v\} \in E\},$$

i.e., the vertices are the neighbours of v, and the edges are those edges of the original hypergraph H which contained v, but with v removed. Clearly $|E(L_v)| = d_H(v)$. Let $F \subset V(L_v)$ be

$$F := \{u \in N(v) : \sigma(u) < \sigma(v)\},$$

i.e., the vertices in the neighbourhood of v which occur before v in the ordering σ. We want $L_v(F)$ to have at most d hyperedges. The vertices occurring before v can be ordered arbitrarily amongst themselves, and similarly for the vertices occuring after v. So we get that the probability that v is selected in I is given by:

$$\Pr\left[v \in I\right] = \sum_{J \subset V(L_v) : \; |E(L_v(J))| \leq d} \frac{(|J|)! \, (|V(L_v)| - |J|)!}{(|V(L_v)|+1)!}$$

$$= \frac{1}{|V(L_v)|+1} \sum_{J \subset V(L_v) : \; |E(L_v(J))| \leq d} \frac{1}{\binom{|V(L_v)|}{|J|}}$$

For $k = 2$, the link hypergraph is a 1-graph i.e. a set of vertices, each vertex being a 1-edge. Hence the summation in the RHS evaluates to $d+1$ (counting 1 for each case when there are exactly $0, 1, \ldots, d$ vertices before v, in the random ordering). Therefore

$$E\left[|I|\right] = \sum_{v \in V} \Pr\left[v \in I\right] = \sum_{v \in V} \min\left\{1, \frac{d+1}{d(v)+1}\right\},$$

and we get the theorem of Alon-Kahn-Seymour (Theorem 2).

For general k-uniform hypergraphs, observe that if $d_H(v) \leq d$, then $\Pr[v \in I] = 1$. However, if $d_H(v) > d$, then we need to look at the link hypergraph which can be an arbitrary $k-1$-uniform hypergraph. In this case, we shall prove the following general lemma, (which may be of independent interest).

Lemma 3. *For each k-uniform hypergraph $H = (V, E)$, such that $|V| = n$, $|E| = m$, we have*

$$\sum_{J \subset V(H) \,:\, |E(J)| \leq a} \frac{1}{\binom{n}{|J|}} \geq c'_k n \left(\frac{a+1}{m}\right)^{1/k}.$$

where $c'_k = 2^{-(1+1/k)}$.

Indeed, we get that the probability that v is selected in I is given by:

$$
\begin{aligned}
\Pr[v \in I] &= \frac{1}{|V(L_v)| + 1} \sum_{J \subset V(L_v) \,:\, |E(L_v(J))| \leq d} \frac{1}{\binom{|V(L_v)|}{|J|}} \\
&\geq \frac{c'_{k-1}|V(L_v)|}{|V(L_v)| + 1} \times \frac{(d+1)^{1/(k-1)}}{|E(L_v)|^{1/(k-1)}} \quad \text{(from Lemma 3)} \\
&\geq \frac{c'_{k-1}|V(L_v)|}{|V(L_v)| + 1} \times \frac{(d+1)^{1/(k-1)}}{d_H(v)^{1/(k-1)}} \quad \text{(as $|E(L_v)| = d_H(v)$)} \\
&\geq c_k \left(\frac{d+1}{d_H(v)+1}\right)^{1/(k-1)}, \quad\quad\quad\quad\quad\quad\quad\quad (5)
\end{aligned}
$$

where

$$c_k = 2^{-\left(1+\frac{1}{k-1}\right)} \left(1 - \frac{1}{k}\right) = 2^{-(1+o_k(1))}.$$

Note that Inequality (5) follows from the fact that since $d_H(v) > d \geq 0$, we must have at least $k - 1$ vertices in the hypergraph L_v, i.e., $|V(L_v)| \geq k - 1$.

It only remains to prove Lemma 3, which we will prove using a *local density* argument.

Proof (of Lemma 3). For all $1 \leq s \leq n$, we define

$$\rho_s := \mathrm{E}_{|S|=s}[|E(H(S))|] = \frac{\sum_{S \subseteq V, |S|=s} E(H(S))}{\binom{n}{s}}.$$

Note that the expectation is taken over all subsets of V of size s, and $E(H(S)) = \{e \in E(H) : e \subseteq S\}$.

Counting the number of pairs (e, S), where $e \in E(H)$, and $S \subset V : |S| = s, e \in S$, in two ways, we get the average local density of sets of size s is

$$\rho_s = \frac{m\binom{n-k}{s-k}}{\binom{n}{s}} = \frac{m(s)_k}{(n)_k} \leq \frac{ms^k}{n^k}.$$

(Here $\binom{a}{b} := 0$ if $b < 0$). This is as follows: let

$$z := \#\left\{(e, S) : e \in E, S \in \binom{V}{s}, e \subset S\right\}.$$

Then each of the $\binom{n}{s}$ sets of size s contributes, on average, ρ_s-many entries to z. On the other hand, each edge $e \in E(H)$ belongs to $\binom{n-k}{s-k}$-many sets of size s. Equating the two summations gives the claimed average local density.

Now, we use the above observation to prove the lemma. Partition the summands on the LHS into n parts, depending on the size of the set J (i.e. the number of neighbouring vertices which precede v in the random ordering):

$$\sum_{i=1}^{n} \left(\sum_{J \subseteq V : |J|=i, |E(H(J))| < a} \frac{1}{\binom{n}{i}}\right).$$

When $i < k$, it is easy to see that the inner summation is 1. The main idea of the proof is the following: first, observe that for any $i \in [n]$, the inner summation is just the probability that a randomly picked set of exactly i vertices has fewer than $a+1$ edges. Then for small enough i, the expected number of edges is upper bounded by ρ_i, and is much smaller than $(a + 1)/2$. So the probability that a random i-set contains more than twice the expected number, is at most half. Therefore for all such i, the contribution to the outer sum is at least $1/2$. The number of such terms in the outer sum, then gives the claimed lower bound.

Formally, let X_i be a random variable giving the number of edges contained in a randomly chosen set on i vertices. By Markov's inequality:

$$\Pr[X_i \geq 2.\mathrm{E}[X_i]] \leq \frac{1}{2}.$$

We have that $\mathrm{E}[X_i] = \rho_i$. Therefore, the LHS becomes:

$$\sum_{i=1}^{n} \Pr[X_i < a + 1] = 1 - \sum_{i=1}^{n} \Pr[X_i \geq a + 1]$$

With foresight, we split the above sum into two parts, when $i \leq t^* := \frac{n(a+1)^{1/k}}{(2m)^{1/k}}$, and when $i > t^*$. Observe that when $i \leq t^*$, we have that $\frac{mi^k}{n^k} \leq \frac{a+1}{2}$. We get

$$\sum_{i=1}^{n} \Pr[X_i < a + 1] = \sum_{i=1}^{t^*} \Pr[X_i < a + 1] + \sum_{i>t^*} \Pr[X_i < a + 1]$$

$$\geq \sum_{i=1}^{t^*} \left(1 - \frac{\mathrm{E}[X_i]}{a + 1}\right)$$

$$= \sum_{i=1}^{t^*} \left(1 - \frac{m(i)_k}{(a + 1)(n)_k}\right)$$

$$\geq \sum_{i=1}^{t^*} \left(1 - \frac{mi^k}{(a+1)n^k}\right)$$

$$\geq \sum_{i=1}^{t^*} \left(1 - \frac{1}{2}\right)$$

$$\geq \frac{n(a+1)^{1/k}}{2(2m)^{1/k}}$$

where in the second step we used Markov's inequality on the first summation, and in the penultimate step we used the observation on t^* noted above.

Remark 1 (Regarding the constant "c_k").

1. Observe that

$$c_k = 2^{-\left(1+\frac{1}{k-1}\right)} \left(1 - \frac{1}{k}\right) \geq \frac{1}{8},$$

   for all $k \geq 2$, and $c_k \to \frac{1}{2}$ as $k \to \infty$.
2. For simplicity of exposition we did not try to optimize c_k.

4 Conclusion

Our randomized algorithm for finding d-degenerate subgraphs of k-uniform hypergraphs inherits the analysis of Srinivasan-Shachnai [SS04] for independent sets:

(i) The RandPermute algorithm runs in RNC, as long as d is polylogarithmic in the number of vertices and edges.
(ii) Our proof technique generalizes to non-uniform hypergraphs.
(iii) All our results generalize to the vertex-weighted scenario, where we want an induced d-degenerate subgraph of maximum weight.

It is interesting to ask if the RandPermute algorithm can be used to obtain lower bounds for the generalized degeneracy of Zaker, or to solve the conjecture of Beame and Luby [BL90], which asks whether iterating the RandPermute algorithm always yields a maximal independent set.

Acknowledgement. Kunal Dutta and Arijit Ghosh are supported by European Research Council under Advanced Grant 339025 GUDHI (Algorithmic Foundations of Geometric Understanding in Higher Dimensions) and Ramanujan Fellowship (No. SB/S2/RJN-064/2015) respectively.

Part of this work was done when Kunal Dutta and Arijit Ghosh were Researchers at Max-Planck-Institute for Informatics, Germany, supported by the Indo-German Max Planck Center for Computer Science (IMPECS).

References

[AKS80]   Ajtai, M., Komlós, J., Szemerédi, E.: A note on Ramsey numbers. J. Comb. Theory, Ser. A **29**(3), 354–360 (1980)

[AKS87]   Alon, N., Kahn, J., Seymour, P.D.: Large induced degenerate subgraphs. Graphs Comb. **3**(1), 203–211 (1987)

[AS08]   Alon, N., Spencer, J.H.: The Probabilistic Method. Wiley, Hoboken (2008)

[BL90]   Beame, P., Luby, M.: Parallel search for maximal independence given minimal dependence. In: Proceedings of the 1st Annual ACM-SIAM Symposium on Discrete Algorithms, SODA 1990, pp. 212–218 (1990)

[Car79]   Caro, Y.: New results on the independence number. Technical report, Tel Aviv University (1979)

[CT91]   Caro, Y., Tuza, Z.: Improved lower bounds on k-independence. J. Graph Theory **15**(1), 99–107 (1991)

[DLR95]   Duke, R.A., Lefmann, H., Rödl, V.: On uncrowded hypergraphs. Random Struct. Algorithms **6**(2/3), 209–212 (1995)

[DMS12]   Dutta, K., Mubayi, D., Subramanian, C.R.: New lower bounds for the independence number of sparse graphs and hypergraphs. SIAM J. Discrete Math. **26**(3), 1134–1147 (2012)

[PP12]   Pilipczuk, M., Pilipczuk, M.: Finding a maximum induced degenerate subgraph faster than 2^n. In: Thilikos, D.M., Woeginger, G.J. (eds.) IPEC 2012. LNCS, vol. 7535, pp. 3–12. Springer, Heidelberg (2012)

[PW13]   Payne, M.S., Wood, D.R.: On the general position subset selection problem. SIAM J. Discrete Math. **27**(4), 1727–1733 (2013)

[She83]   Shearer, J.B.: A note on the independence number of triangle-free graphs. Discrete Math. **46**(1), 83–87 (1983)

[She95]   Shearer, J.B.: On the independence number of sparse graphs. Random Struct. Algorithms **7**(3), 269–272 (1995)

[Spe72]   Spencer, J.H.: Turán's theorem for k-graphs. Discrete Math. **2**, 183–186 (1972)

[SS04]   Shachnai, H., Srinivasan, A.: Finding large independent sets in graphs and hypergraphs. SIAM J. Discrete Math. **18**(3), 488–500 (2004)

[ST83]   Szemerédi, E., Trotter, W.T.: Extremal problems in discrete geometry. Combinatorica **3**(3), 381–392 (1983)

[Thi99]   Thiele, T.: A lower bound on the independence number of arbitrary hypergraphs. J. Graph Theory **30**(3), 213–221 (1999)

[Tur41]   Turán, P.: On an extremal problem in graph theory. Math. Fiz. Lapok **48**, 436–452 (1941). (in Hungarian)

[Wei81]   Wei, V.K.: A lower bound on the stability number of a simple graph. Technical Memorandum TM 81-11217-9, Bell Laboratories (1981)

[Zak13]   Zaker, M.: Generalized degeneracy, dynamic monopolies, maximum degenerate subgraphs. Discrete Appl. Math. **161**(1617), 2716–2723 (2013)

Erratum to: Packing and Covering Immersion Models of Planar Subcubic Graphs

Archontia C. Giannopoulou[1]([⊠]), O-joung Kwon[2],
Jean-Florent Raymond[3,5], and Dimitrios M. Thilikos[3,4]

[1] Technische Universität Berlin, Berlin, Germany
archontia.giannopoulou@tu-berlin.de
[2] Institute for Computer Science and Control, Hungarian Academy of Sciences,
Budapest, Hungary
ojoungkwon@gmail.com
[3] AlGCo Project-Team, CNRS, LIRMM, Montpellier, France
sedthilk@thilikos.info
[4] Department of Mathematics, National and Kapodistrian University of Athens,
Athens, Greece
[5] University of Warsaw, Warsaw, Poland
jean-florent.raymond@mimuw.edu.pl

Erratum to:
Chapter "Packing and Covering Immersion Models of Planar Subcubic Graphs" in: P. Heggernes (Ed.): Graph-Theoretic Concepts in Computer Science, LNCS, DOI: 10.1007/978-3-662-53536-3_7

The original version of this chapter contained an error. The affiliation of the author was incorrect in the original publication. The original chapter was corrected.

The updated online version of this chapter can be found at
http://dx.doi.org/10.1007/978-3-662-53536-3_7
A.C. Giannopoulou—The research of this author has been supported by the European Research Council (ERC) under the European Union's Horizon 2020 research and innovation programme (ERC consolidator grant DISTRUCT, agreement No 648527) and by the Warsaw Center of Mathematics and Computer Science.
O.-j. Kwon—Supported by ERC Starting Grant PARAMTIGHT (No. 280152)
J.-F. Raymond—Supported by the (Polish) National Science Centre grant PRELUDIUM 2013/11/N/ST6/02706.

© Springer-Verlag GmbH Germany 2017
P. Heggernes (Ed.): WG 2016, LNCS 9941, p. E1, 2016.
DOI: 10.1007/978-3-662-53536-3_26

Author Index

Alrasheed, Hend 145
Angel, Eric 97

Bampis, Evripidis 97
Böhmová, Kateřina 282
Bonnet, Édouard 233
Brandstädt, Andreas 38
Brettell, Nick 233

Cai, Leizhen 62
Cao, Yixin 13
Chalopin, Jérémie 282
Chitnis, Rajesh 133

Dębski, Michał 1
Dourado, Mitre C. 25
Dragan, Feodor F. 145
Dusart, Jérémie 121
Dutta, Kunal 295

Escoffier, Bruno 50, 97

Fomin, Fedor V. 171

Ghosh, Arijit 295
Giannopoulou, Archontia C. 74
Goedgebeur, Jan 109
Golumbic, Martin Charles 121
Gözüpek, Didem 195

Kamma, Lior 133
Kleist, Linda 158
Köhler, Ekkehard 145
Kolay, Sudeshna 245
Kou, Shaowei 220
Krauthgamer, Robert 133
Kwon, O-joung 74, 233

Lampis, Michael 97
Lonc, Zbigniew 1

Maffray, Frédéric 85
Marx, Dániel 233

Mathew, Rogers 121
Mihalák, Matúš 282
Montealegre, Pedro 183
Mosca, Raffaele 38

Özkan, Sibel 195

Pandurangan, Ragukumar 245
Panolan, Fahad 245
Pastor, Lucas 85
Paul, Christophe 195
Penso, Lucia D. 25
Pergel, Martin 207
Proietti, Guido 282

Rajendraprasad, Deepak 121
Raman, Venkatesh 245
Ramanujan, M.S. 269
Rautenbach, Dieter 25
Raymond, Jean-Florent 74
Rząӟewski, Paweł 1, 207

Sau, Ignasi 195
Schaudt, Oliver 109
Shalom, Mordechai 195
Strømme, Torstein J.F. 171
Suchý, Ondřej 257

Tale, Prafullkumar 245
Thilikos, Dimitrios M. 74
Todinca, Ioan 183

Wang, Jianxin 13
Widmayer, Peter 282

Xiao, Mingyu 220

Ye, Junjie 62
You, Jie 13

Ziedan, Emile 121

Printed in the United States
by Bookmasters

Printed in the United States
By Bookmasters